# Philosophy and Engineering: Reflections on Practice, Principles and Process

# Philosophy of Engineering and Technology

VOLUME 15



For further volumes:
http://www.springer.com/series/8657

Diane P. Michelfelder • Natasha McCarthy
David E. Goldberg
Editors

# Philosophy and Engineering: Reflections on Practice, Principles and Process

*Editors*
Diane P. Michelfelder
Macalester College
St Paul, MN, USA

Natasha McCarthy
Royal Academy of Engineering
London, UK

David E. Goldberg
ThreeJoy Associates, Inc.
Douglas, MI, USA

ISSN 1879-7202 ISSN 1879-7210 (electronic)
ISBN 978-94-007-7761-3 ISBN 978-94-007-7762-0 (eBook)
DOI 10.1007/978-94-007-7762-0
Springer Dordrecht Heidelberg New York London

Library of Congress Control Number: 2013957884



Printed on acid-free paper

Springer is part of Springer Science+Business Media (www.springer.com)

# Foreword

## Prospects in the Philosophy of Engineering: An Exchange between the Editors and Carl Mitcham

1. From the outset, we editors (Diane P. Michelfelder, Natasha McCarthy, and David E. Goldberg) have thought of this volume as a sequel to *Philosophy and Engineering: An Emerging Agenda* (edited by David E. Goldberg and Ibo van de Poel). To what extent does the new volume make a contribution to this emerging agenda?

*Philosophy and Engineering: Reflections on Practice, Principles, and Process* is clearly a companion to *Philosophy and Engineering: An Emerging Agenda*. The *Agenda* volume was divided into sections dealing with philosophy, ethics, and reflection. In the new volume, reflections have become primary and are addressed to practice, principles, and process. There is thus less reliance on the categories of past discourse and more of an effort to develop appropriate categories for the future.

Complementary differences in the spectrum of authors also deserve notice. *Agenda* has 32 authors in 28 chapters. *Reflections* has 38 authors in 30 chapters. What is more significant is that 31 of the *Reflections* authors did not contribute to *Agenda* and thus bring new perspectives; 21 *Agenda* authors are not repeat contributors to *Reflections*. Whereas what might be called the usual suspects dominate in *Agenda*, new suspects play a major role in *Reflections*, expanding representation from three new countries (Ireland, Italy, and Sri Lanka) beyond the old *Agenda* representations (Canada, China, Germany, Netherlands, New Zealand, United Kingdom, and United States). Obviously, many countries still wait to be included. *Reflections* also includes three times more women than *Agenda*.

Finally, it can be observed that whereas two-thirds of the *Agenda* chapters have "engineer" or "engineering" in the titles, closer to only half of the *Reflections* chapters do so. *Reflections* appears slightly more interested in drilling into the particulars of engineering.

2. Our second question is a direct follow-up to the first. In the concluding session of the fPET 2012 meeting in Beijing, Pieter Vermaas made a comment to the effect that the philosophy of engineering, as a sub-discipline, has already emerged. To what extent do you agree with Pieter's remark? To what extent has the philosophy of engineering successfully emerged as a sub-discipline? To what extent is it still emerging? How does this volume contribute to establishing the philosophy of engineering?

Vermaas's comment may be more rhetorical than substantive. Let us consider some possible meanings of "sub-discipline" and "emergence."

First, sub-disciplines come in different granularities. Philosophy as a whole is commonly divided into the branches of logic, ethics, epistemology, and metaphysics. A different branching occurs with regionalizations such as philosophy of art, of religion, and of science. But since the mid-twentieth century, the philosophy of science itself has been sub-divided into philosophy of physics, of biology, of chemistry, and more. The philosophy of technology has sometimes been treated as a related sub-division (as in Gabbay, Thagard, and Woods, eds., *Handbook of the Philosophy of Science*); at other times and more commonly (as in some of the basic introductions to philosophy of technology) as on a par with philosophy of science as a whole. So the first question is, at what level of granularity is the philosophy of engineering emerging?

There is a debate—which is perhaps reflective of its emergence—with regard to whether philosophy of engineering should be conceived as a sub-discipline of the philosophy of technology or as its own regionalization. Among the general introductions to philosophy of technology, only my *Thinking through Technology: The Path between Engineering and Philosophy* (1994) gives engineering any prominence. Don Ihde's *Philosophy of Technology: An Introduction* (1993), Frederick Ferré's *Philosophy of Technology* (1995), and Val Dusek's *Philosophy of Technology: An Introduction* (2006) all give engineering short shrift. The Olsen, Pedersen, and Hendricks, eds., *A Companion to the Philosophy of Technology* (2009) includes engineering in the titles of only two ("Engineering Science" and "Engineering Ethics") of 98 chapters. Both the *Agenda* and *Reflections* volumes give technology more attention (in three and four chapter titles, respectively) than the Ihde, Ferré, and Dusek books give engineering.

Second, there is the issue of emergence, which takes place in at least two different forms: as a self-conscious pursuit among a group of like-minded scholars and as a discourse or research program that is acknowledged by non-participant scholars and even the non-scholarly public. On the basis of the *Agenda* and *Reflections* volumes themselves it is reasonable to affirm emergence in the former sense but not necessarily in the latter.

Finally, it is possible to conceive of philosophy of engineering as less a sub-discipline of whatever type and more as a field of interdisciplinary interaction: philosophy *and* engineering rather than philosophy *of* engineering. As an interdisciplinary field, interactions can be traced to the eighteenth century origins of engineering itself. The founders of engineering in the modern sense drew on the work of philosophers to conceptualize their new endeavor. In formulating what has become the classic definition of engineering as "the art of directing the great sources of power in nature for the use and convenience of man," British engineer Thomas Tredgold (1788–1829) implicitly referenced the thought of David Hume (1711–1776) and other Scottish Enlightenment philosophers, although this is not generally appreciated.

3. As you look at this volume, you can see that one of its recurring themes has to do with the role that philosophy of engineering plays in the rethinking of

engineering education. In what ways has this way of putting the philosophy of engineering to use been successful? In what ways does it need to be rethought?

From the beginning, engineers, especially in the United States, have been concerned with education to a greater extent than is the case with any other learned profession. Much more than physicians and professors of medicine or lawyers and law professors with regard to their educational programs, engineers and engineering professors have debated the proper content and structure of the engineering curriculum. Questions about the technical core and the proper balance between mathematics, science, and practical design experience have been hotly contested, as have concerns about the proper roles of the humanities and the social sciences. Arguments have been made for making engineering a graduate program after the manner of medicine and law, which generally require a bachelor degree of some type prior to admission to their respective schools that lead to doctorates.

Engineers have probably worried about the proper character of engineering education even more than philosophers have debated philosophical education. But any critical examination of education necessarily engages philosophical issues, from questions of the relation between knowing and doing to the anthropological and political implications of learning, although not always explicitly. Insofar as the philosophy of engineering attempts to make the implicit more explicit it cannot help but deepen discussions of engineering education.

A number of chapters in *Reflections* also give education explicit attention. In Part I, Chaps. 6 and 7 address pedagogical aspects of engineering education. In Part II, Chaps. 14 and 17 propose relevant new content for engineering curricula. Indeed, Charles E. Harris, Jr's concept of aspirational ethics and W. Richard Bowen's ideal of peace engineering complement each other in calling engineers to think more idealistically and imaginatively in regard to their professional self-understandings. Chapter 24, in Part III, is also relevant; Bruce A. Vojak and Raymond L. Price's epistemological analysis of the innovation process advances a wide-spread concern to make innovation a more conscious aspect of engineering education.

With regard to education, however, it might be helpful to make more conscious use of another regionalization of philosophy, that of the philosophy of education. Philosophical discourse about education has not yet played a significant role in engineering discussions of engineering education. Instead the focus has been largely on the extent to which the teaching and learning of philosophy itself might benefit engineering practice and the professional engineering life.

4. Our fourth question follows directly from the one before. Another way of looking at this same theme is to say that philosophy is being used instrumentally in order to repair engineering education and to bolster the status of engineering as a practice. Do you think this is a legitimate use of philosophy, or not?

There is nothing wrong with the instrumental use of philosophy—although this is not all that philosophy is. Rocks have both extrinsic or instrumental and intrinsic value. The trick with using rocks (or philosophy) instrumentally is not to allow such usage to occlude their (or its) intrinsic reality. Too much of a focus on the use of rocks to build bridges or as a source of oil and gas can obscure their intrinsic complexity and beauty and the wonder of rocks. The same goes for philosophy.

But the question could also be turned around. Does philosophy run the danger of using engineering only instrumentally, as just another phenomenon on which to deploy its analytic skills and reflective resources? Philosophy needs to be aware of this danger and to work to respect the intrinsic complexity and wonder that is engineering.

Here again it is possible to make a few (necessarily incomplete) references to relevant chapters in *Reflections*. Philosophers Hans Poser (Part I, Chap. 1), Joseph C. Pitt (Part I, Chap. 8), Peter Simons (Part II, Chap. 12), and Ibo van de Poel (Part III, Chap. 20) all pay philosophical attention to engineering in ways that respect its own inherent complexity.

In this regard I want to call special attention to Wang Guoyu's examination of feasibility from a Chinese perspective (Part III, Chap. 28). This is a penetrating analysis of a philosophically important but neglected engineering concept in a way that combines insights from the English, German, and Chinese languages for the potential benefit of engineering. It thus models one path for the philosophy of engineering in a globalizing world.

5. Part of the purpose of this volume and the meetings that created it has been to encourage practicing engineers to think and write philosophically. What are the challenges for engineers in thinking this way? Do you think philosophical thinking is more natural to engineers than we might imagine?

Yes and no. There is certainly a school of philosophy—pragmatism—that has deep affinities with engineering and for which engineers also appear to have a natural attraction. However, there is more to philosophy than pragmatism.

This yes-and-no is also well illustrated by some engineers who have a natural affinity for more than pragmatism, who are more generally philosophical inclined (although whether this is because they are engineers or more than engineers, human beings, is an open question). Byron Newberry (Part II, Chap. 13) and David E. Goldberg (Part III, Chap. 30) stand out in this respect, although they are by no means the only ones. Both Newberry's and Goldberg's contributions have to be counted as high quality philosophy. Indeed, I cannot help but acknowledge Goldberg's deft philosophical criticism of my own work. There is no more careful reader or insightful critic of my argument about the philosophical weakness of engineering; I want to acknowledge my need to re-think things as a result of his analysis.

But the question here may also be considered in conjunction with the next, concerning the influence of engineering on philosophy.

6. This volume represents the voices of philosophers, other humanists, engineering faculty, and engineering practitioners. To what extent do you think this mix of voices has generated new themes or possibilities for changing the self-understanding or self-perception of philosophy: for example, has it contributed to looking at philosophers as practitioners rather than theorists?

This is a provocative question and relates to new philosophical efforts to take engineering as a model for philosophy. A leading example of such an effort is William Wimsatt's *Re-engineering Philosophy for Limited Beings: Piecewise Approximations to Reality* (Cambridge, MA: Harvard University Press, 2007). Wimsatt proposes the possibility of something that might be called the philosophy of an engineered and engineering nature.

In reaction against his father's messy work in biology, Wimsatt initiated his professional life in physics, moved from there into engineering and philosophy, and finally circled back into biology. Along the way he worked as an engineer in the adding machine division of NCR and picked up a B.A. in general studies and philosophy (Cornell, 1965) and a Ph.D. in philosophy (University of Pittsburgh, 1971), before joining the faculty at the University of Chicago. At Chicago he is Professor Emeritus of philosophy and a member of two interdisciplinary research committees, one on the Conceptual Foundations of Science and another on Evolutionary Biology. Although most well known as a philosopher of biology, one of his basic arguments is that what goes on in biological evolution is fundamentally like what happens in engineering. To quote from the glossary of *Re-engineering Philosophy*:

> ENGINEERING PERSPECTIVE. A cluster of theses derived from the assumption that theory has much to learn from practice and application. Teleological: Design is design for an end. View scientific activities as functional, and evaluate their designs for that supposed end .... Relation to practice: Focus not only on theory and *in principle* arguments, but on the practical implications of a view of science, how to apply it, and how it must be adjusted or qualified to do so. The central role of heuristics as fallible inferential tools, rather than sources of certainty. Applied not only to our theories and methods as instruments, but also to our mental capabilities and inferences. Most engineering is re-engineering, recognizing that we rarely start from scratch, but will use what comes readily to hand, as quicker, cheaper, more convenient. This has two consequences: (1) [H]istory matters; to understand our methods we must understand where they came from and how. The genetic fallacy is not a fallacy. (2) There is no "perfect adaptation" *ex nihilo*: adaptation commonly co-opts something else to a new role, so exaptation is common. This view is profoundly instrumental, but denies any necessary tension between instrumental usefulness and truth or realism. (p. 354)

Wimsatt's understanding of engineering design sees both technology and the natural environment as manifesting fundamentally similar processes. As Wimsatt argues earlier in the book with regard to genetic engineering, what in a 1976 paper he had first called the "engineering paradigm," does not design from scratch. Genetically engineering molecules "are not examples of *ab initio* constructions, but rather examples of the conversion of naturally occurring organic factories to the production of other products." "There is some assembly to be sure, but it is assembly of the jigs on the production line and sometimes rearrangement and redirection of the line—not construction of the factory" (p. 202). What engineering does is to assemble "complex systems out of simpler parts, a process that can be iterated" (p. 206).

What is true of engineering and the natural environment, Wimsatt further suggests, is equally true of philosophy. It does not begin from scratch. It takes previously occurring ideas and arguments and re-assembles them in new ways. What philosophy does is to assemble complex systems of thought out of simpler parts in a process that is historically iterated. All philosophizing is re-philosophizing.

What are the implications of this understanding of engineering—and philosophy—as opportunistic modular construction and nature as unconscious engineering? Responses remain to be worked out. Any such working out would include reflection on the implications for basic questions in the philosophy of engineering concerning ontology, epistemology, and ethics in ways to which the *Agenda* and *Reflections* volumes both contribute.

7. Some of the contributions contained in this volume focus on the construction of engineering identity through narrative or other forms of reasoning. One of the "sparks" for the creation of the Workshop on Philosophy and Engineering was concern on the part of many with the engineer's place in the world and the connection of that with the formation of engineering identity. To what extent do you believe philosophical perspectives are helpful in the construction of engineering identity? To what extent does this volume make a contribution to this identity?

When you mention contributions focused on the construction of engineering identity, I assume you have in mind at least those by Andrew Chilvers and Sarah Bell (Part I, Chap. 5) and by Priyan Dias (Part II, Chap. 11). The Chilvers-Bell story of Ove Arup describes one quite remarkable engineering identity and at one end of an identity studies spectrum; the Dias analysis is a more abstract contrast between two identity types and thus at another end of the spectrum. The fact that both are included in *Reflections* enriches the volume.

Given the truth that all human beings are, by virtue of being human, to some degree philosophers, philosophy cannot help but play a role in both types of reflection (particular and abstract) and in the construction of any engineering identity. Yet as Gary Downey and Juan Lucena (among others) have argued at length, engineering identity is not some one thing. Engineers are different in the United States, in France, in Germany, in Japan, in China, and so on. Of course, there are also some commonalities, so that one challenge in the philosophy of engineering is to explicate the ways engineering identity is both same and different across national borders and cultures. By bringing together engineers and philosophers from different national traditions of engineering and of philosophy both the *Agenda* and *Reflections* volumes stimulate precisely this kind of analysis. Cultural anthropology and ethnography can also make important contributions in this regard.

8. We've asked you seven questions. What question did you anticipate we might ask you that we haven't asked yet? What would your response to this question be?

No expectations preceded the questions, all of which have significantly stimulated my own thinking—and no doubt will continue to do so. Dialogue, listening to the questions of others, is one of the core methods of philosophy. Your questions and the questioning that necessarily takes place in one form or another in all the chapters in this book are multiple pathways into the philosophy of engineering.

At the same time, I would propose that there seem to be two basic pathways. One begins in engineering and uses philosophy to try to improve or enhance engineering. The other begins outside engineering and uses philosophy to try to better understand what must increasingly be recognized as an aspect of the human equal in importance to politics, to religion, and to art—and perhaps even to philosophy.

However, from the perspective of an engaged outsider, engineering is promoting the creation of a world of artifice that is historically unprecedented in its breadth and depth. It is important that the philosophy of engineering seeks to reflect on the implications of this project—implications that range from confidence in achievements to intimations of fragility and risk. Surely at some point this too must become a basic aspect of the interdisciplinary encounter between philosophy and engineering or engineering and philosophy.

In this regard, I venture to call attention to three more contributions to *Reflections*: those by Jon Alan Schmidt (Part I, Chap. 9), by Scott Forschler (Part III, Chap. 21), and by Zachary Pirtle (Part III, Chap. 29). In Schmidt's chapter, an engineer draws on the transcendental Thomism of Canadian philosopher Bernhard Lonergan to develop an account of the volitional dimension of engineering. Forschler draws on one of the philosophical founders of modern economics, Adam Smith, to raise basic questions about the kind of volition than can distort engineering practice. And Pirtle, an engineer-philosopher embedded in a major government engineering agency reflects on how to adjust efforts that are ultimately volitional for societal benefit. Such complementary and mutually stimulating chapters are a hallmark of this *Reflections* collection.

9. Do you have any final reflections for the readers of this volume?

Only an invitation to take seriously the initiative presented here and then to contribute to furthering a reflection on practice, principles, and process that engages with the ways in which all of us, engineers and non-engineers alike, are increasingly embedded in an engineered and perhaps engineering world.

# Preface

If the word "and" in the title of this volume, *Philosophy and Engineering*: *Practice, Principles, and Process*, is anything more than a mere grammatical marker, it signals the ongoing opening up of a conversational space whose dimensions and potential are still in fairly early stages of development. For this reason, those who join the conversation within this space—philosophers, engineers, practitioners, and others in the humanities and social sciences—participate in a risky business. Because it is risky, it is also intellectually exciting. We hope the set of papers appearing in this volume brings out a sense of this exciting conversation and might stimulate others both to reflect on and to become participants in it in the future.

These papers—the "voices" of this volume, if you will—represent a highly select group of papers originally presented in three different conferences. One group of papers was solicited from the 2008 Workshop on Philosophy and Engineering (WPE-2008), held at the Royal Academy of Engineering. Contributions to this volume also came from a track on reflective engineering at the 2009 meeting of the Society for Philosophy and Technology at the University of Twente in the Netherlands. A third group of papers were drawn from an outgrowth of the WPE: the 2010 Forum on Philosophy, Engineering, and Technology (fPET-2010), held in Golden, Colorado at the Colorado School of Mines. Some of the contributors represented in this volume have made thinking about engineering their life's work; other contributors are in the early stages of their careers. All but one of the papers appearing here are previously unpublished.

In a broader sense, this volume traces its origins to a meeting held at MIT in Fall 2006 in which a small group of philosophers and engineers met to discuss the possibility of ways in which engineers and philosophers could meet and exchange views in a series of intimate, reflective workshops. The first of those workshops was held in 2007 at the Technical University of Delft (TUDelft), leading to a volume published in 2010: *Philosophy and Engineering: An Emerging Agenda,* with Ibo van de Poel and David E. Goldberg as editors. It is also fair to say that that the original meeting and the continuations were helpful in demonstrating demand for published works at the intersection of philosophy and engineering, and the Springer Series on Philosophy of Engineering and Technology, in which the present volume as well as its predecessor is published, came about through the current editor-in-chief's efforts during this same time period.

The present volume continues in the same spirit as *Philosophy and Engineering: An Emerging Agenda*, as well in the spirit of WPE 2008 and fPET 2010. In hosting WPE 2008, The Royal Academy of Engineering was interested in providing an international forum to prompt discussion and debate over the nature and purpose of engineering, and the role and impacts of engineering within society. Similarly, the mission of fPET is to encourage reflection on engineering, engineers, and technology by philosophers and engineers alike and to build bridges between existing organizations of philosophers and of engineers. In both cases, there are real-world, change-related implications to the dialogue. Without a greater understanding of the issues involved here, the ability of engineering to address global societal challenges is seriously compromised.

This volume aims to sustain this spirit in four specific ways. One is to continue to move forward the emerging field of the philosophy of engineering, to add both to its conceptual scaffolding as well as to its substantive content. Another is to advance the development of reflective engineering and to encourage a culture of reflection among engineering practitioners. A third is to show how reflective engineering can assist in the process of the construction of engineering identity: what it is to be an engineer. A fourth is to show how integrating engineering and philosophy might lead to innovation in engineering design and curricula. These motivations cannot be easily disentangled from one another. Similarly, the division of this volume into reflections on practice, principles, and process is fairly porous. Even so, we believe this distinction among subjects of reflection to be a useful one in pointing to areas of inquiry that have emerged as significant as conversation among philosophers, engineers and other researchers and practitioners about the issues just mentioned has intensified.

Reading through the essays presented here, one can find yet another theme tying together considerations of practice, principles, and process: that of challenging prevalent assumptions and commitments within engineering and philosophy alike. Exploring the ontological and epistemological dimensions of engineering challenges the notion that engineering is simply the application of scientific knowledge to problem solving, a challenge that has deep implications for the design of engineering curricula. This exploration also presents challenges to the basic philosophical assumption that theoretical knowledge is superior to practical ways of knowing. Considering how engineering ethics might be refocused on bringing about social change challenges its current dominant, "do no harm" approach. Above all, the perspectives collected here ought to challenge any lingering beliefs that a conversation between philosophy and engineering is bound to be unproductive because the two disciplines do not have enough in common for a substantive dialogue to take place.

Do these perspectives—these reflections on practices, principles, and process—add up to a unified vision of what the philosophy of engineering or the practice of reflective engineering can be said to be? The answer from the essays comprising this volume is a clear "no." Just as there are key themes to be found here, there are also conflicting voices, debates, and disagreements. Since the publication of this volume, fPET-2012 has taken place in Beijing, and a track on reflective engineering and the ethics of complex, sociotechnical systems, sponsored by fPET and the Council of Engineering Systems, has also been held at SPT-2013 in Lisbon. As other opportunities such as these develop to draw engineers and philosophers into further conversations with one another, it is hoped that the issues and questions raised here will be continually revisited. We hope such exchanges will help shape the way engineering is presented and taught to the engineers of the future, ensure that those engineers are fully engaged with the pressing issues at the center of global society, and contribute to making the voices of engineers a more audible part of society's self-understanding.

It would not be appropriate to bring this brief overview of the volume to an end without pausing to express our gratitude to the host institutions, organizers, and sponsors of WPE-2008, SPT 2009, and fPET-2010—without these conferences this volume would simply not have come to be. We are indebted to Carl Mitcham for his willingness to participate in the "exchange" that forms the foreword to this volume, and for his comments, marked by his characteristic vitality of insight and depth of knowledge. As general editor of the Philosophy of Engineering and Technology book series, Pieter Vermaas has been generous with advice, adept with guidance, conscientious, patient, and ever supportive of our project. It has indeed been a pleasure for us to work with him.

Our thanks go to all the members of the Springer team, especially Christi Lue and Sridharan Asanimshi, who helped move this volume through the publication process. We are also grateful to Denise Carlson of North Coast Indexing for her exemplary work and to our respective institutions: Macalester College, the Royal Academy of Engineering, and the University of Illinois at Urbana-Champaign, for providing the support for indexing this volume.

Diane P. Michelfelder
Natasha McCarthy
David E. Goldberg

# Contents

## Part II Reflections on Principles



## Part III Reflections on Process

# Part I
# Reflections on Practice

# Chapter 1
# The Ignorance of Engineers and How They Know It

**Hans Poser**

> *The very concept of research presupposes conscious ignorance about the object of research at the outset; otherwise there is nothing to research.*
>
> (Smithson 2008: 218)

**Abstract** An engineer starts his design from a problem, i.e. from ignorance as non-knowledge. This corresponds to a question and indicates a direction towards an aim. Therefore the engineer needs knowledge concerning *means* as a *functional* compliance for an *aim*, knowledge of *how to gain and to use* such a means, knowledge concerning *values* behind the aim, and knowledge of how to *modify the aim* in the light of values, if necessary. This is connected to epistemological presuppositions not only as theoretical and practical rationality, but much more as teleological reasoning by a reflective power of judgment.

**Keywords** Ignorance • Creativity • Knowledge • Reflective judgments • Teleology

## 1.1 Introduction

In his famous book *What Engineers Know and How They Know It*, Walter G. Vincenti (1990) analyses the way of problem solving in engineering design as an epistemological problem. But even his lucid undertaking in describing the steps involved in problem solving ignores that all problem solving starts from ignorance in the sense of non-knowledge or rational ignorance, or knowledge about the limits of

H. Poser (✉)
Institut für Philosophie, Technische Universität Berlin, Berlin, Germany
e-mail: hans.poser@tu-berlin.de

D.P. Michelfelder et al. (eds.), *Philosophy and Engineering: Reflections on Practice, Principles and Process*, Philosophy of Engineering and Technology 15,
DOI 10.1007/978-94-007-7762-0_1, 

knowledge: There would be no problem at all if we already had the necessary knowledge (including know-how, etc.). Therefore we need a further epistemological step back; namely, we need to take a look at that kind of rational ignorance or non-knowledge from which the technological problem originates.

Throughout the last two decades 'ignorance' has become a topic of investigation, starting from Michael Smithson (1989, 1990, 1993, 2008) through the interdisciplinary collection of Robert N. Proctor and Londa Schiebinger (2008), where *agnotology* is introduced as a new area of research. Knowledge management has been extended by ignorance management, sociologists and psychologists study phenomena of ignorance. Nearly all these studies, though, take ignorance as manipulated, suppressed, overlooked, but nonetheless still existing knowledge. And secondly, technology ignorance is nearly ignored, and not only in the collection on *Agnotology*. Elsewhere it is addressed only under the heading of uncertainty (e.g. Banse et al. 2005; Gamm and Hetzel 2005) or manipulation (Magnus 2008).

Lying in the background of all of this is the permanent struggle of human beings with contingency: Our life world is full of uncertainties, imponderability, unforeseen accidents; and we as human beings try to overcome this situation through the sciences, which impose necessity – in the first place and ever since Plato, as timeless mathematical truth (*a priori* necessity); followed by Galileo's ideal of the book of nature written in numbers explored by empirical research (physical necessity); and then in a further step by the installation of rules of action within a society to establish stable social structures, fixed by laws and punishments, which allow behaviour to be anticipated and predictions to be made concerning actions (social and ethical necessity). But one of the most important elements banning contingency is technology, which we suppose works properly (technological necessity), i.e. in a foreseeable way, may it be a knife or a car or a whole industrial plant (Poser 2009). However, technology might fail, since it does not work properly in many cases. So, our problem of ignorance concerning engineering is a central one thinking of our understanding of technology.

The guiding idea of this chapter is that there are at least four different types of an engineer's ignorance in the sense of non-knowledge:

1. Ignorance is the starting point of each design and its development by marking a *problem*.
2. Problem solving often needs *creativity*, which excludes predictions – a hard case ignorance.
3. R & D departments need to *communicate* about ignorance, namely concerning the guiding problem, which has to be solved.
4. Unknown possible consequences of technology – i.e. hard cases of ignorance – have to be evaluated by means of the methods of technology assessment.

So, the engineer's ignorance is characterized by a *problem* or a *question* demanding a missing *solution* to the problem. This is the reason for understanding ignorance here in the first place as a state of non-knowledge or nescience state – what Robert Proctor called the "native state," to differentiate it from other states e.g. "ignorance as lost realm, or selective choice" in neglecting other possibilities, or "ignorance as strategic ploy" in keeping knowledge a secret (Proctor and Schiebinger 2008: 4–10).

Moreover, in what follows ignorance will not be discussed from a sociological viewpoint, but rather from the side of epistemology. To understand ignorance as an epistemological and not as a sociological problem needs an approach which asks for the conditions of possibility of knowledge – or in this particular case: What are the conditions, which allow us to conclude from ignorance as non-knowledge, what the question is which has to be answered.

All this presupposes dealing first with *knowledge* and with *ignorance* as the absence of knowledge. As a second step one has to clarify the *fundamental limits of knowledge*, and hence the limits of the problem solving capacities of engineers. The third step deals with the important point that problems depend on an *evaluation* of the given or expected actual social situation, as well as of the types of solution at hand – which presupposes a knowledge of norms and values (e.g. functioning, efficiency, safety, sustainability, etc.). This is connected with a further problem shift, since we have to deal with complex systems and complexity reduction. This causes new kinds of ignorance and problems – not only for technological reason in a narrow sense, but for ethical reasons as well. Ignorance here becomes an essential challenge not only for engineering and for philosophers of technology, but possibly also for the survival of humankind, because we have to find the way between the Scylla of knowing the impossibility of predictions in complex systems and the Charybdis of probably insufficient parameter reductions.

## 1.2 Knowledge and Ignorance

Ignorance as non-knowledge is human to the core – and always subject-related. Therefore ignorance concerning the lack of knowledge has been a topos of human reflection since Socrates and the Sceptics via Nicholas of Cusa up to Emil du Bois-Reymond. For two decades, management theories concerning knowledge management have been enlarged via risk management by ignorance management; but this is only partly adoptable to technology, because economic considerations are only a small element within the broad scale of reflections in engineering.

Now, in order to talk about ignorance as non-knowledge in an epistemological perspective, it first needs to be clarified what non-knowledge would mean: Epistemology as a theory of (positive) knowledge is a well-known discipline since Plato and Aristotle – but what about ignorance? Let me start from non-knowledge; it corresponds to a problem, which can be put as a question of the form "Do you know *that and that*?" – and this indicates a *that and that* as a content of the non-knowledge. It is by no means sufficient to consider ignorance only as something not yet known. An epistemology of ignorance, postulated and roughly sketched by Nancy Tuana (2004), is no pure gap of knowledge and no simple negation, but a quite distinctive kind of gap and, as a logical term, a type of privation. It cannot consist in an analysis of *social* practices as causes, since doubt, trust and uncertainty have to be taken as *cognitive* terms; but it might be necessary to extend epistemology by including elements from the side of sociology as has been done in approaches to social epistemology (e.g. Goldman 1999), since criteria for knowledge have changed in history.

In each case it has to be clarified what 'knowledge' means. Concerning technology, at least four kinds of knowledge and its corresponding ignorance have to be taken into account:

- *Knowing that* as factual knowledge;
- *Knowing why* as theoretical and causal knowledge;
- *Knowing how* as practical action knowledge; and
- *Knowing wherefore* as normative value knowledge.

To speak of ignorance presupposes to know what knowledge is in all these cases. Sociologists take it as factual compliance, as *communis opinio* in a given society at a given time, such as in Smithson (1985). This is not a criterion of truth, but the basis of actions as well as of jurisdictions depending on the state of the art – and therefore it is the way in which it is used in engineering sciences. From a philosophical standpoint one would prefer at least intersubjective agreement – but even that demands the Platonic definition of knowledge as justified true belief (forget the tricky criticism from analytic philosophy known as Gettier's Problem). But since Karl Popper we are confronted with the insight that knowledge is a methodologically justified proposition as a hypothesis; and Thomas Kuhn made clear that the criteria of justification depend on history and culture. However, thinking of technological knowledge is not a question of the truth of propositions, but of the functioning of an action rule – which limits the SCOT-approach of a social construction of technology (Pinch and Bijker 1987). But back to knowledge and ignorance as non-knowledge.

Forms of ignorance corresponding to the above forms of knowing seem to be immediately visible:

- *Unawareness* concerning facts;
- *nescience* concerning theoretical reasons;
- *inability* to achieve something; and
- *blindness* concerning norms and values.

Yet these terms are misleading, because with respect to engineering, knowledge as well as its deficiency is a state of consciousness. The distinction Willem H. Vanderburg (2002: 90) draws between 'useful' and 'harmful' ignorance indicates that we are searching for the first type. In fact, Smithson (2008: 214ff.) starts from a differentiation, which includes that kind of non-knowledge as ignorance. He speaks of four "different kinds of accounts that focus on ignorance." Two of them are important for us (quotations shortened and supplemented by an 'S'):

S 1 *Ignorance as encountered in the external* [non-social] *world*… These accounts make strong claims about meta-knowledge and explain ignorance in exogenous (and usually non-social) terms.

S 4 *Managing ignorance*: How people think about ignorance … and how they act on it.

Smithson attributes the first account to the sciences, but clearly it has to be our starting point for an engineer's ignorance: it is important for each R&D undertaking. But there is a remarkable difference between the sciences and engineering, as Mario Bunge's formulation of the well-known difference between their aims brings out well: Scientists are seeking for the most general laws, whereas engineers are looking for better ends. This has far reaching consequences for the ignorance in question. According to Smithson (2008: 209), this S1-ignorance seems not to be "socially constructed" and to be independent from the "sociocultural origins" of ignorance. This might be the case for the sciences, since the problems they deal with originate with the sciences themselves. But for engineering at least, the situation differs completely from the outset, since its aims stem from the needs of individuals or of society. And since all intentions as well as ways of taking something as a problem depend on a cultural background, it will be necessary to accept that even the S1 account has in part sociocultural origins. Moreover, in this account ignorance as "meta-knowledge" is meant as a meta-language predicate like 'truth' or 'knowledge': It indicates that we know that we do not possess an answer concerning the content of the question.

Now, Smithson's S4 account is important, too. Vanderburg (2002: 91) speaks in this connection of two kinds of ignorance. The first is "related to the fact that, as specialists, we cannot know everything there is to know," whereas what we know is embedded in a second kind of ignorance, as "[w]e forget that any human knowledge is relative to a vantage point determined by our professional experience, formal education, life experience, convictions, values, and, last but not least, the culture of our society." This makes clear that there is no knowledge at all without Vanderburg's second kind of ignorance – which can be turned into a useful kind of ignorance, "if its existence is clearly recognized" (Vanderburg 2002: 91). Thus it will be necessary to include Smithson's fourth account, and with this an element of social construction – e.g. when we ask about risks in regions, where scientific answers are either not yet possible as in nanophysics or impossible for formal reasons as in cases mapped by systems of complexity, since chaotic or dissipative structures exclude predictions for purely mathematical reasons. In these cases the questions which constitute our ignorance depend primarily on values, expectations and fears, which have sociocultural origins and are as such part of our life world.

Therefore, concerning our question Smithson's well-known taxonomy scheme of ignorance (Smithson 1989: 9, 1990: 211) is misleading already in its starting point "error" versus "irrelevance": The ignorance we are looking for means: *I have a problem, but I do not know its solution!* Remember Popper's "All life is problem solving": Hence an engineer's ignorance is neither irrelevant nor erroneous – it is highly relevant, but a knowledge concerning a specific point is missing.

Briefly summing up the results so far, one can say that an engineer's ignorance has a typical structure depending on epistemological connections, since this ignorance is a knowledge of non-knowledge, i.e. a meta-knowledge. It has a content, it leads to a problem, and it can be formulated as a question. Therefore an engineer's ignorance has both a structure and a content. Thus our task will be first to exclude impossibility, and subsequently to analyse the cognitive presuppositions and its epistemological conditions.

## 1.3 Ignorance as Knowledge of the Fundamental Limits of Knowledge

A look at the history of science and technology shows that there have been questions throughout the centuries, which indicate ignorance as a lack of knowledge in cases where we today know that Emil du Bois-Reymond's *Ignoramus – Ignorabimus –* We do not know now and in future – (1872/1912) is insurmountable, even if what he took to be the points of the limits of knowledge are not the same as today. We know about limitations as e.g. the impossibility of deducing the Euclidian parallel axiom from the other axioms; or designing a perpetual mobile; or Gödel's proof of the impossibility of a complete axiomatization of second-order predicate logic and by this of mathematics. Einstein and Heisenberg showed the limits of human experience, since we cannot enter the region inside the uncertainty relation and outside the light cone.

All this seems to be irrelevant for engineering, although there are borderline cases, which we cannot discuss here. But one of the most relevant limitations of knowledge causing an inevitable Ignorabimus consists in the mathematical properties of complex systems beginning from systems of non-linear differential equations (deterministic chaos) via dissipative structures up to autopoietic ones: Even in the case of deterministic chaos it is impossible to derive a closed function as a result. We may reach approximate solutions, but they depend in a highly sensitive way on initial conditions and constraints. Now, it is precisely these complex structures which are indispensable in bio-technology, in communication technology and its networks, and in simulations of technology assessment including the social consequences of a projected technology. This is a new element of known ignorance – and it includes as a further new element norms and values as the basis of the evaluation of possible technological solutions as well as their influence on society and the environment. This enforces the need to introduce cultural traditions and consequently historicity, where our intention is to analyze the epistemological side of the engineer's ignorance. All this demands a new kind of ignorance management in technology, not restricted to economy.

This points to a further Ignorabimus, because there is no absolute or rational foundation of ethics, of norms and values. Moral rules are needed, though neither Kant's categorical imperative nor John Rawls' approach nor any other one warrants an absolute foundation. We have to admit that a universal ethics or a universal theory of norms and values is impossible, since all of them depend on history and culture. This kind of Ignorabimus is not only a challenge for philosophers. It also affects engineering in an essential way, thinking of Bunge's 'better ends': There is no theory fixing once and forever what better ends are.

Altogether, we must be aware of the fact that there are inevitable kinds of ignorance as an Ignorabimus, stemming from formal, namely logical and mathematical, limitations, from physics as well as from foundational problems of ethics.

## 1.4 Ignorance as Knowledge of a Problem to Be Solved

Knowledge concerning technology and engineering sciences has to be specified, because it has its characteristic elements within the broad scope of knowledge mentioned at the beginning. An engineer's ignorance means: There is a problem to be solved. And "a problem to be solved" means: There is an aim to be reached. Therefore the engineer needs

1. knowledge concerning the *means* as a *functional compliance* for an *aim*;
2. knowledge of *how to gain and how to use* such a means;
3. knowledge concerning *values* behind the aim; and
4. knowledge of how to *modify the aim* in the light of values, if necessary.

All this is *problem solving knowledge*. The first is causal knowledge related to a purpose, the second refers to the given situation, the third depends on the cultural horizon of values, whereas the last presupposes knowledge of how to deal with values and aims with respect to needs and intentions. These kinds of knowledge are at the same time the foundation of engineering science, because otherwise they would lose their applicability.

The forms of ignorance corresponding to these kinds of knowledge are immediately visible and are connected to a typical epistemological background. Because the first kind of non-knowledge does not mean the causal laws of nature as hypotheses, but rather reproducible effects in an aims-means relation fulfilling a function, none of these concepts belong to an observational language, but depend on a view from the side of acting: They are *interpretations* of real and possible facts and causal connections. Ignorance, therefore, is not the missing knowledge of nature, but of functions transforming a given situation into an intended end. Therefore ignorance in this field, taken as a problem, demands an extension of knowledge beginning from a new combination of already existing technological knowledge up to new creative solutions.

This is not trivial, because here we meet a hard epistemological problem: How can I know what I am looking for, if the starting point is the knowledge of my ignorance, and by this of a problem to be solved. Furthermore, how do we get from a problem to an aim as an interpretation of a possible state, and from there to a means as an interpretation of a function which itself depends on an interpretation as well? Ignorance, seen from an epistemological viewpoint, presupposes two structural elements: (1) the *direction of a question* oriented at an imagined aim, and (2) the possession of the cognitive ability to develop *heuristic methods* of solution and/or to develop a creative and up to now completely unknown solution.

In the *second* case, the one of knowing how, the corresponding ignorance is related to the absence of an ability: it indicates that there is something to be learned or to be organized. In fact, this is a dominating problem, even if engineers would not say so; but the trickiest technology would be senseless if we were not able to actualise it: Actualizability is a *conditio sine qua non* from the very beginning of each engineering design. But in contrast to the first case, it must be possible to *learn* the know how to overcome this ignorance. Now, even learning has been a classical epistemological problem since Plato, who argued: "Learning is nothing but re-memorization" (Plato, *Phaidon*, 72e) of something which exists already in the soul. In the tradition of the philosophy of technology, this led Friedrich Dessauer to the Platonic presumption that all technological solutions are part of a "fourth empire" of ideas (Dessauer 1956: 155). No one would accept this metaphysical thesis today as a solution of the epistemological problem of learning and of technological creativity to overcome ignorance – but it shows that we understand ignorance in this case as *presupposing the human faculty of learning and creating something entirely new*.

The *third* case has gained substantial weight throughout the most recent decades, since it has become apparent just how complex the area of values in technology is – values which partly bear a great tension, as e.g. economic efficiency and security. All these values and their corresponding norms depend on culture and history. Moreover, normative and epistemological problems are interwoven, which is obvious in all trials to predict future consequences not only concerning the possible results of a new technology, but also its influence on social structures and newly developed values, including an evaluation of all these steps and of the outcome. Ignorance, in this case, includes as a part of its content not only the knowledge of what is unknown, but also and at the same time the *knowledge of values*. Otherwise the aim-oriented question, associated with this kind of ignorance, would be impossible.

The *fourth* case is highly important for our problem, because it would be too simple to presuppose that the non-knowledge ignorance fixes the aim completely. This might be the case when there is a clear-cut task – but normally the problem and its corresponding question adumbrates an aim and sketches a direction in connection with values attributed to imagined ends. So *ignorance means an open structure*. The kind of knowledge in this case presupposes a *knowledge of value hierarchies*, since when thinking of the needs to be fulfilled it might be necessary to substitute a specific value *a* by a differing one *b* which fulfils the same more general value as *a* does, or even to substitute an end for a different one fulfilling the same function. This is well known from the perspective of the practical syllogism as a scheme of action explanation, because there are always infinitely many possible means to bring about an intended end. But the same holds for ends and aims, and finally for the values behind them.

Looking at the four cases all together and asking not only about the knowledge which is presupposed as the content of an engineer's ignorance, but also about the epistemological conditions of its possibility, we reach a deeper level of presupposition. First of all, it is essential that the human being (or to say it with Kant – the transcendental subject) is capable of *imaginations* independent from the actual situation. Moreover, this has to include:

1. *thinking about possibilities* (which might as theoretical reasoning correspond to the conditions of Kant's *Critique of Pure Reason*),

2. *thinking about norms and values* (as practical reasoning corresponding to Kant's *Critique of Practical Reason*), and
3. *thinking teleologically about means and ends* (as teleological reasoning *by means of the reflective power of judgment,* corresponding to Kant's *Critique of Judgment (KdU)* – as we will see now).

Kant did not really deal with technology, not to speak of ignorance as a form of non-knowledge, but he does speak of the arts, in the sense of distinguishing the mechanical arts from the liberal arts (*KdU*, § 43; AA V.303). Since the ignorance we are discussing here presupposes knowledge, it is helpful to pick up some of Kant's points, especially concerning the teleology of nature compared to the teleology of artefacts. In the case of the causation of an object depending on free will, Kant speaks of an intentional technic (technica intentionalis) (*KdU*, § 72, AA V.390). Its principles do not so much depend on causality, which he calls "technically practical" (technisch-praktische Prinzipien), but on "morally practical" (moralisch-praktisch) principles – and he adds that the technical ones belong to "theoretical philosophy," the latter ones to "practical philosophy." He goes on: "All technically practical rules (i.e. of arts and of skilfulness) so far as their principles are based on concepts, have to be seen only as corollaries to theoretical philosophy" (*KdU*, Einleitung, I Einteilung der Philosophie, AA V.172) But if the rules depend on free will, their principles do not depend on the knowledge of nature, but as morally practical ones on moral principles. These are just both transcendental areas, mentioned above, which characterize the cognitive and the normative element of ignorance.

All this is only the first step. Kant's substantial new approach is expressed when he writes that a "teleological (technical) method of explanation" belongs to the "reflective judgement" – or nearer to the original text – to the "reflective power of judgement" (*KdU*, § 71, AA V.389), since it is a faculty or ability of the transcendental subject to think in terms of means and ends. Now, this new and essential concept of '*reflective judgment*' is explained already in the *Introduction* to the *Critique of Judgment*:

> Judgment in general is the faculty of thinking the particular as contained under the universal. If the universal (the rule, the principle, the law) be given, the judgement which subsumes the particular under it [...] is *determinant.* But if only the particular be given for which the universal has to be found, the judgement is merely *reflective*. (*KdU*, Introduction, IV On Judgment, AA V. 179)

This can be taken as a very clear conceptualisation of the cognitive situation of an engineer. Since there are no universal laws which would allow the deduction of a special technological solution, he has to start from the particular – in his specific case from his singular problem and its corresponding question in order to reach not a universal, but an actualizable solution (which, since it is not yet realized, is a universal, namely conceptual one, but not a law – yet it is remarkable that engineers speak of the 'solution principle'). Here, teleological reflection finds its adequate expression as an *a priori* faculty: It presupposes the categories of knowledge, it presupposes the moral principles, but it adds intentional technique as the teleological element, which makes all the difference between an artefact and a natural object.

**Table 1.1** Engineering ignorance

| Missing knowledge | Solution | Epistemic condition |
|---|---|---|
| I. Adaptation of given methods | Heuristic | Imagination |
| *Know how* | Teleological reasoning | Reflective judgment |
| II. New method | Development | Creativity |
| III. Basic theoretical knowledge | Research | Theoretical rationality |
| *Know why* | Empirical and theoretical reasoning | Social epistemology |
| IV. Moral consequences | Ethical reasoning | Practical rationality |
| *Know wherefore* | | |
| V. Complex system Combination of I–IV | To avoid Ignorabimus: parameter reduction in feasibility studies | Theoretical and practical rationality, reflective judgment |

These short remarks might explain why the engineer's ignorance is really an epistemological problem opening a wide horizon of reflection.

Here we need to include an additional point. Kant in his theory of reflective judgments thinks of the problem of a teleology of nature – but we need a teleology independent from nature, and related to artefacts and artificial processes depending on human aims. When it comes to aims as better ends, norms and values play an essential role, since they are at the same time warranting the openness on the side of the aim, because it can be substituted by a different one which actualizes the same value. This indicates that openness is already a constitutive part of ignorance. It is the direction of the possible solution, which is indicated as an epistemic content of ignorance.

Nevertheless, two further capacities have to be added, namely our ability to *learn* and to be *creative*. Both of them presuppose free will. The latter one, *the* fundamental category of Whitehead, breaks open the Kantian scheme of categories, since creativity allows the development of new schemes of ideas in history. Therefore, a Whiteheadian enrichment of our tools allows including elements of the social theory of knowledge. Both elements fit into a better understanding of ignorance as non-knowledge, because it is already an act of creativity as an openness to develop new imaginations and to be aware of a new problem as an element of ignorance. This allows us to understand the background of the awareness of non-knowledge as a cultural element of learning, knowledge transmission, and traditions of methods.

All this might be put together as steps and as conditions in the following scheme (Table 1.1).

It is necessary to say that all this is far from a complete list or a complete disjunction – in fact, what has been called a 'kind of ignorance' here is only a demarcation of a focal point within overlapping phenomena. Yet it is important that ignorance is no blindness, but a highly structured content depending on a broad and differentiated knowledge as well as on extended human cognitive capabilities. It is this which allows for communication with others on missing elements of knowledge and to indicate the direction of creative problem solutions. Therefore ignorance of this kind is the precondition of development as well as of technological creativity.

## 1.5 Conclusion

Let me assemble in a few sentences the elements we have found.

The engineer's ignorance has both structure and content:

- It is a kind of meta-knowledge (knowledge of non-knowledge);
- it characterizes a problem (knowing the direction of an aim);
- it leads to a question (asking for a means to an end);
- it has as a background explicit technological and normative knowledge.

In at least two cases the engineer's ignorance is characterized by an Ignorabimus:

- Creative solutions are never predictable;
- complex developments, mapped in simulations and feasibility projects, are never predictable.

We need the knowledge, activity, and creativity of engineers. Their ignorance is not at all an error or something irrelevant – and by no means nothing which one ought to force them to give up. No creative solution is predictable, but creativity as expression of human freedom, in most cases and especially in engineering, is aim-directed and no sheer hazard. And complex technology developments – even if they are not predictable – need a diligent handling, where ethical principles demand careful reflections on decisions to be made. This is an essential part of human life; thus life experience might be the guide where formal procedures fail. Therefore we need experienced engineers. And therefore, the engineer's ignorance remains an interesting epistemological problem.

All this has to be seen at the same time as a problem of knowledge and of the conditions of the possibility of knowledge. Naturally, one of the central presuppositions is free will – but it is highly important to see which elements and capacities come into play. These fall under the heading of reflection – first of all in a very Kantian sense, namely to have *imaginations* – not in the sense of a picture, but as part of *thinking in possibilities and necessities*. As we saw, we need not only pure reason, but practical reason as well, namely in thinking of norms and values. This is the precondition of *teleological reasoning*, namely thinking of and reflecting on means and ends. But beneath all that one has to include further abilities, which are at least partly non-Kantian ones, namely the *hermeneutic ability* of interpreting facts as means or ends, and of attributing a function, a value and/or a need to facts. When asking how this might be possible we are directed back to life experience in a phenomenological mode. This is not astonishing since technology, its development and, consequently, its kind of ignorance, belong to the most essential preconditions of human life.

## References

Banse, G., Hronsky, I., & Nelson, G. (Eds.). (2005). *Rationality in an uncertain world*. Berlin: Edition Sigma.

Dessauer, F. (1956). *Streit um die Technik*. Frankfurt a.M.: Knecht Vlg (Rev. and ext. ed. of *Philosophie der Technik. Das Problem der Realisierung*. Bonn: Cohen 1927, $^{2}$1928, $^{3}$1932).

du Bois-Reymond, E. (1872/1912). Über die Grenzen des Naturerkennens, 1872. In E. du Bois-Reymond (Ed.), *Reden von Emil du Bois-Reymond* (Vol. 2, pp. 441–473). Leipzig: Veit.

Gamm, G., & Hetzel, A. (Eds.). (2005). *Unbestimmtheitssignaturen der Technik. Eine neue Deutung der technisierten Welt*. Bielefeld: Transcript.

Goldman, A. (1999). *Knowledge in a social world*. Oxford: Oxford University Press.

Kant, I. *Gesammelte Schriften* [=AA] (Preussische Akad. der Wissenschaften, Ed.). Berlin: de Gruyter, vol. 5, 1913.

Magnus, D. (2008). Risk management versus the precautionary principle. In R. Proctor & L. Schiebinger (Eds.), *Agnotology: The making and unmaking of ignorance* (pp. 250–265). Stanford: Stanford University Press.

Pinch, T. J., & Bijker, W. B. (1987). The social construction of facts and artifacts. In W. B. Bijker, T. P. Hughes, & T. J. Pinch (Eds.), *The social construction of technological systems* (pp. 17–50). Cambridge, MA: MIT Press.

Poser, H. (2009). Technology and necessity. *The Monist, 92*(3), 441–451.

Proctor, R., & Schiebinger, L. (Eds.). (2008). *Agnotology: The making and unmaking of ignorance*. Stanford: Stanford University Press.

Smithson, M. (1985). Towards a social theory of ignorance. *Journal for the Theory of Social Behaviour, 15*(2), 151–172.

Smithson, M. (1989). *Ignorance and uncertainty: Emerging paradigms*. New York: Springer.

Smithson, M. (1990). Ignorance and disasters. *International Journal of Mass Emergencies and Disasters, 8*(3), 207–235.

Smithson, M. (1993). Ignorance and science: Dilemmas, perspectives, and prospects. *Science Communication, 15*(2), 133–156.

Smithson, M. (2008). Social theories of ignorance. In R. Proctor & L. Schiebinger (Eds.), *Agnotology: The making and unmaking of ignorance* (pp. 209–229). Stanford: Stanford University Press.

Tuana, N. (2004). Coming to understand: Orgasm and the epistemology of ignorance. *Hypatia, 19*(1), 194–232 (Repr. in Proctor & Schiebinger 2008, 108–145).

Vanderburg, W. H. (2002). *The Labyrinth of technology: A preventive technology and economic strategy as a way out*. Toronto: University of Toronto Press.

Vincenti, W. G. (1990). *What engineers know and how they know it*. Baltimore: John Hopkins University Press.

# Chapter 2
# Rules of Skill: Ethics in Engineering

**Wade L. Robison**

**Abstract** Rules of skill tell us how to achieve a particular end: to bake a cake, do such-and-such; to buttress a girder, do so-and-so. They are the tools of the trade, so to speak, for any profession. Surgeons learn how to cut out cancerous tissue; software engineers learn how to write code. There are also the norms of the profession, and when failing to follow the right rule of skill leads to significant harm, they carry ethical weight: a professional ought, ethically, to do such-and-such. They also serve in engineering in another way. The intellectual core of engineering is the solution to design problems, and any solution is a rule of skill: to solve this problem, do so-and-so. At a minimum such solutions should not cause unnecessary harm. That is a moral injunction, and so ethics enters into the core of engineering both through its tools, the standing rules, and through design solutions. Engineers are ethically obligated to use the right rule of skill and to provide design solutions that cause no unnecessary harm. We should like them to provide design solutions that produce more benefits than harms as well, but satisfying the minimal condition of causing no unnecessary harm is sufficient to show how ethics enters into the heart of engineering.

**Keywords** Rules of skill • Professional norms • Morality • Design solutions

## 2.1 Introduction

A special set of skills is an essential feature of a professional. Physicians and surgeons learn how to identify various body parts, but only surgeons need learn how to extract cancerous tissue, for example. Lawyers must learn how to marshall

---

W.L. Robison (✉)
Department of Philosophy, Rochester Institute of Technology, Rochester, New York, USA
e-mail: wlrgsh@rit.edu

D.P. Michelfelder et al. (eds.), *Philosophy and Engineering: Reflections on Practice, Principles and Process*, Philosophy of Engineering and Technology 15,
DOI 10.1007/978-94-007-7762-0_2, 

reasons for and against a particular legal claim. Engineers need none of these skills, but engineers within the various engineering disciplines do need to learn how to brace a building so that it will not succumb to high winds (Morgenstern 1995) or write software that will not fail at a crucial point of execution (*New York Times* 1996; Manes 1996).

Every profession has its failures, some coming from not properly understanding or executing one of its rules of skill. Engineering is no different. "Use the same unit of measurement throughout any single project" is a simple rule, but because a subcontractor used imperial units of measurement while NASA used metric units, the Mars Climate Orbiter came into the Martian atmosphere at the wrong angle and burned up (Orbiter Report 1991). "Take into consideration all the variables" is another rule, somewhat more difficult to follow because it is sometimes not clear what all the variables are. Yet even some obvious variables are sometimes ignored. The Hubble telescope failed to work properly because the engineers forgot to compensate for how zero gravity would affect the curvature of the lens (BBC 2000). These sorts of problems ought to resonate for engineers. No professional is immune from such professional failures and mistakes.

Every engineering artefact – from space orbiters to software – is the result of engineers using sets of rules, the tools of the trade, so to speak – about how to calculate loads, determine the density of materials, measure trajectories, devise a consistent set of commands for a computer, and so on. Kant calls these rules of skill. They tell us how to achieve a particular determinate end. Yet the rules within a profession become so natural to its practitioners, so much second-nature, that those within the profession may not even realize they are following rules. This sort of problem is commonplace. Ask a group which way to turn the knob to open a door, and many will raise a hand, turn it the way they would to open a door, and then report the result: "Clockwise!" "To the right!" Their behavior is so hidden in a habit that giving the answer requires paying attention while replaying the habit.

When such rules are brought to consciousness, they seem purely practical: if you wish to do so-and-so, do such-and-such. Indeed, Kant argues that rules of skill have no moral import. One reason he gives is that they tell us how to achieve a determinate end without regard to whether the end is good or bad.

> The precepts to be followed by a physician in order to cure his patient and by a poisoner in order to bring about certain death are of equal value in so far as each does that which will perfectly accomplish his purpose. (Kant 1969)

But Kant is mistaken – and in two different ways:

- Rules of skill articulate norms and their use, and misuse, can have ethical import. They tell us how we ought to achieve a particular end, and
- when the particular end is itself a good, or necessary to achieve that good end – the health of patients, the safety of an engineering artefact, the defense of an accused – the norms they articulate clearly have ethical import.

We will need to examine some features of rules to put us into a position to understand how they can carry ethical weight and thus, as I shall argue, how ethics enters through them into the core of such professions as engineering.

## 2.2 The Nature of Rules

When I play Monopoly, I throw the dice, pick up my designated piece, and move it along a row of boxes representing properties until I reach the number on the dice – neither more nor less. I make these moves in the order given, as mandated by the rules of Monopoly. The rules preclude any other ordering of these moves, turning what could be a random set of activities or events into a play in a game – a sequenced order.

Rules give coherence as well as order. First, a rule sets the beginning of an activity and an end: I throw the dice first and stop after counting off the correct number of properties on the board. Second, it separates out those features of an activity which "belong together by virtue of the rule, and are set off from other activities which may be accidentally associated with it,..." (Wolff 1963). A tennis player may blink to clear an eye of dust while preparing to serve, for instance, but blinking is not part of serving. The serve is an activity that begins at some point prior to tossing the ball and ends with the follow through after the racket has hit the ball. Anything occurring before or after is not part of the activity specified by the rule, and we ignore what happens during the toss that lacks standing because it is not marked out as part of what it is to serve.

Besides providing coherence and order, rules also constrain us – to stop at stop signs when no traffic is about, for instance, or to open a door by following a set of activities. But when we teach a child how to open a door, we are also liberating the child by showing the thread of effective actions. "Grab the knob, turn it to the right – this way – and then pull. The door will open!" The rule for opening doors describes how to open doors, and also how we ought to open doors. Both constrain us to a certain coherent sequenced order of steps and free us from experimentation whenever we go to open a door.

One benefit of a rule is thus security. If we do what the rule tells us we ought to do, we are as well positioned as we can possibly be to achieve the end the rule is designed to achieve. That is why, when a lawyer is asked to defend a physician against a malpractice suit, the lawyer's first question must be, "Did you follow the standard procedure?" A history of use hones a rule. We have a rule, use it, and discover a problem. So we correct the rule, use it again, find another problem, and correct it again. Eventually we have a rule we can use with the security that comes from knowing it will not have any of the common problems that arise because earlier versions met those problems and the rule was modified to ensure that those problems would not arise. The standard procedure is the standard for a reason, and that is one reason we ought to use the rule. If the physician answers the lawyer's question by saying, "No. I thought I'd try something different," the lawyer knows that he and the physician have a problem. It will be the physician at the dock, not the medical profession. The lawyer can no longer defend the physician by saying, "This physician did what any physician would, and should, have done. You need to sue the profession, not my client."

A physician or an engineer adopting a new rule thus risks a blot on the profession: "A professional would do that?!" But we want professionals to push the envelope of the rules of skill of their profession. We want innovation and do not want professionals stuck doing what has always been done.

Where to draw a line between a new rule that properly builds on old ones and a new rule that puts a professional and the profession at risk is no doubt a delicate issue, determined in part by the details of particular cases. In any event, besides providing a sequenced and coherent order to a set of activities, rules set norms for how we ought to engage in those activities, norms sanctioned by the profession. We would admonish those who wrote software however they wished, or braced a skyscraper with two by four's. These activities have rules for our engagement, and implicit in these admonitions is the normativity that marks all rules. They tell us what we ought to do, and we open ourselves to criticism if we fail to do what we ought to do. But this fourth feature – a sequenced, coherent, normative order – has different aspects we need to distinguish.

## 2.3 Following the Rules

We need to distinguish several ways in which we can fail to follow a rule:

(a) We may use the proper rule, but fail to do all the steps it entails.
(b) We may use the proper rule, but fail to do one or more of the steps properly.
(c) We may use the proper rule, but do one or more of the steps in the wrong order.
(d) We may use the wrong rule.
(e) We may fail to have any rule at all.

In making fudge, for instance, we could just make things up as we go, following no recipe, or somehow use the recipe for chocolate sauce by mistake and wonder why our "fudge" fails to set, or put in vanilla at the beginning rather than the end, or put in too little chocolate, or completely forget the chocolate and wonder why our "fudge" fails to turn brown. And, of course, we can fail in any one of these standard ways in a variety of ways. Think of all the ways we can fail to do a step properly – too much flour, too little, too old, the wrong kind, and so on. These ways of failing are standard ways to make a mistake in trying to follow a rule – as the example of following a recipe is meant to illustrate.

We can find a multitude of examples of each kind of failure in engineering practice, but we shall look at only one, an example of failing to have any rule at all.

Determining everything that went wrong in the 2010 BP oil spill in the Gulf and assessing the relative weights of the various causal factors would be a daunting task, but we know that one contributing factor concerned the rule adopted for ensuring that the well was properly closed off so that it could be "temporarily abandoned," as the oil industry says. The "basic sequence" was laid out in an "Ops Note" that Brian Morel, a BP engineer, sent out at 10:43 a.m. on April 20th, the morning of the blow-out (*Deep water* 2011: 98, 104). The crew responsible for performing the procedure first saw the sequence at a meeting at 11 a.m. We already have one red flag: if this procedure was the standard operating procedure, why would the crew even need to be presented with it – as though it were new? In any event, here is what they saw:

1. Perform a positive-pressure test to test the integrity of the production casing;
2. Run the drill pipe into the well to 8,367 feet (3,300 feet below the mud line);
3. Displace 3,300 feet of mud in the well with seawater, lifting the mud above the BOP and into the riser;
4. Perform a negative-pressure test to assess the integrity of the well and bottom-hole cement job to ensure outside fluids (such as hydrocarbons) are not leaking into the well;
5. Displace the mud in the riser with seawater;
6. Set the surface cement plug at 8,367 feet; and
7. Set the lockdown sleeve.

This rule has seven steps to be performed in sequence. But when we look at its history immediately prior to the blowout, we discover at least one reason why the crew needed to see it.

> 'BP's Macondo team had made numerous changes to the temporary abandonment procedures in the 2 weeks leading up to the April 20 "Ops Note." For example, in its April 12 drilling plan, BP had planned (1) to set the lockdown sleeve before setting the surface cement plug, and (2) to set the surface cement plug in seawater only 6,000 feet below sea level (as opposed to 8,367 feet). The April 12 plan did not include a negative pressure test. On April 14, Morel sent an e-mail entitled "Forward Ops" setting forth a different procedure, which included a negative-pressure test but would require setting the surface cement plug in mud before displacement of the riser with seawater. On April 16, BP sent an Application for Permit to Modify to MMS [the Mineral Management Service] describing a temporary abandonment procedure that was different from the procedure in either the April 12 drilling plan, the April 14 e-mail, or the April 20 "Ops Note." There is no evidence that these changes went through any sort of formal risk assessment or management of change process' (*Deep water* 2011: 104).

We have more red flags. The sequence was different in the April 12th plan, and that plan left out one step. The April 14th plan added that step, but changed the sequence. The April 16th plan was different from both former plans and different from the final plan listed above. No wonder the crew needed to see the procedure. There were four different plans floating about – from the 12th, the 14th, the 16th, and the 20th.

The "basic sequence" lays out a procedure for "temporary abandoning" a well, for locking it down so it can be left without any fear that the pressure of the oil and gases will blow up the well pipe. That is the end to be achieved. The sequence lays out what we are to do to achieve that end. Do it correctly, and we achieve the end. There may indeed be four or more different ways to achieve the same end, and so all the plans may be proper: each may, if followed correctly, succeed in achieving the end in question.

But things do not appear that way. When changes keep getting made in some recipe, or rule, or procedure, we presume that those proposing the changes are still getting clear on what needs to be done. The reason is that we presume reasons for the sequence embodied in a rule. A baserunner is to touch first base before touching second base. A runner who fails to touch first before touching second will be called out. Baseball imposes the rule to ensure that base runners do not cheat by cutting the corners of the diamond and shortening their run. So we would presume that the seven steps in the "basic sequence" were each necessary and that the sequence in which they were to be done mattered. Yet the four different sets of basic procedures laid out from April 12th to the 20th certainly make it appear that the particular order of the procedure of the 20th did not matter and even that at least one of the procedures, the negative-pressure test, was not really necessary. Which was the "correct" rule – if any?

It is not that the engineers failed to follow the proper rule. They failed to have a "basic sequence" in place about how to proceed. It appears they proceeded haphazardly over several days in designing a "basic sequence" – a presumption that would need to be proven, but disconcerting in the extreme if true. Indeed, given the number of changes suggested, we can make no presumption about what rule, if any, the crew followed in trying to lock down the well. Despite the 11 a.m. meeting where the last iteration of the rule was presented, the crew had no time to train themselves to that rule and so had no chance to ensure that every crew member understood how the new iteration differed from the other three – if they were aware of the other versions. We cannot presume therefore that every crew member was using the same version of the rule. Even if they had had the same "rule" in front of them as they worked, we cannot know that they did not change the sequence on the job, on the fly, as the disaster unfolded. The obvious inference to draw from all the changes in the rule is that no one had a fix on what would work; so if things began to go awry in following whatever rule was being used, it would not be unreasonable to improvise.

In brief, with no standard basic sequence to which we can appeal, we can have no idea whether the sequence used was proper or improper. It may be that the April 20th version was the correct way to go, or the version of the 16th, or the 14th, or the 12th – or, indeed, none of those at all. It may be, that is, that no rule was followed in temporally abandoning the well.

We have a second problem here as well. Even if a rule was used, it lacked authority for the engineers. We distinguish between rule-directed activities and articulating a rule, and that distinction allows us to make another distinction between rule-directed activities and activities directed at determining the rules themselves. In creating a rule for some end or in ensuring that any changes in an existing rule determinative of an activity will improve the rule, improve, that is, the likelihood of achieving the particular end, some rule or set of rules needs to be followed. Rules of skill come to have an authority for professionals within a profession, that is, only after these rules have been vetted in some way.

We can appeal to the standard practice to justify what we do only because the practice has become standard – the way things ought to be done. This can happen in at least two ways: a rule may be honed by experience or vetted by some body authorized to examine a rule and assess and approve or disapprove it. It can be tested, that is, or approved, or, better, both.

We can find a paradigm of how rules are honed by experience by looking at the development of common law. A judge decides a case one way, creating a precedent for similar cases, and when new cases arise, that judge or another either appeals to the precedent as settling the matter or modifies the precedent to account for changed circumstances. Through a series of cases, the rule by which the first case was decided is honed to the point where it would take a very unusual situation for the rule to be changed: through its application in a variety of different situations, the rule becomes the standard for how to handle the particular kind of situation to which it applies. That is why a lawyer for a physician accused of malpractice would be delighted for the physician to have followed standard practice. The development of the common law ensures, over time, that changes that handle problems are incorporated into the original rule and, in the best of cases, that the ends to be achieved by the rules are more readily achieved.

We can also imagine a body authorized to examine proposed rules of skill. But, as the report indicates, there is no evidence that the "changes [in the rule for temporarily abandoning a well] went through any sort of formal risk assessment or management of change process."

A rule gains authority within a profession through being honed by experience so that it becomes standard or through its adoption by some body authorized to adopt it. So if the engineers did use one of the variants of the rule articulated between the 12th and the 20th, that rule had no authority: they cannot claim that they were obligated to do what the rule told them to do. Indeed, they lacked the security experience with a rule and vetting of a rule give us: they had no idea, and we can have no idea, whether what they did was adequate in those circumstances, let alone the best thing to do.

## 2.4 How Ethics Enters

So how does ethics enter into this? Rules of skill lay down procedures for achieving particular ends: bake a cake by doing this, throw a spiral pass in a football game by doing that. So a failure to follow a rule of skill will prevent our achieving the end or achieving it as fully as we had hoped. For many rules of skill not achieving the end in question raises no ethical issues. Failing to follow all the steps for baking a cake will result in, well, a half-baked cake – not a good culinary end, but no great ethical problem in normal circumstances. And Kant is right, of course, in saying that rules of skill tell us how to achieve an end independently of whether the end is good or bad. So if the end is bad – poisoning a rival, say – a failure of the culprit to follow the proper rules would be occasion for applause, not moral condemnation.

But clearly some failures have consequences that are so harmful they rise to the level of being morally wrong. The failures in the BP oil spill resulted in 11 deaths and 17 injured as well as significant damage to the ecosystem and economic damage in the billions to the fishing and tourist industries on the Gulf. When a failure involving a rule of skill results in death, we have a moral problem. Leaving out a step in a rule of skill, or taking the steps out of sequence, or failing to follow a rule vetted by past experience and/or approved through a formal process are each morally wrong if, as a result, the proper end is not achieved and significant harm results instead. Engineers have an ethical obligation to use proper rules of skill and to use them properly.

We have examined only one kind of failure here – the failure to have any clear rule. But it does not take imagination to provide examples of the other sorts of failures. We need only examine various engineering failures to see how not having a rule, or not following properly one that is in place, has led to significant enough harm to rise to the level of being morally wrong.

Intentions are irrelevant. All that is relevant is whether an engineer used, or failed to use, the proper rule and used it properly. If my dentist drills completely through a tooth while daydreaming – "Drill, baby, drill!" – it does not matter that he did not intend the harm he caused. What matters is that as a professional, he needs to pay attention to what he is doing. Whatever the intentions of the engineers who approved the epoxy that was used to hold in place the three-ton concrete slabs on the ceiling of the Boston tunnel, they failed to ensure that the slabs would stay in place – despite a warning from a safety officer for the construction company that the slabs were at risk of falling over time (Zezima 2006). We would also need to question why three-ton slabs that might fall were put on the ceiling in the first place and also why the bolts that were used to fasten the slabs were stored outside where they rusted before being put in place. In any event, it is not the intentions of the engineers involved that matter, but the failure to do what they ought to have done to achieve the end they were trying to achieve. If we were to sort through the problems with the Big Dig in Boston, we would find, time and again, failures tied to rules of skill.

In examining the rules of skill of engineering, we are defining the role of morality within engineering. Among all that someone must master to become a professional within a discipline are rules that are essential to the practice of those within that discipline – both because they sort out disciplines one from another, the rules of skill for lawyers being different from those for surgeons, for instance, and because they tell someone within the discipline how to achieve the ends appropriate to the discipline. Ethics thus enters into the heart of a discipline in two ways.

A person must work within the boundaries set by those rules to work as a professional within that discipline. A general practitioner who decides to amputate a leg had better have a very good reason for not calling on a surgeon. A charge of unprofessional conduct will be difficult to rebut otherwise. A patent lawyer could not be acting as a patent lawyer and amputate a leg any more than a dentist could be acting as a dentist in drawing up a will. Lawyers learn how to make out a will, a seemingly simple matter that unfortunately can go wrong in many ways and requires mastery of a complex set of laws and procedures to get right. Surgeons master a variety of instruments for cutting, and a delicacy and sureness of touch is as crucial a set of skills as a mastery of where to cut and in what order and how deeply. Working outside one's area of professional expertise as though it were an area of professional expertise is deceptive and morally wrong. It is also likely to lead to mistakes, of course, but even without the possibility of failure, we are morally wrong to misrepresent ourselves as being professionals of a certain sort when we are not.

In addition to working within the boundaries of our profession, we ought to follow the rules essential to the profession. The rules articulate the norms of the profession – how someone within that professional field ought to achieve an end in question – and so when the failure to use them or to use them properly causes harm, they have ethical import. Ethics thus enters into the heart of every profession, not just engineering, but it certainly enters into engineering where a failure to use a rule of skill of the profession or use one of its rules of skill properly can result in enormous harms. A failure of those sorts is not just a mark of incompetence, but an ethical failure.

But we have so far understated how it is that ethics enters into engineering through its rules of skill. In considering whether the rule for temporarily abandoning a well had been honed by experience or vetted by an authoritative body, we were asking whether the rule was a standing rule of the profession, a rule such as that for converting from the metric to the decimal system, for example: a rule that is a constant in the lives of engineers working in an area where the rule is relevant.

Yet that understanding of the role of a rule of skill misses its most important use in engineering and so misses the most important way in which ethics enters into the core of engineering. The intellectual core of engineering is the solution to design problems of a certain sort. That solution has a form Kant would recognize as a rule of skill: if we wish a bridge to go from this place to that, we must do such-and-such and so-and-so. The end is a determinate particular – this bridge, that vehicle, this switch, that software – and the rule states the conditions for producing that end. In short, the intellectual core of engineering is the creation of rules of skill to solve particular problems. Those creations are constrained by the standing rules of the profession, among other things, but they are also constrained at a minimum by a simple ethical principle: do no unnecessary harm. We would like engineers to solve their design problems with a creative genius that produces a solution which we can all applaud for being so good. But the minimal condition for producing a good solution is producing a solution which causes no unnecessary harm. Whatever our criteria for a solution's being good – whether we appeal to cost, or aesthetics, or any other variable we find in engineering – we will forfeit our claim to have a good solution if it causes unnecessary harm.

## 2.5 Creating Rules of Skill

We distinguish what we do from how we do it. A player may mistakenly pass a ball to an opposing player, but do so with such a deft touch as to provoke admiration. A surgeon may amputate the wrong leg, but do it well. A dentist may fill the correct molar, but do it so poorly it will not last. We can judge both what we do and how we do it, and the two judgments need not coincide. What is right may be done badly, and what is wrong may be done well. Cicero's last words are said to have been, "There is nothing proper about what you are doing, soldier, but do try to kill me properly" (Wikipedia).

If we presume a bell curve of professional competence, we can get a sense of what is at issue here in solving design problems. Suppose your primary care physician recommends surgery to save your life. You do not ask, "Can you recommend the worst surgeon you know?" You do not say, "I'll settle for someone mediocre." The same is true for any professional. An engineering firm that advertises that its engineers are all "pretty much below average" is not going to get many customers. Rather obviously, we prefer the best over the worst – even if what needs to get done gets done by the worst. Why is that?

As we saw, rules of skill are conditional. They tell us that if we are to achieve such-and-such, we must do so-and-so. A design problem calls for the creation of a rule of skill: "This is what you need to do to achieve a solution!" But no design problem determines its solution. No matter how detailed, a statement of the problem to be solved will not necessitate any particular conclusion the way 2+2 necessitates 4. A design solution is a contingent outcome of a design problem, and many solutions are possible for the same problem. Engineers can satisfy all the constraints posed by the problem as well as the constraints of the standing rules and end up with radically different solutions to a design problem. We humans innovate, and so some solutions are significantly better than others.

That is how the bell curve of professional competence plays itself out in engineering, with one engineer producing a brilliant solution to a design problem and another a mediocre one. Both solve the problem, we will assume, but one solution is significantly better in all the ways that engineers measure – easier to use, less expensive to manufacture, longer-lasting, easier to repair, recyclable, and so on. And one value among these standards of measurement is ethical. Any design solution, once realized in an artefact, will have its causal effects, some beneficial, some not. A solution which causes unnecessary harm raises an ethical red flag. It is significantly worse than than a solution which does not cause unnecessary harm.

Cup holders in vehicles, shower controls, door handles – we all have our favorite examples of design solutions that have caused or could easily cause harm. I have been scalded when I accidentally backed into a shower control that jutted out from the wall and moved very freely. At a conference I once attended, participants were unable to get into a lecture room because the door mechanism, as we later discovered, required simultaneously pressing in on two little levers hardly discernible between the two doors, which slid apart far enough to let your fingers in easily to press the levers only after the levers were pressed. We could barely get our fingers into the gap where the levers were, and we could only operate the levers by setting down whatever we were carrying. There was no serious harm on that occasion, but that particular design solution could easily cause harm. We would have been hard pressed to figure out how to open those doors had there been a fire: and even if we had known how to open them, we would have been hard pressed to do so.

We can cause more than several hundred dollars damage to some Cadillacs simply by closing the trunk. The trunk lids are designed to be pushed down to about a foot from where they would latch, where a motor takes over, latching the trunk securely. Push the trunk lid down all the way, as we do for other cars, and we break the mechanism. The trunk will then neither latch nor close, and fixing it requires taking the back seat out of the car to get at the motor. That trunk is an accident waiting to happen – especially because there is no warning on the trunk itself not to treat it the way we treat all other trunks.

We can each no doubt generate our own lists of engineering design solutions that are less than optimal and some that are harmful, but we need to consider how less than optimal design solutions raise ethical questions. They do so in at least three ways that I can only briefly sketch here.

First, an engineer can correctly follow all the rules, but still come up with design solutions that cause unnecessary harm. A part that could be recyclable may be designed in such a way as to make recycling impossible without, say, a huge expense. A part that could be made of something readily available and not at all harmful to us or the environment may be made with something toxic. Car manufacturers made use of over 36 million trunk lights containing mercury prior to 2000, over half of them GM products. The collection and safe disposal of that mercury remains a problem. There is no need to detail the problems having mercury in our environment poses to our health – a presumably unnecessary harm.

Second, professionals have a moral imperative to strive to be the best that they can be. We would find it morally obtuse for budding engineers to say, in response to queries about their life's ambitions, "We want to be mediocre." It seems inevitable that varying talents and drives and places of education will produce differing levels, and we rank those in the professions by how well they do professionally. Some surgeons are better surgeons than others; some engineers are better engineers than others. Yet it is part of the drive of an engineer – one of the animating principles of the profession – to improve things, to ferret out ways to make things work better – more efficiently, simpler, with fewer parts, and so on. Lacking that drive is a character fault, and that criticism carries moral overtones.

Third, each profession serves a social purposes or set of purposes; and the state recognizes and regulates a profession to ensure that the purpose or purposes are properly realized, giving those within the profession a monopoly in return. It sets standards for membership in the profession, requires that individuals meet those standards to become a practicing member of the profession, and can, generally, remove professional certification should a member fail in a significant way to meet those standards. Anyone entering into a profession thus comes into a new set of moral relations – to the state and to others in the profession. One obligation members have is to push the envelope of development, strive always to make things better. Someone satisfied with things as they are – "It works. What's the problem?" – is not going to help the profession achieve the goals for which the state gives it a monopoly. That is not just a practical problem for those in the profession who must work with someone who is not concerned to improve matters, but a moral problem as well because that person is a drag on the profession's achieving the purposes for which the state gives it a monopoly.

These are sketches of arguments that would need to be given at far greater length to be fully persuasive, but in combination with the problems that can arise from less than optimal design solutions, we can see how ethics enters into engineering in this second way – through how engineers provide design solutions which, at a minimum, are not to cause unnecessary harm if they are to be good solutions.

## 2.6 Summary

We are all familiar with engineering successes – roads that survive the rigors of traffic and weather, bridges that seemingly float above deep chasms or over deep water despite high winds, software that works seamlessly. When things go well, we can thank the engineers who fashioned creative rules of skill in response to the design problems they faced and who used properly the standing rules of the profession.

Unfortunately, we are also all familiar with engineering failures – door knobs that stick when they should not, shower controls that move only with great effort, or with too little effort, or not very smoothly. The list is long. Mediocre engineering is one of the banes of modern civilization: we have design solutions that are less than optimal, and, in some cases, solutions which are positively harmful.

It is in these failures that the ethical aspects of all engineering can best be seen. Either a standing rule of skill was not followed or, if followed, not followed properly, or the rule of skill created by an engineer to solve a particular design problem introduced harms that could have been avoided without harming the benefits an alternative solution would bring. These failures raise ethical red flags if only because they produce unnecessary harms. So Kant was mistaken. Rules of skill have ethical weight.

## References


BBC News (2000, February 10). Hubble's painful birth.

*Deep water: The Gulf oil disaster and the future of offshore drilling*. (2011). Washington, DC: National Commission on the BP Deepwater Horizon Spill and Offshore Drilling, p. 104.

Kant, I. (1969). *Foundations of the metaphysics of morals* (trans: Beck, L. W.). Indianapolis: Bobbs-Merrill, p. 32.

Manes, S. (1996, September 17). When trust in "data" is misplaced. *New York Times*.

Mars Climate Orbiter Mishap Investigation Board Phase I Report, 6. (1991). ftp://ftp.hq.nasa.gov/pub/pao/reports/1999/MCO_report.pdf. Accessed 10 November 2011.

Morgenstern, J. (1995, May 29). The fifty-nine-story crisis. *The New Yorker*, pp. 45–53.

New York Times. (1996, August 24). Pilot's wrong keystroke led to crash, airline says.

Wikipedia. http://en.wikipedia.org/wiki/Cicero. Accessed 19 January 2011.

Wolff, R. P. (1963). *Kant's theory of mental activity* (p. 123). Cambridge: Harvard University Press.

Zezima, K. (2006, July 27). Boston papers say memos question tunnel safety. *New York Times*.

# Chapter 3
# Engineering as Performance: An "Experiential Gestalt" for Understanding Engineering

**Rick Evans**

**Abstract** There is a growing interest in exploring engineering practice, especially as it reveals that which might be considered essential or distinctive. However, such an exploration often constructs a dichotomous view that artificially separates science from non-science, the technical from the social; and thereby distorts what engineering actually is and what engineers really do. In this paper, I propose an alternative to that dichotomous view – engineering as performance. Like engineering practice, engineering as performance highlights the everyday activities of engineers, although the focus changes from what is essential or distinctive about those activities to the "performative accomplishment." Consequently, the actual work of engineering and the real performances of engineers can now be viewed as a genuine ensemble that includes both science *and* non-science, the technical *and* the social.

**Keywords** Engineering practice • Performance theory • Communication

## 3.1 Introduction

Engineering practice has long been a topic of interest. More recently, some of that interest has focused on exploring engineering practice with the aim of defining "the [essential] nature of engineering and engineering beliefs, values, and knowledge" (Pawley 2009). The primary motivation seems to be the belief that a better understanding of the nature of engineering will suggest better approaches to teaching engineering, e.g., problem-based/project-centered learning (Sheppard et al. 2009) or "the CDIO approach" (Crawley et al. 2007). In a 2008 study entitled,

R. Evans (✉)
College of Engineering, Cornell University, Ithaca, NY 14865, USA
e-mail: rae27@cornell.edu

D.P. Michelfelder et al. (eds.), *Philosophy and Engineering: Reflections on Practice, Principles and Process*, Philosophy of Engineering and Technology 15,
DOI 10.1007/978-94-007-7762-0_3, 

*Changing the Conversation: Messages for Improving Public Understanding of Engineering*, the National Academy of Engineering offers a few additional motivations: first to attract and retain more young people, especially women and members of underrepresented populations; also to offer those young people as well as the general public a more accurate understanding of engineering and of the professional identities available to engineers; and finally to encourage interaction with and the participation of a general public more informed about what engineering is and what engineers can and should do (National Academy of Engineers 2008).

## 3.2 Engineering Practice: A Dichotomous View

Indeed, Sheppard et al. (2006) in their initial attempt to answer the question, "What is engineering practice?" seem to embrace a dichotomous view of this essence. They claim that "[e]very professional engineer . . . is called on not only to achieve a certain degree of intellectual and technical mastery, but also to acquire a practical wisdom that brings together knowledge and skills that best serve a particular purpose for the good of humanity" (Sheppard et al. 2006). In effect, they attempt to identify (and to some extent describe) the two core elements of engineering practice. On the one hand, engineering practice involves intellectual and technical mastery or knowledge. On the other hand, they acknowledge the relevance of practical wisdom or the necessity of certain so-called skills.

Before I suggest some of the problems that such a dichotomous view creates, I would like to elaborate, very briefly, on the differences that exist between these two elements or sides. Sheppard et al. (2006) propose that as knowledge, engineering practice is specialized. It is knowledge that is both unusual and particular, i.e., available only to engineers. As specialized knowledge, it is dynamic or changing, always becoming more comprehensive, complex, and complete. And as specialized, dynamic knowledge, learning is constant – it becomes a "highly desirable secondary product" (Sheppard et al. 2006). Conversely, as practical wisdom, engineering practice requires skills common to everyone, skills not only generally available, but also discrete or generally available *apart* from engineering. And, since practical wisdom is both common and discrete, the skills are unvarying and therefore generalizable. And finally, because those skills are common, discrete, unvarying, and generalizable, they suggest that practical wisdom can be learned once and for all. In *Educating Engineers*, Sheppard et al. (2009) articulate the relation of these two elements, the two sides of the dichotomy. While "learning how to communicate," "learning to work in teams," or "learning to acquire attitudes of persistence, healthy skepticism, and optimism," and so on are critically important; the primary concern is (and should be) to develop "professionals who are . . . ***technically competent*** [italics my own] because being technically competent today and tomorrow is a natural outcome of the conception of the engineer as professional" (Sheppard et al. 2009).

As I stated above, I believe that such a dichotomous view of engineering practice creates a whole host of problems (those mentioned below are only a few) related to understanding what engineers do and who engineers are. First, it actually misrepresents engineering practice. For example, Sheppard et al., certainly consider communication a skill and the ability to communicate as something common, discrete, unvarying, and generalizable. However, one cannot do engineering, cannot be an engineer, without using language that is scientific, or certainly technical in ways that are established and conventional within the relevant engineering discourse communities. Indeed, one cannot become an engineer unless one enters and becomes a participating member of one or more of those discourse communities (Winsor 1996). Consequently, language use or communication in engineering contexts – just like all the other practices/actions that constitute technical competence – is simultaneously and inextricably technical *and* social. It involves both knowledge *and* skilled action.

Second, such a view potentially re-inscribes longstanding stereotypes associated with engineering and with who can and should be engineers. Lisa Frehill et al. (2009) make reference to various messaging efforts related to the science and technical side of the dichotomy that are simply off-putting, "especially for girls." For example, since many girls (as well as boys) understand or are unfortunately told that "engineering is hard" or that it is only "a great field if a student 'loves' mathematics or science," they then self-select other professional and career directions. Instead, Frehill et al. (2009) maintain that those girls (and boys) should be told about the actual work that engineers do, be presented with a more complete understanding of what that work involves – "the excitement associated with solving problems or working in teams." Again, in engineering practice, solving problems and working in teams are simultaneously and inextricably technical *and* social. Again, they involve both knowledge *and* skilled action.

Lastly, Gary Downey (2005) in an article entitled, "Are Engineers Losing Control of Technology" states that "[e]ducators in chemical engineering around the world are working hard to re-imagine the field in response to rapid technological change." He further suggests that "[r]real concern exists about the possible loss of cohesion and identity for the field and the profession" (Downey 2005). If, as the dichotomous view suggests, technical competence is to be the single, "natural outcome" of engineering education; while, as seems to be the case, technological innovation and what it means to be technically competent is and will continue to change faster than schools and colleges of engineering can respond; how can engineering educators and students of engineering keep pace? Certainly not, according to Downey (2005), by suggesting as this dichotomous view does, that "breadth . . . [while relevant] is supplementary" or that "the human dimensions . . . [are] extraneous." I agree with Downey (2005) when he says that "[w]orking as an engineer would [and should] mean both that one brings engineering technical knowledge . . . and appropriate and sufficient non-technical knowledge" – both knowledge *and* skilled action simultaneously and inextricably to bear in solving human problems, in preparing "students for what has always counted as quality work by the best engineers."

Dichotomies are a distinctive feature of western thought – mind versus body, nature versus humanity, or idealism versus materialism (Prior 2006). However, this dichotomous view of engineering practice as science versus non-science, technical/ technology versus the social provides not only an overly determinative lens through which to see engineering and what engineers actually do, but it also defines only one half of that dichotomy as *real* engineering. Currently, there are a growing number of qualitative and/or ethnographic-like studies that are investigating real world engineering practice. And, rather than highlighting simply the technical competencies per se, they reveal the significance of social relationships within a range of different engineering contexts (Bucciarelli 1994; Downey 1998; Vinck 2003).

In this regard, James Trevelyan is doing some very interesting research. Using interviews and direct observations, he offers an understanding of engineering practice as "technical coordination" (Trevelyan 2007). According to Trevelyan (2007), "[t]he engineers we interviewed devoted little of their attention to hands-on technical work. . . . The evidence showed that engineering work was coordinated and driven by engineers, but the end results were delivered through the hands of other people. The link between engineers and the ultimate production and service delivery was a complex series of social interactions." He claims that such an understanding facilitates the important recognition that "engineering is [both] a technical and a social discipline . . . [and that] the social and technical are inextricably intertwined" (Trevelyan 2007). And, in a later paper and apropos of a more inclusive understanding of engineering practice, he claims that there is a "fundamental misunderstanding" of communication (2009). This misunderstanding is perhaps best illustrated in *Communication Patterns of Engineers* by Carol Tenopir and Donald W. King (2004). Theirs is the dominant yet limited view that communication in engineering is simply "a one way information transfer" (Trevelyan 2009). However, Trevelyan (2009) suggests that such a view belies the "realities of [authentic engineering] practice . . . [and] the means by which complex interactions are sustained." As a sociolinguist, I certainly agree that a one way information transfer understanding of communication seriously lacks both descriptive and explanatory power, and thereby trivializes the role of communication in engineering. More about what communication is and its role in engineering practice later on.

This dichotomous view of engineering practice – potentially emphasizing on the one hand either science and the technical or on the other non-science and the social – strikes me as similar to Bucciarelli's (1994) characterizations of the *savant* and the *utilitarian*. For the savant-like students of engineering practice, technical knowledge is determinate. Whether that knowledge is applied through problem-solving or through design (or some combination of both or other means) matters less than that it is technical and applied in some systematic way because that is the natural outcome of the conception of the engineer as professional. However, for the utilitarian-like student of engineering practice, social process appears to be determinate. While the technical is certainly inextricably intertwined with the social, the emphasis now falls on the communal process, that which seems at least to the experience of engineers to be "uncertain," "ambiguous," [and maybe even] "nonrational" (Bucciarelli 1994). Bucciarelli (and I agree) criticizes both the savant and the utilitarian perspectives as being abstracted from engineering practice itself and more than a little tautological.

In part, what has led me to propose the metaphor of performance as an alternative to practice (itself also a metaphor, by the way) is that the latter seems to maintain the dichotomy of science versus non-science, technical versus social, indeed to privilege science and the technical almost in opposition to non-science and the social. However, I believe understanding engineering as performance will free us from that dichotomy, and allow for a more open-ended investigation, conversation, and reflection on what engineering actually involves and what engineers really do. I believe about engineering and about being an engineer something similar to what Judith Butler believes about gender – that "[it] is in no way a stable identity or a locus of agency from which various acts proceed; rather it is an identity tenuously constituted in time – an identity instituted through a *stylized repetition of acts*" (Butler 1990a). Consequently, I believe that if we can study the "performative accomplishment" that is engineering and that is being (and becoming) an engineer; then, perhaps we can also develop not only a genuine appreciation for all the ways that engineering and engineers can and do make a difference in the world, but for the best ways to prepare them to make that difference (Butler 1990b).

Next, I offer an understanding of performance, "an essentially contested concept" and borrow very eclectically from just a few of the possible fields/disciplines – sociology (Goffman 1959, 1974), anthropology (Turner 1974, 1982), linguistics (Hymes 1974, 1975; Bauman 1977, 1986, 1992), literary and rhetorical studies (Burke 1945), theatre and/or performance studies (Schechner 1977, 2002), even philosophy (Butler 1990a, b) – to describe it. Then, since my particular interest is language use in engineering, I discuss the ways that performance helps us to better understand communication. Communication, in conjunction with other ways of doing in engineering, is an ever varied and variable collection of situated and recurring actions relevant to purpose. Understanding communication in this way not only helps us to appreciate the real role of communication or language use, but by extension the real role of other collections of situated and recurring actions – ethics, aesthetics, politics, culture – all similarly relevant to purpose. Finally, I suggest that the metaphor of performance represents a better "experiential gestalt," or "a structured whole within our experience," one that will allow us to explore the many and various possible constructions of engineering and being an engineer all in terms of ***doing***, ***re-doing***, and ***showing doing*** (Lakoff and Johnson 1980).

## 3.3 Performance: "An Essentially Contested Concept"

Marvin Carlson, in his seminal book, *Performance: A Critical Introduction*, begins his concluding chapter stating that "[s]o much has been written by experts from such a wide range of disciplines, and such a complex web of specialized critical vocabulary has been developed . . . that a newcomer seeking a way into the discussion [about performance] may feel confused and overwhelmed" (Carlson 1996). Certainly, in the limited space that I have to introduce performance, I do not expect to eliminate that confusion. Rather, I intend a simple (and inevitably somewhat simplistic) introduction, attempting to distil from this essentially contested concept a few key ideas that I believe are especially relevant to the understanding of engineering as performance.

Performance or performing is ***doing***; it is ***re-doing***; and it is ***showing doing*** (Carleson 1996; Butler 1990a, b; Schechner 1977). To say that performance is doing emphasizes the importance, indeed, the primacy of action and acting. It highlights someone, often a performer (although sometimes not recognized as such) but quite possibly (and more often) an assembly of performers, who in some context engage(s) in activities associated with some endeavour for some purpose(s). Ultimately, according to Kirshenblatt-Gimblett (1998), "[i]t is about getting something done." Richard Bauman (1992) suggests that there are two general kinds of performances: "aesthetically marked" and "aesthetically neutral." Aesthetically marked performances are heightened modes of action. They are "set up and prepared for in advance" (Bauman 1992). They are temporally and spatially bounded. There is a structured sequencing of actions or an established process. And finally they have the feel of an occasion, an event that is "open to view by an audience and to collective participation" (Bauman 1992). Aesthetically marked performances are also sometimes referred to as "cultural performances" (Kirshenblatt-Gimblett 1998). A formal paper presentation at the Forum on Philosophy, Engineering & Technology (fPET) is such a performance. Aesthetically neutral performances most surely involve actions, but unlike an occasion or event, they are not nearly as scheduled, bounded, or programmed. And, if they have a feel, it is that of the mundane. An aesthetically neutral performance is "all the activity of a given participant on a given occasion which serves to influence in any way any of the other participants" (Bauman 1992). Conversation following from such a presentation over coffee or dinner is an example of an aesthetically neutral performance.

Clearly, we all can imagine different examples, as well as examples in which it would be difficult to separate an aesthetically marked from a neutral performance. Consequently then, it is important that we attend to Richard Schechner's (2002) suggestion that there is actually a continuum of "various kinds of performing" that extends, similar to Bauman's kinds of performances, from the "large-scale public events and rituals . . . to the great and small roles of everyday life." All of which are, to reiterate, about getting something done.

Performance is also a re-doing. A performer who engages in particular activities never does so apart from a history of like activities or the present-in-time conventions that guide them. Rather, that history and those conventions, while they may not wholly determine what practices and activities are possible, certainly provide a conceptual framework that suggests which are appropriate, effective, and even efficacious. Schechner (2002) names re-doing "restored behaviour." Restored behaviours are "routines, habits, and rituals; the recombination of already behaved behaviours" (Schechner 2002). He claims that there are no *new* or *original* performances. There is never a "first time" (Schechner 2002). However, because the activities that make up a performance are never new or original; they are marked, can be identified, and therefore can be "worked on . . . played with, made into something else . . . [even] transformed" (Schechner 2002). Re-doing is acting with an appreciation of the history of past action and of the conventions that direct current action and the understanding that made that history and formed those conventions. Re-doing both allows the rituals of the past and the routines of the present to direct, and yet, allows for variation as well – wandering in doing. So, just as the activities related to giving papers at fPET are always a re-doing – they most certainly involve routines, habits, and rituals – so the activities related to performing within one's particular profession – as an engineer, lawyer, doctor, teacher, and factory worker – are always a re-doing. Indeed the different actions, their histories and conventions are what separate those professions from one another.

Finally, performance is showing doing. Showing doing is a kind of display of our awareness (Kirshenblatt-Gimblett 1998). First, showing doing is a display of our awareness of our own distinctive agency – that certain activities constitute a particular way of doing. Second, it is a display of our awareness that that doing is a re-doing. Showing doing acknowledges the understanding, reveals an appreciation of the history and the conventions related to doing – that certain practices and activities have preceded ours, and that certain other practices and activities surround and are contemporary with and influence ours. And third, showing doing is a display of our awareness of our selves as actors, or better, performers engaged in doing and re-doing. It is a display of our awareness that our identity as a particular kind of performer is constructed and represented through those very activities. I suspect that presenters at fPET are aware of themselves as performers, are aware of how this performance is enacted, and are aware (or at least hopeful) that their performances reveal agency – contribute to getting something done. However, underlying the notion of display is the presumption of an audience for that display, someone else who attends, who through attending to that display in some way participates. The nature of that participation can be various: observational (spectator), experiential (participant), evaluative (critic), and so on. So, while showing doing is a display of one's awareness of doing, re-doing, and through doing and re-doing one's identity; it is also always a display for someone else. That it is a display for someone else makes showing doing reflexive, or it creates the opportunity for those who participate to think about what a particular performance has to do with their professional lives as they choose and continue to choose to live them. It even encourages participants – the often-stated aim of academic conferences like fPET – to explore the extent and the limits of their own awareness.

There is a common misunderstanding of performance that, in turn, might have an impact on how useful performance is in helping us to better represent engineering, what it means to become an engineer, even the teaching and learning of engineering. The misunderstanding is that performance is often thought of as *a mere show, something of a spectacle, a simple demonstration.* Something not really work. Nothing could be further from the truth. At an academic conference, for example, performance is always purposeful. The performers are always doing, re-doing, and showing doing. And their performances are, after all, about getting work done, whatever the work may be.

## 3.4 Engineering as Performance and Communication

Earlier I stated that understanding communication as a one way information transfer (otherwise referred to as the conduit or process model of communication) has neither descriptive nor explanatory power. In fact, while information is typically transferred in communicative interactions, there has been a growing consensus that communication is more, much more. Instead, communication – reading and writing, speaking and talking, or our many other ways of using language – is actually a collection of activities – individual and social actions – that are as foundational, as fundamental to any professional (as well as personal) performance as are any other. Indeed, in the College of Engineering at Cornell University, I have long been advocating for an understanding of communication as action, as always and everywhere situated, as learned through processes of participation, and as sometimes instrumental, representative, and even constitutive of doing the real work of engineering. This way of understanding communication can be generally labeled as the "genre perspective" (Bhatia 2004).

There is a vast literature relating to this genre perspective. Much of it is very interesting and applicable. However, there is one approach that is perhaps immediately relevant to my particular focus on engineering as performance or doing, re-doing and showing doing. It is what Charles Bazerman (1999) calls "the North American approach to genre." According to Bazerman (1999), this "North American approach to genre directs our attention to the typification of rhetorical action – that is, the repeated communicative actions people do with each other, the repeated forms by which they do it, and the interpretive practices by which they recognize what they are doing." In other words, genre refers to those particular actions related to communication that are typically, routinely, and (I would argue) necessarily part of professional work or, more narrowly, part of the work of engineering.

Further, he suggests that this approach also "directs our attention to the historical emergence of. . . [communicative action], the current social organization of communication, and [engineers'] strategic use of [conventionalized] forms to participate in socially organized activities" (Bazerman 1999). In other words, genres have histories and present-in-time conventions that relate to communicative action in context. That history and those conventions provide a scaffolding for engineers' participation as language users, for enacting those genres in ways that are appropriate, effective, and efficacious. Bazerman (1999) goes on to suggest that this approach attunes engineers "to the particularity of [the] processes" [of their participation] . . . by showing [them] how specific texts [examples of particular genres] functionally mediate the socially organized . . . [work] of engineering."

Finally, he concludes that a genre-based . . . approach toward communication or language use in context not only helps engineers develop an understanding of communicative activities necessary for the conduct of their professional work, but also provides them with analytic tools and a framework to recognize and adapt to "the changing genre landscapes that their professional lives will travel across" (Bazerman 1999). In other words, once engineers understand that genres perform particular and necessary actions – literally do engineering work; once they learn to appreciate the history and conventions that inform how that work gets done – can take advantage of the traditional as well as the current scaffolding for doing that work; then those engineers can begin to understand their own agency in the field and identity as a performer – as engineers engaged in doing and re-doing in evolving and new contexts. In addition to understanding themselves as engineers, they are also representing – performing, if you will – themselves as engineers to others.

It is that performance for others that encourages reflexivity, to choose and continue to choose how they might realize engineering through being an engineer.

Clearly, I am claiming communication is performance and that, in ways particular, it is a part of and not apart from the performative accomplishment that is engineering. It is ***doing*** through the genres that engineers use to get things done. It is ***re-doing*** in that all those genres emerge from a history of use in engineering and adhere to conventions that relate the form of that communication to an engineering context. And it is ***showing doing*** in that both doing and re-doing are revealed along with the agency and identity of engineers to others and even to themselves. Communication is not science or non-science, technical or social, knowledge or skill. Rather, communication is action, always and everywhere situated in engineering contexts, learned through processes of participation as an engineer, and sometimes instrumental, representative, and even constitutive of doing the real work of engineering. And, if communication can be so understood – why not then those other things non-science and social – ethics, aesthetics, politics, culture? After all, all can be understood as doing, re-doing and showing doing. Again, to understand engineering as performance, as a performative accomplishment, allows us to consider all of the above as well as science and the technical as as much a part of the real work of engineering as anything else.

## 3.5 Engineering as Performance: An Experiential Gestalt

In their book, *Metaphors We Live By*, George Lakoff and Mark Johnson (1980) dismiss the idea that metaphors are "just a matter of language, and can at best only describe reality." To accept such a point of view is to conflate the study of reality with that of the physical world, in effect to leave out the "human aspects of reality, in particular real perceptions, conceptualizations, motivations, and actions that constitute most of what we experience" (Lakoff and Johnson 1980). Instead, they suggest that metaphors provide an "experiential gestalt," or "a structured whole within our experience" (Lakoff and Johnson 1980). Metaphors help us to find coherence. Metaphors help us to impose meaning, literally, to make sense. However, in a chapter entitled, "New Meaning," Lakoff and Johnson (1980) admit "that new metaphors make sense of our experience in the same way that conventional metaphors do . . . highlighting some things and hiding others." The actual usefulness of a metaphor resides in what it highlights and in what it hides.

Practice and performance both highlight the everyday activities of engineering and engineers. Yet, practice attempts to highlight the distinctive, the essential; and to hide that which seems marginal, not perhaps unnecessary, but certainly ancillary. So science, those activities that are considered technical, that which is considered knowledge – they are engineering. They define what it means to be an engineer. Practice hides those activities that seem marginal, that which is considered not to be science, whatever is considered to be social and to involve so-called skill. Further, in ways that I believe are false and certainly exclusive, practice highlights being technically competent as if being technically competent is the real "natural outcome of the conception of the engineer as professional" (Sheppard et al. 2009).

While it also highlights the everyday activities of engineering and engineers, performance does so as a genuine ensemble. Again, as an ensemble, whatever is doing, re-doing, and showing doing – like communication or ethics or aesthetics or politics or culture – in an engineering context is engineering; and, along with all the other ways of doing, re-doing, and showing doing can define what it means to be an engineer. There is no dichotomy of science versus non-science, of the technical versus the social. Both knowledge and skilled action are united in the ensemble. Performance, however, does hide the essential, that which is engineering and nothing else. The identity of engineering and the identities of engineers, to refer again to Judith Butler, are "in no way a stable;" they are "tenuously constituted in time;" they are a "*stylized repetition of acts*" [italics my own] (Butler 1990a, b). Further, through this identity and/or these identities so constituted or, better, continuously constructed; performance highlights, in ways now more complex and I would argue more inclusive, exactly how engineering and engineers can and do make a difference in the world. The focus changes from what is distinctive about engineering and about the individual engineer to the "performative accomplishment" that is engineering and the actual performances of engineers (Butler 1990a, b). To massage the phrase of Barbara Kirshenblatt-Gimblett – engineering as performance is about the doing and what gets done in order to make a difference in the world!

## References

Bauman, R. (1977). *Verbal art as performance*. Rowley: Newbury House.

Bauman, R. (1986). *Story, performance, event: Contextual studies in oral narrative*. New York: Cambridge University Press.

Bauman, R. (1992). Performance. In R. Bauman (Ed.), *Folklore, cultural performances, and popular entertainments: A communications-centered handbook* (pp. 41–49). New York: Oxford University Press.

Bazerman, C. (1999). Introduction: Changing regularities of genre. *IEEE Transactions on Professional Communication, 42*(1), 1–2.

Bhatia, V. (2004). *Worlds of written discourse: A genre-based view*. New York: Continuum.

Bucciarelli, L. (1994). *Designing engineers*. Cambridge, MA: MIT Press.

Burke, K. (1945). *A grammar of motives*. Berkeley: University of California Press.

Butler, J. (1990a). *Gender trouble: Feminism and the subversion of identity*. New York: Routledge.

Butler, J. (1990b). Performative acts and gender constitution: An essay in phenomenology and feminist theory. In S.-E. Case (Ed.), *Performing feminisms: Feminist critical theory and theatre* (pp. 270–282). Baltimore: The Johns Hopkins University Press.

Carlson, M. (1996). *Performance: A critical introduction*. London: Routledge.

Crawley, E., Malmqvist, J., Ostlund, S., & Brodeur, D. (2007). *Rethinking engineering education: The CDIO approach*. New York: Springer.

Downey, G. (1998). *The machine in me: An anthropologist sits among computer engineers*. New York: Routledge.

Downey, G. (2005). Are engineers losing control of technology: From 'problem solving' to problem definition and solution' in engineering education. *Chemical Engineering Research and Design, 83*(A6), 583–595.

Frehill, L. M., Carolyn Brandi, M., Amanda L., & Frampton, F. (2009). Women in engineering: A review of the 2009 literature. *Society of Women in Engineering Magazine*. http://www.nxtbook.com/nxtbooks/swe/summer10/#/50
Goffman, E. (1959). *The presentation of self in everyday life*. Garden City: Doubleday.
Goffman, E. (1974). *Frame analysis*. Garden City: Doubleday.
Hymes, D. (1974). *Foundations of sociolinguistics: An ethnographic approach*. Philadelphia: University of Pennsylvania Press.
Hymes, D. (1975). Breakthrough into performance. In D. Ben-Amos & K. S. Goldstein (Eds.), *Folklore: Performance and communication* (pp. 11–74). The Hague: Mouton.
Kirshenblatt-Gimblett, B. (1998). *Destination culture: Tourism, museums, and heritage*. Berkeley: University of California Press.
Lakoff, G., & Johnson, M. (1980). *Metaphors we live by*. Chicago: The University Of Chicago Press.
National Academy of Engineering. (2008). *Changing the conversation: Messages for improving public understanding of engineering*. Washington, DC: The National Academy Press.
Pawley, A. L. (2009). Universalized narratives: Patterns in how faculty members define engineering. *Journal of Engineering Education, 98*(4), 309–319.
Prior, P. (2006). A sociocultural theory of writing. In C. MacArthur, S. Graham, & J. Fitzgerald (Eds.), *The handbook of writing research* (pp. 54–66). New York: The Guilford Press.
Schechner, R. (1977). *Essays on performance theory*. New York: Drama Book Specialists.
Schechner, R. (2002). *Performance studies: An introduction*. London: Routledge/Taylor & Francis.
Sheppard, S., Colby, A., Macatanggay, K., & Sullivan, W. (2006). What is engineering practice? *International Journal of Engineering Education, 22*(3), 429–438.
Sheppard, S., Macatangay, K., Colby, A., & Sullivan, W. (2009). *Educating engineers: Designing for the future of the field*. San Francisco: Jossey-Bass (Wiley).
Tenopir, C., & King, D. W. (2004). *Communication patterns of engineers*. Piscataway: IEEE Press.
Trevelyan, J. (2007). Technical coordination in engineering practice. *Journal of Engineering Education, 96*(3), 191–204.
Trevelyan, J. (2009). *Steps toward a better model of engineering practice*. Paper presented at the Research in Engineering Education Symposium, Palm Cove, Australia.
Turner, V. (1974). *Dramas, fields, and metaphors: Symbolic action in human society*. London: Cornell University Press.
Turner, V. (1982). *From ritual to theatre: The human seriousness of play*. New York: Performing Arts Journal Publication.
Vinck, D. (Ed.). (2003). *Everyday engineering: An ethnography of design and innovation*. Cambridge, MA: The MIT Press.
Winsor, D. (1996). *Writing like an engineer: A rhetorical education*. Mahwah: Lawrence Erlbaum.

# Chapter 4
# The Formulation of Engineering Identities: Storytelling as Philosophical Inquiry

**Russell Korte**

**Abstract** Along with the development of knowledge and skills, professionals in the early stages of their careers strive to formulate a sense of their worthiness or fit for a profession. Developing this sense of 'fitting in' involves a process of identification whereby an individual categorizes him or herself and others for purposes of answering the questions: *Who am I?* and *Who are we?* A common means of making sense of one's experiences is to organize experiences into a story or narrative. The self-reflective formulation of one's identity into a narrative draws upon individual and social data to develop a coherent view of oneself, similar to the process of philosophical inquiry, which attempts to create an orderly and coherent account of the world.

**Keywords** Social identity • Narratives • Storytelling • Professional socialization • Identity theory

## 4.1 Introduction

One answer to the question of why people do the things they do can be grounded in the concept of identity, which contends that individuals do what they do, at least partially, because of who they believe they are. Psychology defines personal identity as an internal cognitive construct of the self that is essentially relational and self-referential (Erikson 1968; James 1891/1952; Mischel 2004). Social psychology and sociology conceptualize social identity as an emergent property of a group that is adopted by members, and informs and guides the behaviors of members of the group

---

R. Korte (✉)
School of Education, Colorado State University, Fort Collins, CO, USA
e-mail: russ.korte@colostate.edu

D.P. Michelfelder et al. (eds.), *Philosophy and Engineering: Reflections on Practice, Principles and Process*, Philosophy of Engineering and Technology 15,
DOI 10.1007/978-94-007-7762-0_4, © Springer Science+Business Media Dordrecht 2013

(Abrams and Hogg 1990; Haslam 2004; Stets and Burke 2000). Thus, whether the result of individual or social processes, identity exists as a construction, typically in the form of a narrative, that answers questions about who one is and how one fits into society as a whole (Lawler 2008).

The purpose of this chapter is to briefly review the essence and development of professional identities (a form of social identity) and the narrative methods often used to formulate and represent them. Two key philosophical issues discussed are the conceptualization of identity as a social construction and the relation of narratives to identity. Illustrating this review are statements taken from a study of the experiences of recent graduates of engineering in the early stages of their professional careers explaining what it means to them to be an engineer. Over the course of this study, the researcher interviewed nearly 120 new engineers working in four different organizations. The interviews were recorded and transcribed verbatim, then analyzed according to the qualitative data analysis methods prescribed by Matthew Miles and A. Michael Huberman (1994) and Anselm Strauss and Juliet Corbin (1998). The chapter draws upon a small sampling of these interviews and closes with some thoughts about the implications of identity development and narrative construction for aspiring engineers.

## 4.2 Conceptualizing Identity

Identity is a broad, and somewhat vague term, used widely across different disciplines (Erikson 1968; Stets and Burke 2000). Identity has been defined in various ways focusing on one's psychological orientation, interactional role, or social affiliation (Bastos and Oliveira 2006). The concept appears in psychoanalysis, psychology, education, and sociology, and is tapped in many applied disciplines and professions as a necessary and carefully cultivated facet of a person. Internalizing the appropriate characteristics of the members of a profession is an important process for becoming or identifying oneself as a professional.

The concept of identity is grounded in the notion that people have the capacity for self-reflection, whereby they can perceive themselves as an object (person) separate from others and in comparison to others (Ryle 1949; Stets and Burke 2000). While the veracity of these perceptions is contested, it is generally assumed that conscious and introspective beings cannot help but be aware of and analyze their thoughts (Ryle 1949).

William James (1891/1952) divided the self into three facets: the constituents of the self, the feelings drawn from the self (esteem or despair), and the actions prompted by the self (self-seeking and self-preservation behaviors). He further divided the constituents of the self into four categories: the material self, the social self, the spiritual self, and the Ego. He claimed that an individual had as many social selves as there were relevant social groups that recognized him or her in some manner. James described the strong influence that one's feelings of self exert on other feelings, perceptions, and even physical characteristics. He concluded that our sense of our identity, including all four constituents, coalesces into a more or less unitary concept based on the continuity of and similarities among the constituents and our interpretations of our self-experiences.

Jan Stets and Peter Burke (2000) argued as well for an integrated view of various concepts of identity. Specifically comparing personal identity to social identity, they found that the differences in the concepts of identity derived more from categorical language (labels) rather than conceptual differences. Common views of identity conceptualize this concept as a continuum, which is multi-faceted and variable depending on the context. At one end of the continuum is personal identity and the other is social identity (Hogg et al. 1995; Stets and Burke 2000; Turner and Onorato 1999). Depending on the context or one's focus of attention, the conceptualization of the self can emphasize either the personal or the social qualities of one's identity.

Edward Sapir (1927/1995) reminded us that the distinct categorizations of phenomena as either individual or social are not grounded in the essence of the focal phenomena, but are imposed by the interests of the observer. Proponents of a social identity, by definition, focus on the social environment that motivates people to perceive their interactions as *we do this* in contrast to *I do this* (Haslam 2004). This distinction focuses one of the major philosophical issues debated in the social sciences, i.e., are social phenomena the aggregation of individual phenomena or are they an emergent factor that is not reducible to individuals (Bishop 2007; Rosenberg 2008)? For most practical purposes, Sapir's (1927/1995) comment that it all depends on the perspective of the observer turns the debate from the concept of identity to the important philosophical underlying assumptions about the way phenomena are observed and the conclusions to be drawn from such observations.

In a similar manner, Richard Jenkins (2004) characterized the interdependence and mutual constitution of personal and social identities as a reciprocating process between the individual and the group. He claimed that it is best to conceptualize identity as a process rather than as an entity. Furthermore, he described identity as a dynamic process comprised of a relatively enduring core process (as personal identity) and less durable, peripheral processes (as various social identities). For purposes of explanation, conceptualizing identity as a dynamic, mutually constituted process complements the entitative view of identity as a characteristic of an individual or group.

The rather singular notion of a professional identity belies the complexity and dynamism inherent in identity phenomena at the individual level. Consequently, the professional identity of a person depends, to some extent, on the situation and the relative salience of internal and external categorizations in effect at any particular time (Jenkins 2004). Categorizing oneself as a professional is largely based on a dynamic flux of contextual, social, and personal factors.

### *4.2.1 Self-categorization*

John Turner and Rina Onorato (1999) articulated a theory of self-categorization that emphasized the processes by which individuals developed their social identities. They proposed that individuals had varying opportunities to join a group, opportunities that depended on their personal and perceived readiness and fit for membership in the group, as well as the group's accessibility. Some groups easily accept new members, while other groups resist outsiders. Obviously, individuals cannot join just any group and groups do not admit just anyone—especially professional groups. New members are expected to adopt the norms of the group and construct an identity aligned with its ideals.

Individuals identify with a group for a sense of pride, involvement, stability, and meaning (Hogg and Grieve 1999). The power of social identity varies, but research has found that it is generally more powerful than personal identity (Hogg and McGarty 1990). Hence, the general tendency of people to go along with the group with which they identify.

The transition from a personal identity to a social identity tends to depersonalize the individual in favor of becoming a group member (Tajfel 1981; Turner and Onorato 1999). Depersonalization does not insinuate a negative connotation in the sense of dehumanization, for it is not a loss of personal identity but rather the acquisition and accentuation of an additional identity. Enhancing self-esteem is one of the basic tenets of social identity theory. The benefits of a social identity help individuals reduce uncertainty and lighten their cognitive load through categorizing and stereotyping. On the downside, this tendency to categorize the self and others fosters rigidity, conflict, and prejudice. Research has found that the bias in favor of one's group (favoritism) and the denigration of others in out groups (discrimination) is pervasive, implicit, and easily enacted (Tajfel 1982; Turner et al. 1987) The ubiquity of categorization in society and the dependence of individuals on groups to function continually reinforce the importance of group membership.

### 4.2.2 Limitations of Identity Theory

Despite the utility of identity as a concept for explaining part of the power of groups, there is widespread debate about its formulation. The three most common questions are: what is it, where is it located, and why is it important? First, the difficulties surrounding the definition of identity stem from confusion and crossover with other related concepts in different disciplines. In many ways, it is an issue of semantics. Anthropologists discuss identity as an artefact of culture, sociologists define identity as the set of social roles, and psychologists define identity as a set of norms (Stets and Burke 2000; Hogg et al. 1995). Despite differences in the construct, overall similarities point to the presence of an important concept for understanding cognition and behavior in the social environment. Theorists do not dispute the utility of the concept as much as the details of its construction.

A second controversy related to where identity exists has disciplinary biases. A key philosophical question highlights the reification of group-level phenomena. It is common in discussions of social identity to jump back and forth between individual and group levels of analysis. Much of this debate focuses on the transferability of the concept between levels of individual and group phenomena. Some theorists locate social identity in the individual, and others construct a supraindividual entity out of the group (Jenkins 2004; Tajfel 1981). Etienne Wenger (1998) discounts this debate as unproductive, claiming that the interaction between the individual and the group is the important point. And, as mentioned earlier, Sapir (1927/1995) locates this debate in the interests of the observer, not the subjects or situation.

Another controversy is the theory's disconnection between explanation and prediction. Social identity theory makes coherent explanations of past individual behavior in social settings from which it is difficult to predict future behavior (Hogg and McGarty 1990). This difficulty in predicting human behavior is not exclusively the weakness of social identity theory, but a characteristic of the social sciences in general.

Identity has become a popular lens to view individual and social phenomena. There seems to be little debate about the existence of a socially influenced identity, i.e., a socially influenced answer to the question of who am I or who are we. Realizing that the concept of social identity is not crystal clear, the following section discusses one of the primary means used to analyze the concept of social or professional identity. Constructing and analyzing narratives is a common method for investigating elusive, interpretive social concepts—a method that depends on important, and contested, epistemological and ontological assumptions.

## 4.3 Narratives Representing a Process of Philosophical Reasoning in the Formulation of Identities

Philosophy is one of the means we have for trying to understand the universe and our place in it (Solomon and Higgins 1996). Typically, this quest for understanding is left for the 'big questions' (Rescher 2010), the ones that science has not answered yet—or cannot answer (Rosenberg 2008). While the answers to these big questions might seem quite distant from the practical concerns of everyday living, there are direct links between the abstract questions concerning our understanding of the world and the ways we live our lives—even if we rarely make the links explicit. Today, there are hotly debated issues about the nature and value of knowledge (Bishop 2007; Rosenberg 2008; Solomon and Higgins 1996). How we make sense of these issues forms much of our worldview and the belief systems we use to guide our lives.

Philosophy addresses the need we have to make sense out of the complex, chaotic, and incoherent experiences we encounter throughout our lives. Thus, the aim of philosophical reasoning is to develop a more consistent, coherent understanding of reality by systematically estimating this reality from the data and information available at the time—data in the form of the facts afforded by science, experts, and authorities; the lessons from history; our everyday experiences; common sense beliefs; and the wisdom of our culture. Given the limitations of our cognitive and rational abilities, and the shortcomings of our data, our efforts to make sense of our experiences can only strive for the best available answer rather than the best answer

(Rescher 2001, 2010). This sense-making process is an imperfect, dynamic process that is continually co-constructed between others and ourselves.

Through reflection we use the data available to us and attempt to construct coherent systems of beliefs that make sense of our lives and formulate our identities. Nicholas Rescher (2001) identified three questions addressed through the process of philosophical reasoning and rational reflection: (a) informative questions related to determining what is the situation, (b) practical questions related to figuring out how to do things or achieve one's aims, and (c) evaluative questions related to deciding what aims to pursue. In this process of reasoning, John Dewey (1938) stated that we start with doubt and end with belief or knowledge. The tension that arises from inhabiting a doubtful situation is reduced by the formulation of beliefs or knowledge. This process of the formulation of beliefs and knowledge is what Dewey calls rational or logical. And he emphasizes that initially, a process or means that is typically considered rational or logical is not so *a priori*, but becomes generally accepted as a rational process out of habit and because it achieves a status over time as a warranted means to achieve desired ends.

The rational construction of a coherent system of beliefs, or narrative, often seems to be a tenuous bridge between objectivity and subjectivity. The use of narratives as sources of data in the social sciences often encounters skepticism or criticism because of biases against the subjective or interpretive nature of qualitative data. Typically, these biases come from those holding to a more positivist or objectivist view of the world (Abell 2004; Bishop 2007; Polkinghorne 1988; Rosenberg 2008; Searle 1995). A closer interrogation of these biases will find them grounded in the philosophical debates focused on the relation of the social sciences to the natural sciences—and indeed on the philosophical questions grappling with the nature and role of science in general. Presenting the depth of arguments on both sides of this issue is definitely beyond the scope of this chapter, however there is compelling evidence that the social world, at least in part, is irreducible to the physical world, as we comprehend it.

An important task of philosophy then, is the systematic formulation of a coherent view of the world that can serve as a guide. From a vast array of experiences, thoughts, beliefs, ideas, and habits one formulates such a view of the world by identifying, evaluating, and synthesizing ideas into a more or less coherent system (Rescher 2001, 2010; Searle 1995). Developing a coherent understanding of reality is the aim of philosophical reasoning and of the ongoing formulation of one's life story and identity.

In constructing a coherent system of beliefs, people most often rely on developing such a system in the form of a story or narrative (Fisher 1989). People tend to characterize their lives as stories that unfold over time (Bishop 2007). Life stories express a person's sense of self—who one is and how one got that way. Constructing narratives is an ongoing activity of making meaning or making sense out of chaos, and strives for a level of coherence and logic that is culturally mandated (Linde 1993). Donald Polkinghorne (1988) identified three suppositions about narratives: (a) that human experience was embedded in personal and cultural meanings and thoughts, (b) that human experience is cognitively constructed from the interaction between personal schema and the influence of the external environment, and (c) that human experience is organized around poetic meaning, not the technical logic used in the natural sciences.

Specifically regarding the formulation of identities as narratives, people piece together a more or less orderly sequence or story from the rather chaotic mix of experiences and narratives they encounter in living. This ordered sequence is a key characteristic of narratives known as the plot. The narrator's 'emplotment' of events into a story is a cognitive process that directs, and is directed by, the structure of the narrative in a mutually constitutive manner. The emplotment process begins with the conscious naming of experiences and the identification of the relationships between the experiences (Linde 1993; Polkinghorne 1988). Which events the individual selects and how they become related depends on the relative influences of individual and cultural factors.

Narrators strive for coherence based on rhetorical logics that differ from formal or technical logics. Rhetorical logics follow some of the strictures of formal logics, but differ significantly in that they are contingent, context-bound, and based on probability (Fisher 1989). Narrators strive to make their life stories coherent, something that will strongly influence how the narrator interprets experiences. In addition to the importance of the plot for narrative construction, narrative coherence is the "structural glue" that gives meaning and sense to lived experiences (Bamberg et al. 2007, p. 5).

Regarding the belief systems that individuals construct as narratives, Rescher (2001) claimed that some are better than others and identified four dimensions for evaluating them: (a) *contextual coherence* (an interpretation that fits within a larger context), (b) *comprehensiveness* (the broader the scope of an interpretation the smaller the number of plausible, competing interpretations), (c) *sophistication* (the more substantial an interpretation, the more complex and ramified it becomes), and (d) *imperfectability* (an interpretation is only plausible up to a point because simplicity and plausibility are negatively correlated).

These narratives also serve as important means by which one communicates and negotiates one's sense of self with others (Ayometzi 2007; Linde 1993; Polkinghorne 1988). The life story is also used to make a claim for membership in a group, as well as to negotiate the conditions of membership in a group (Linde 1993). Becoming a member of a group often means identifying with and adopting the 'master narrative' of the group (Ayometzi 2007). This is obvious in many of the master narratives recounted by members of a profession in response to the question: *What do you do?*

## 4.4 Formulating an Engineering Identity: Adopting the 'Master Narrative'

For early career engineers, the formulation of a professional identity is a form of narrative development involving, at best, the rational reflection upon their status as novice engineers in the workplace. The developing professional identities of new engineers are reflected in the narratives they construct regarding who they are and how they fit in to the profession and their work.

In the field of engineering, some of the big philosophical questions concern the juxtaposition of the natural and social worlds. Larry Bucciarelli (1994) claimed that engineering effectiveness requires the ability to operate in two different worlds: the object world and the social world. Engineering work in the object world is typically reductionist—based on the application of scientific principles (e.g., the principles of mathematics and the natural sciences). By comparison, the work in the social world is based on communication, negotiation, and consensus building (Bucciarelli 1994; Korte et al. 2008; Trevelyan 2010). Bent Flyvbjerg (2001) characterized these two worlds as complementary and mutually exclusive. From the stories told by new engineers, it is apparent that many of them grapple with the tensions between the technical and social worlds. Some deliberatively choose to identify more with the technical side of the profession as shown by the following statement.

> Technical things are more interesting to me than people issues. But when I turn older maybe things will change. I have supervised people in the past, and you know, it's just… At the moment that's not what I'm interested in. I can probably be a program leader and work in that scope, but not a manager or… I'm not striving to become a supervisor right away. I'm working on my technical niche and become the expert.

For others, formulating a coherent synthesis of these two worlds is an important task in their work. For example, as one novice engineer explained:

> It's [engineering] very people dependent. Certain people in the group that I'm working with right now do not like me telling them what I thought they should do. They like me to more present it as a suggestion so they can figure it out or they can do the approval on it. … and vice versa, some of the people you'll kind of have to say to them—I think this is how you should do it because my data shows that this is how you should do it. So that it really depends on how people are. So nowadays I'm more likely to suggest the things than to say things.

The typical model of engineering taught to students in schools is criticized by some as over-emphasizing the natural scientific perspective (the object world) at the expense of the social scientific perspective, yet again and again the experiences of engineers in practice recount the predominance of social influences on their work (Korte 2009; Trevelyan 2010).

By narrating their experiences new engineers transformed a relatively disorganized set of experiences into a more meaningful and coherent series of events (Abell 2004; Ochs and Capps 1996). The following statement reported by an early career engineer shows that individuals not only make sense of the past and present, but construct plausible futures with which they identify as well.

> I guess eventually I have to figure out a personality, which basically represents [*company*], it's like—yeah, this is a [*company*] person. He is talking like a [*company*] person. This is probably where I need to get at. I'm not sure. That's what I figure out. It's like—okay what is a typical [*company*] person? Even if somebody is really doing really, really great, what is different about that person? I don't think it's the technical expertise. It's more than that or it's something else. Actually I take this as a challenge for me, because I know there are a lot of things at stake.

Another example told the story of the move from outsider (a contract person) to insider (company employee) and the change in identity entailed in that move.

> Well, when I started as a contract engineer, you know, your badge is different, it's yellow, I don't know why they chose yellow, but… I always felt like an outsider. … Nobody ever said that to me—well, you're an outsider or whatever. But carrying the yellow badge,

everybody else has got a blue badge, you definitely felt outside. I think it's supposed to be that way. I remember that as soon as I got the blue badge [*company badge*] we were in a meeting with one of the guys from advanced development, a big scientific person there, and I was expressing an opinion. When I stood up, he goes—oh, you're an employee now. So I don't know if he was filtering that through something, like—here's the contract guys from outside. ... When I was a contractor I felt sometimes like people thought I was a salesman for the software, that there are competing products in this industry. And so when I presented results, you know, and there a was question on those results, when I had the yellow badge sometimes I felt like a salesman, like, you know—our product, this is... I have to uphold the integrity of the product. Whereas now I use a variety of tools and nobody ever—I don't think anybody ever questions unless they have some preconceived idea about which one's better. I don't feel like I have to be a salesman, I just have to do what I'm supposed to do. That's all contract stuff and that's all behind, but I was never made to feel overtly like I'm second-class. I think it was just more—I got the yellow badge [*contractor badge*] and so I'm different. I mean that's probably just all personal.

The social environments from which new engineers draw their data heavily influence the series of events that describe and connect these stories. Their backgrounds prepared them to reasonably interpret technical data, but they found it more difficult to integrate the social data from their experiences. In fact, some stated that they had learned to be suspicious of social data suspecting that it was invalid and irrational—it was 'noise' in the system.

In construction you always have the contractor that's going to yell at you, but it's never personal. You know, it's like, don't yell at me. I don't care, you know. Yelling at me is not going to solve the problem. Come on, get past the yelling. OK, we're past the yelling. Good job.

The categorization of social experiences, however extreme or mundane, as interference is one of the vestiges of the logico-scientific view that has come to characterize engineering. This view only makes it more difficult for new engineers to practice their profession in a more coherent manner.

Integrating the disparate and often conflicting aspects of professional reality is aided by reasoning through the inconsistencies and incoherencies one encounters through experience (Rescher 2001). John Searle (1995) described a more integrated view of social reality as a mix of institutional facts (those requiring human agency for their existence) and brute facts (those that exist independently of human interpretation). Working out a reasonable synthesis of this combination of facts affects our views of reality, our identities, and how we perceive and consequently make sense of our lives.

## 4.5 Conclusions

The view that the world appears to people and is interpreted and understood by them as a narrative is the foundation of efforts to better understand the meaning that people make of their lives (Abell 2004; Bishop 2007; Czarniawska 2004; Ochs and Capps 1996; Searle 1995). Thus for engineers and their work, an informative analysis of their development might be found by perceiving engineering as a narrative that is often interpreted and understood in the form of a story. As such, the narrative becomes a useful tool encompassing a broader scope of the institutional or social world in which engineers work. Peter Abell (2004) and others supported narrative as a means of sense making (Ochs and Capps 1996; Polkinghorne 1988) and the sense that people make of their experiences informs their identities and guides their future actions in a mutually constitutive manner.

The narratives reported by new engineers generally began within the technical realm of work—grounded in the sciences and a systematic way of thinking about problem solving. Over time, this boundary might expand to include the social dynamics among the set of actors with whom they interacted in their professional settings. Organizational, industrial, and societal factors also appeared in their stories as they expanded and redefined what it meant to them to be an engineer. Thus, engineering became a more complex and more socially integrated profession for some of these individuals, and the stories they constructed to make sense of their experiences reflected this development. Others clung more tightly to the scientific paradigm that characterized the social world as a disturbance in the system.

The development of coherent, rational, and logical narratives that guide people in their work is an important foundation of developing well-rounded professionals. Using a narrative perspective for understanding engineers and engineering in organizations helps deepen our understanding of the practice of engineering. Furthermore, it helps inform and refine the institutional definitions and practices of engineering education.

The language of engineering is not math, as some contend, but the language of engineering is language (Goldberg et al. 2010). And the philosophy of language and linguistics, while certainly not conclusive in its contention that language controls thought, makes a strong case for the power of language to influence thought. This also relates to our beliefs in the objectivity of science and the real possibility that language and master narratives (paradigms) configure our notions and beliefs about science and math, as well as our personal and professional identities. The attempts to include more of the social world into engineering bumps up against paradigmatic obstacles that relegate the social and language phenomena to a lesser status. The insights of linguistics, philosophy, and narrative analysis indicate that providing more attention to and emphasis on narrative development in the education of engineers will better prepare students for the world in which they work and live.

**Acknowledgments** An earlier version of this paper was presented at the Forum on Philosophy, Engineering, and Technology (fPET-2010), Colorado School of Mines, Golden, CO, USA, May 9–10, 2010. Support came from the Center for the Advancement of Engineering Education, funded by NSF grant no. ESI-0227558, and the Department of Management Science and Engineering at Stanford University.

## References

Abell, P. (2004). Narrative explanation: An alternative to variable-centered explanation? *Annual Review of Sociology, 30*, 287–310.

Abrams, D., & Hogg, M. A. (1990). An introduction to the social identity approach. In D. Abrams & M. A. Hogg (Eds.), *Social identity theory: Constructive and critical advances* (pp. 1–9). New York: Harvester Wheatsheaf.

Ayometzi, C. C. (2007). Storying as becoming: Identity through the telling of conversion. In M. Bamberg, A. De Fina, & D. Schiffrin (Eds.), *Selves and identities in narrative and discourse* (pp. 41–70). Amsterdam: John Benjamins.
Bamberg, M., Anna De Fina, A., & Deborah Schiffrin, D. (2007). Introduction to the volume. In M. Bamberg, A. De Fina, & D. Schiffrin (Eds.), *Selves and identities in narrative and discourse* (pp. 1–8). Amsterdam: John Benjamins.
Bastos, L. C., & de Oliveira, M. C. L. (2006). Identity and personal/institutional relations: People and tragedy in a health insurance customer service. In A. de Fina, D. Schiffrin, & M. Bamberg (Eds.), *Discourse and identity* (pp. 188–212). Cambridge: Cambridge University Press.
Bishop, R. C. (2007). *The philosophy of the social sciences: An introduction*. London: Continuum.
Bucciarelli, L. L. (1994). *Designing engineers*. Cambridge, MA: The MIT Press.
Czarniawska, B. (2004). *Narratives in social science research*. London: Sage.
Dewey, J. (1938). *Logic: The theory of inquiry*. New York: Henry Holt.
Erikson, E. H. (1968). *Identity: Youth and crisis*. New York: W. W. Norton.
Fisher, W. R. (1989). *Human communication as narration: Toward a philosophy of reason, value, and action*. Columbia: University of South Carolina Press.
Flyvbjerg, B. (2001). *Making social science matter: Why social inquiry fails and how it can succeed again*. Cambridge: Cambridge University Press.
Goldberg, D. E., Somerville, M., Kerns, S. E., & Korte, R. (2010). A war of words: The role of language in transforming engineering education. Special session. *Frontiers in Education Conference, 2010,* Washington, DC.
Haslam, S. A. (2004). *Psychology in organizations: The social identity approach*. London: Sage.
Hogg, M. A., & Grieve, P. (1999). Social identity theory and the crisis of confidence in social psychology: A commentary, and some research on uncertainty reduction. *Asian Journal of Social Psychology, 2*, 79–93.
Hogg, M. A., & McGarty, C. (1990). Self-categorization and social identity. In A. Dominic & M. A. Hogg (Eds.), *Social identity theory: Constructive and critical advances* (pp. 10–27). New York: Harvester Wheatsheaf.
Hogg, M. A., Terry, D. J., & White, K. M. (1995). A tale of two theories: A critical comparison of identity theory with social identity theory. *Social Psychology Quarterly, 58*(4), 255–269.
James, W. (1981). The consciousness of self (chapter X). In R. M. Hutchins (Ed.), *Principles of psychology* (Great books of the Western world) (pp. 188–259). Chicago: Encyclopaedia Britannica (Original work published 1952).
Jenkins, R. (2004). *Social identity* (2nd ed.). London: Routledge.
Korte, R. F. (2009). How newcomers learn the social norms of an organization through relationships: A case study of the socialization of newly hired engineers. *Human Resource Development Quarterly, 20*(3), 285–306.
Korte, R. F., Sheppard, S., & Jordan W. C. (2008). *A study of the early work experiences of recent graduates in engineering*. In Proceedings of the American Society for Engineering Education Conference, 2008, Pittsburgh, PA.
Lawler, S. (2008). *Identity: Sociological perspectives*. Cambridge: Polity Press.
Linde, C. (1993). *Life stories: The creation of coherence*. New York: Oxford University Press.
Miles, M., & Michael Huberman, A. (1994). *Qualitative data analysis* (2nd ed.). Thousand Oaks: Sage.
Mischel, W. (2004). Toward an integrative science of the person. *Annual Review of Psychology, 55*, 1–22.
Ochs, E., & Capps, L. (1996). Narrating the self. *Annual Review of Anthropology, 25*, 19–43.
Polkinghorne, D. E. (1988). *Narrative knowing and the human sciences*. Albany: State University of New York Press.
Rescher, N. (2001). *Philosophical reasoning: A study in the methodology of philosophizing*. Malden: Blackwell.
Rescher, N. (2010). *Philosophical inquiries: An introduction to problems of philosophy*. Pittsburgh: University of Pittsburgh Press.
Rosenberg, A. (2008). *Philosophy of social science* (3rd ed.). Boulder: Westview Press.
Ryle, G. (1949). *The concept of mind*. London: Hutchinson & Company.

Sapir, E. (1995). The unconscious patterning of behavior in society. In: Blount, B. G. (Ed.), *Language, culture, and society: A book of readings* (2nd ed., pp. 29–42). Prospect Heights: Waveland Press. (Reprinted with permission from P. Sapir, Editor-in-Chief, *The collected works of Edward Sapir.* Berlin/New York: Walter de Gruyter. First published in *The unconscious: A symposium*, by E. S. Drummer (Ed.), 1927, New York: Alfred A. Knopf)

Searle, J. R. (1995). *The construction of social reality*. New York: The Free Press.

Solomon, R. C., & Higgins, K. M. (1996). *A short history of philosophy*. New York: Oxford University Press.

Stets, J. E., & Burke, P. J. (2000). Identity theory and social identity theory. *Social Psychology Quarterly, 63*(2), 224–237.

Strauss, A., & Corbin, J. (1998). *Basics of qualitative research: Techniques and procedures for developing grounded theory* (2nd ed.). Thousand Oaks: Sage.

Tajfel, H. (1981). *Human groups and social categories: Studies in social psychology*. Cambridge: Cambridge University Press.

Tajfel, H. (1982). Social psychology of intergroup relations. *Annual Review of Psychology, 33*, 1–39.

Trevelyan, J. (2010). Reconstructing engineering from practice. *Engineering Studies, 2*(3), 175–196.

Turner, J. C., & Onorato, R. S. (1999). Social identity, personality, and the self-concept: A self-categorization perspective. In T. R. Tyler, R. M. Kramer, & O. P. John (Eds.), *The psychology of the social self* (pp. 11–46). Mahwah: Lawrence Erlbaum.

Turner, J. C., Hogg, M. A., Oakes, P. J., Reicher, S. D., & Wetherell, M. S. (1987). *Rediscovering the social group: A self-categorization theory*. Oxford: Basil Blackwell.

Wenger, E. (1998). *Communities of practice: Learning, meaning, and identity*. Cambridge: Cambridge University Press.

# Chapter 5
# Ove Arup: Theoretical and Moral Positions in Practice and the Origins of an Engineering Firm

**Andrew Chilvers and Sarah Bell**

**Abstract** Founded by Sir Ove Arup in 1946, Arup is one of the largest global engineering consultancies offering design services for the built environment. Throughout his career Sir Ove continually reflected on his practice and its role in producing more or less socially robust urban environments. Analysis of documents from his personal and professional archive provides a case study of a practice-based engineer-philosopher. Sir Ove's writings and reflections develop the central elements of his 'Total Design' philosophy: a philosophy that can be characterized as an engineering philosophy of technology as defined by Carl Mitcham (1994), based on an instrumentalist understanding of the nature of technology (Feenberg 2002). Through this case study we see how an influential engineer addressed issues of engineering method, the purpose of engineering, and its role in society, and also developed a framework for the translation of values into practice in engineering.

**Keywords** Sir Ove Arup • Philosophy of engineering • Values • Reflective practice • Organizational discourse

## 5.1 Introduction

Most engineering design for the built environment takes place in large firms, positioned between architects, urban designers and planners who conceptualize buildings and spaces, and construction contractors who build them. Engineering design

A. Chilvers
Department of Science, Technology, Engineering, and Public Policy,
University College London, Gower Street, London, WC1E 6BT, UK
e-mail: a.chilvers@ucl.ac.uk

S. Bell (✉)
Department of Civil, Environmental and Geomatic Engineering, University College London,
Chadwick Building, Gower Street, London, WC1E 6BT, UK
e-mail: s.bell@ucl.ac.uk

D.P. Michelfelder et al. (eds.), *Philosophy and Engineering: Reflections on Practice, Principles and Process*, Philosophy of Engineering and Technology 15,
DOI 10.1007/978-94-007-7762-0_5, 

mediates between creative, scientific, technical, political and practical interests in shaping the built environment. Engineers' own conceptualizations of this role have important implications for understanding economic, social and environmental change in modern societies.

Sir Ove Arup is an important figure in twentieth century British engineering, best known as the founder of the firm which now bears his name. Arup is a global consultancy whose core business is providing engineering design services for buildings, infrastructure and urban development. Throughout his career and his leadership of the firm, Sir Ove recorded his reflections on the role of engineering in society and how to achieve good design in practice. His thoughts were shaped by his experience as an engineer working within the industrial and artistic networks that constituted the built environment of post-war Britain, and were underpinned by his early education in philosophy.

Sir Ove dealt with many conventional engineering considerations for achieving quality design. He was strongly influenced by modernist viewpoints and an instrumentalist conception of science and technology, and maintained a strong interest in incorporating art and aesthetics into structural and urban design. His leadership of the firm focused on the integration of knowledge (both technical and conceptual) across the boundaries within the construction industry. Towards the end of his career he was compelled to articulate his ideas to the growing firm which was structured according to his understanding of the aims and means of good design.

This chapter maps the issues of concern to Sir Ove, a practice-based engineering-philosopher. The case study is intended to illustrate some of the moral, theoretical, organisational and personal concerns of engineers. We characterise Sir Ove's reflections as an example of what Carl Mitcham (1994) has defined as the "engineering philosophy of technology – or analyses of technology from within, and oriented towards, an understanding of the technological way of being-in-the-world as paradigmatic for other kinds of thought and action" (p. 39). We show that Sir Ove's analysis of technology conforms to the instrumentalist view, which Andrew Feenberg (2002) identifies as consistent with dominant policy and engineering approaches. Sir Ove's instrumentalist view of technology does not correspond to an instrumentalist view of engineering. The Arup case also shows how large, modern-day engineering consultancies are underpinned by specific theoretical and moral perspectives.

This chapter begins with an introduction of the core analytical concepts derived from Mitcham (1994) and Feenberg (2002) – engineering philosophy of technology and instrumentalist theory of technology respectively. We then provide a brief biography of Ove Arup before analyzing his speeches and writing in terms of his thoughts on technology and morality, the structure of the building industry, his theory of Total Design, and the 'Aims and Means' of the firm he founded. This material is based on a document archive held at the Arup's London headquarters, which includes papers, conference proceedings, speeches, lectures and addresses, interviews, notes, doodles and other memorabilia. The material analyzed spans a 41 year period of Ove Arup's career from 1942 (just before he set up his own firm) to 1983 (5 years before his death). We conclude by drawing attention to the contribution of practice based engineering-philosophy in understanding the complex relationships between values, technology and society.

## 5.2 Considering Philosophical Positions

Ove Arup's practice-based engineering philosophy is consistent with analyses of philosophies of technology by Carl Mitcham (1994) and Andrew Feenberg (2002). Whilst Sir Ove's contribution to the profession was innovative, it can also be shown to be consistent with accepted understandings of the role of technology in liberal progress, and a tradition of engineering analysis of technology from within.

Mitcham (1994) divides philosophies of technology into two broad categories; 'Engineering Philosophy of Technology (EPT)' and 'Humanities Philosophy of Technology (HPT)'. EPT describes any "attempt by technologists or engineers to elaborate a technological philosophy" (p. 17). EPT is philosophy of technology from 'within' and is pre-conditioned towards a pro-technology stance, often proceeding first with an analysis of the nature of technology – its concepts, methods, cognitive structures etc – and then seeking to explain further aspects of human experience or affairs in these terms (Mitcham 1994). HPT represents "effort by scholars from the humanities...to take technology seriously as a theme for disciplined reflection" (p. 17) and provides a more expansive framework, tending towards more critical accounts of technology and its relation to other aspects of human experience such as art, literature, ethics and politics. Mitcham argues for the primacy of HPT on the basis that humanist aspects of engineering are usually taken for granted in EPT, which "is only one kind of questioning and can itself be questioned" (p. 140).

Feenberg's (2002) schema distinguishes between instrumental and substantive theories of technology. Instrumental theories treat technology as "subservient to values established in other social spheres" (p. 5), and are associated with liberal faith in progress. Substantive theories claim that "what the very employment of technology does to humanity and nature is more consequential than its ostensible goals" (p. 5), and are associated with more critical perspectives, including calls for a retreat to more traditional forms of society.

Engineering theories of technology are most commonly associated with an instrumental viewpoint. Technology is conceived of as tools that engender a universal rationality which is sociopolitically indifferent (i.e. neutrally serving human ends) and hence transferable across every social context. Feenberg shows that such a view focuses discourse on the notion of 'trade-offs' and boundaries. The technical sphere can be limited but not transformed in character by nontechnical values. Since there is a universal rationality underpinning technology, this point of view limits questions to those regarding what extent technological efficiencies should be traded off against culturally mediated considerations such as environmental, ethical or religious ones (Feenberg 2002).

Positioning the reflections of Sir Ove as engineering-philosophy grounded in an instrumentalist view of technology provides a starting point for analyzing his specific concerns with the organization of the construction industry and the role of values in shaping his firm. What follows demonstrates how these broad characterizations of engineering philosophy are enacted in the specific concerns of one of the twentieth century's leading engineers.

## 5.3 Ove Arup and the Firm

Born in England in 1895, Arup took his first degree in philosophy and mathematics before studying engineering, specializing in structures. As a graduate Arup developed an interest in reinforced concrete and joined a specialist contractor in this field, Christiani and Nielsen, designing and constructing structures such as quay walls, bridges, silos, water towers and coal bunkers. Despite becoming chief designer of the firm's London branch, he grew frustrated by the contractor's limited scope for developing new ideas for concrete (Arup 1969a).

Arup became increasingly inspired by the pioneering architects of the Modern Movement such as Walter Gropius and Le Corbusier, who shared a commitment to the functional use of structural materials and an enthusiasm for engineering. Arup's willingness to explore emerging ideas meant that his collaboration as a structural engineer was welcomed. Motivated by this, Arup entered J. L. Kier & Co as a director of designs and tenders. He also joined the Architectural Association and the Modern Architectural Research (MARS) Group, a think tank for modernism in British architecture and began a long association with Tecton, the architectural partnership founded in 1932 by Berthold Lubetkin. With Tecton, he completed works such as the blocks of flats known as "Highpoint I and II" in Highgate, London, the Gorilla House and award winning Penguin Pool at London Zoo, flats for low income families, and the first examples of 'box-frame' construction in Britain (Jones 2006).

In 1946, again seeking increased freedom to provide engineering solutions for the Modern Movement, Arup set up 'Ove N. Arup, Consulting Engineers', which has been known simply as 'Arup' since 2000 (Arup 1969a). As an engineer, Arup is perhaps best known for his work with architect Jorn Utzon on the Sydney Opera House (detailed in Jones 2006).

## 5.4 Technology and Morality

For Arup, making the benefits available from scientific and technological advances through engineering was an inherently moral undertaking. He was vocal in emphasising the imperative of wide and participatory deliberation (to include engineers and scientists) on what the benefits of technology should be and how they should be administered. This call was based on a wholly instrumental definition of engineering as utilising technology to bring natural forces and resources to human advancement, consistent with Robert Treadgold's early definition of Civil Engineering in nineteenth century England (Mitcham 1991). Along with new capabilities stemming from the technological revolution that allowed humans to win their "battle with nature", came a moral responsibility to properly administer the "conquered territory" (Arup 1970a, p. 391).

> ...this is not a technical problem at all. It is not even mainly a problem of organisation... The difficulty is rather one of getting agreement as to what benefit to humanity means ... It becomes therefore a moral or social or political problem.
>
> (Arup 1942, p. 57).

This call for scientists and technicians "as citizens with a social conscience" (Arup 1942, p. 57) to resolve the social problem of agreement on aims is in line with Feenberg's description of the manifestations of instrumental theory. Arup consistently maintains a division between the technical sphere in which a clearly articulated aim can be achieved through rational means, and social and political spheres in which inherently irrational aims must be considered:

> ...to decide what to do next invariably involves value judgments, ethical and aesthetic considerations, and an understanding of human aspirations and behaviours – all of which cannot be logically deduced.
>
> (Arup 1981, p. 1)

Arup calls for scientists and engineers to engage with the arts and humanities in order to contribute to and enliven social and political debates, not to extend their analyses to bear on them. In this regard Arup refrains from an imposition of technological principles to these arenas as one might expect an Engineer-Philosopher to advocate. He does however maintain a seemingly unproblematic relation between aims as defined by such spheres and their rational realisation through engineering; he does not consider, as a substantivist might, that to realise a humanitarian aim through technological means might itself entail a further substantive shaping of either the technology itself or the social context.

## 5.5 The Structure of the Building Industry

In establishing and practicing within his own firm, Arup situated his moral and theoretical concerns within the wider building industry of the time, focusing on three critical themes throughout his career: the architect-engineer divide; divisions between briefing, design and construction; and the limits to the specialization of knowledge.

### *5.5.1 The Architect-Engineer Divide*

Arup was closely aligned with the artistic and functional ideals of modernist architecture, and saw the longstanding division between architect and engineer as outmoded. Rather, he saw two equally valuable perspectives on any one whole design. He envisioned a balanced synthesis of the architect's concern with human reactions to form and space, and the engineer's emphasis on conquering natural forces in a rational way with the aid of science and technology.

In practice, a deep division was embodied in the industry by firms who split themselves between builders working for architects and engineering contractors working for engineers. Arup lamented esoteric practices that reinforced this divide, beginning within professional education. An emphasis on quality and architectural theories in architectural schools neglected the important technical aspects of how to translate these values into real buildings, whilst:

> ...the natural tendency of a designer to care for the appearance of what he creates was actually thwarted rather than encouraged in the education of engineers...
>
> (Arup 1970a, p. 394)

Again, Arup's instrumentalist treatment of this problem focused on trade-offs; an architectural understanding without engineering conceives of buildings and spaces without any regard for the implied trade-offs in efficiencies in structure and method of construction (Arup 1956). Conversely, optimised efficiency does not appropriately prioritise human goals of architectural delight and humane design (Arup 1972a). Arup's 'synthesis' is best thought of as achieving the most appropriate trade-offs between architectural concerns and engineering efficiencies given the human goals. Feenberg (2002) again sensitises us to alternative substantive perspectives that might point to fundamental cultural tensions where the engineering method is applied to the creation of quality spaces for human experience, for which Arup's philosophy does not account.

### 5.5.2 Divisions Between Briefing, Designing and Construction

Arup objected to the rules and norms surrounding a persistent division between design (assigned to the architect and consulting engineers) and construction (the domain of the contractor who was absent from design). Again he argued that constraints on design undermined efficiency and quality:

> You cannot create designs for which the technical and constructional facilities do not exist, yet on the other hand no contractor is interested in creating facilities which are not yet called for by design...The architectural design is very largely the special interpretation of the client's wishes. The client himself does not really know what he wants before the architect has put pencil to paper and has shown the client what could be done ... wise decisions can only be based on a knowledge of facts, and this means that the technical adviser should be brought into the business...at an early stage. It is essential for economy that the design takes into account the method of construction as well as the final structure.
>
> (Arup 1956, p. 2)

The means of construction embody particular knowledge, which must be integrated with the very first architectural design concepts. The transfer of this knowledge was a key problem. Clients were reluctant to collaborate in initial design stages that led to design briefs which were meant to articulate aims, preventing quality design (Arup 1972b, p. 3). Integrating construction considerations into the design briefing process would impose intellectual rigor on architects' responses to briefs, requiring them to "rationalize their purely whimsical predilections by reference to function or structural honesty" (Arup 1954, p. 29).

Arup called for design to become an interactive process involving both client and contractor. The client should formulate their brief alongside an exploration of design possibilities with the designer, and the designer should be closely informed by the contractor's knowledge of construction possibilities, processes and costs.

This might also benefit the technology development process since it was typically down to contractors to develop new plant technology and construction techniques, and they derived their obligation for this from building designs (Arup 1965). Designs thus determined the technological development agenda for new plant and construction techniques. If design activity were more closely informed by construction possibilities, then the development of efficient technology and technique would itself become much more efficient. In Arup's view, the cultural objectives manifested in design briefs might define the character of technological means. That is, these means should be responsive to the human aims of technology expressed through design objectives.

This is how Arup arrived at the view that the design stage must permeate the building process with client, architect, engineer and contractor collaborating together. As the realisation of technical benefits for humanity was a moral imperative, so too was achieving this integration.

### *5.5.3 Specialization and the Limits to Knowledge*

A further barrier to the synthesis of design-pertinent knowledge across the industry was the specialisation resulting from scientific and technological advances. The ever broadening body of knowledge and technique was causing ever greater specialisation in all areas of the industry with no one group covering a wide enough field to discern all design information from often bewildering possibilities. Specialisation was necessary to deal with problems in a manageable way, but for Arup the danger was to forget the connections "so ruthlessly severed" (Arup 1970a, p. 391). Arup's characterisation of specialised views on any design correspond well with Bucciarelli's (2002) 'object worlds' which explain the different knowledge, values and languages of specialists in the design process. These again presented a barrier to the 'synthesis' that Arup sought between quality, form, and safe and efficient functionality.

Arup maintained that while any problem of design could be broken down into specialised parts only the whole or the totality of the parts expressed the ultimate aim, which was both "dream and action" (Arup 1969b, p. 514). In an industry where no individual or group covered a wide enough field to discern all design information, the creation of what he termed the 'composite mind' was key.

## 5.6 Total Design

### *5.6.1 The Total Design Ideal*

Arup's reflections are rich with detail on his efforts and experiments to develop his collaborative, 'composite mind' alternative to the fragmented approach he typically encountered. For the built environment this was 'Total Architecture'; more generally the term used was 'Total Design'.

> The term 'Total Architecture' implies that all relevant design decisions have been considered together and have been integrated into a whole by a well organised team empowered to fix priorities.
>
> (Arup 1970b, p. 1)

A design was the sum of all the decisions recorded and communicated in the form of drawings, sketches, models, prototypes and so on, covering all the facts that needed to be known and processes that needed to be gone through to achieve the aims that had been collaboratively explored. In line with his criticisms of current practices, this had to occur across:

- design perspectives (between both architectural and engineering disciplines *and* emerging sub-specialisms therein); and
- client/designer and designer/builder boundaries.

Arup freely recognized that integrated planning and design of this sort for the whole human environment was sufficiently lofty an aim never to be achieved, nevertheless he still explicitly stated this as the Total Design ideal (Arup 1970a). In any case, if he and his colleagues strived to find what was needed for the best possible result in any single case, then what applied to one entity might well apply to most, as the need for proper integration of parts was a feature of all design (Arup 1970a). Thus, experience gained in working towards any ('locally bounded') total design was valuable for extrapolation to large scales of built environment (Arup 1970a).

This then was Total Design as a moral goal: the instrumental integration of high level aims with the most economical and effective means, which should ideally be extended to all scales of human-mediated environments. In this rationalisation of parts and whole, to be achieved through scientific and engineering method, and partnered with the proposed extension indefinitely across scales, we can see a firming up of Arup's ideas for urban design. We can also see Arup's instrumentalist conceptions of science, technology and design being extrapolated in a way that starts to parallel the tendencies noted by Mitcham (1994) within EPT traditions of thought.

At other times, Arup tackled aspects of what it is to be human, as when for instance he reflected on the nature of 'delight' fostered by architecture, indicating a wider scheme of thought that is conventionally associated with EPT. Furthermore, Arup never denied the social and political complexity of obtaining agreement about the desired character of the 'whole' to be achieved. Whilst he did not devise a sophisticated philosophy of the nature of technology and its implications for humanity, his conceptions about what it means to be human in a technological age underpinned his leadership of a large engineering practice and his formulation of principles for good design in the built environment.

### 5.6.2 Total Design in Practice; Implications for the Firm

The organisational form of Total Design could only mean one thing; achieving committed collaboration and teamwork from the earliest possible stage between the client, the architect, the engineer and the contractor. The expansion of the boundaries of design teams and the overall firm to include other engineering disciplines was essential. Eventually, when the opportunity arose, architecture was also included within the growing 'Arup Group' with the establishment of Arup Associates. Only this approach could eliminate the barriers to quality design presented by the division of practices and responsibilities between architectural and engineering roles, between briefing, designing and constructing processes and between increasingly specialised expert groups.

With his colleagues, Arup sought to experiment with team arrangements for such collaboration. In an address to trustees, Arup compares two approaches to achieving Total Architecture. 'Answer A' involved "small multi-disciplinary teams with stable membership who get to know each other intimately and shed their sectional prejudices" (Arup 1973, p. 2). 'Answer B' consisted of separate, mono-disciplinary firms specialized in a portion of the design and co-ordinated by a project leader with an overarching view of the design, traditionally the architect (Arup 1973). He concluded that "To generalise about the organisation of the team is, however, quite impossible" (Arup 1970a, p. 396). Rather, the firm needed to develop the capability to deliver both approaches to design. In Arup's view, this partly meant continued but carefully considered expansion –

> We are then led to the ideal of 'Total Architecture', in collaboration with other like minded firms or, still better, on our own. This means expanding our field of activity into adjoining fields – architecture, planning, ground engineering, environmental engineering, computer programming, etc. and the planning and organisation of the work on site.
>
> It is not the wish to expand, but the quest for quality which has brought us to this position.
>
> (Arup 1970b, p. 1)

The move by an engineering consultancy to establish an architectural practice received criticism from architectural circles and concerns from Arup members who were worried about alienation of their existing collaborators. Arup's reflection on this again makes it clear that Total Architecture was always to be central:

> ...our ideological commitment – if I may call it that – was to Architecture, and that meant Total Architecture, not just aesthetics. It was not to the architectural profession as such. And we knew that working as structural consultants only, our opportunity to pursue the ideal of Total Architecture would be severely limited. By working with our own architects who shared our ideas we would perhaps be able to make progress towards complete integration...
>
> (Arup 1972c, p. 13)

## 5.7 Aims and Means

Expansion to cover a wide range of specialist knowledge was not in itself synonymous with quality work. The Total Design model also necessitated a particular culture and set of attitudes, and eventually Arup and his partners became concerned over the impact of rapid growth on the core 'Arup values'. Collaboration and the appropriate fixing of priorities, Arup reflected, came only from mutual trust and respect for, understanding of and sympathy toward the work and perspectives of others. As the firm grew in terms of the specialisms and geography covered, Arup was prompted by his partners to make these attitudes explicit.

In the early 1970s Arup delivered a series of organisational addresses to the firm entitled 'Aims and Means', which led to the formulation and delivery of what became known as 'The Key Speech'. It reflects the challenge of devising an organisational form and culture around his Total Design ideal, as well as the usual management concerns associated with running a large and growing organisation. The moral tone is notable:

> By creating a model fraternity, so to speak, we make a contribution to what is almost the central problem of our time: how to overcome social friction and strife… We could become a small scale experiment in how to live and work happily together. This would also have a profound influence on the quality of our work.
>
> (Arup 1969b, p. 514)

Arup explains his continual reference to aims, ideals and moral principles:

> … I do this simply because I think these aims are very important. I can't see the point in having such a large firm with offices all over the world unless there is something which binds us together. If we were just ordinary consulting engineers carrying on business… to make a comfortable living, I can't see why each office couldn't carry on, on its own… unless we feel that we have a special contribution to make which our very size and diversity and our whole outlook can help to achieve, I for one am not interested.
>
> (Arup 1970b, p. 3)

Arup also makes a particular point of de-emphasising the importance of profit. This became embodied most clearly in when the firm was transferred into trust ownership on behalf of its employees in 1977. This was a considered decision to give the staff maximum freedom from short-term commercial pressures in the pursuit of the long-term integration of high level aims (Jones 2006).

This structure of the firm reflects what Michael Davis (1998) characterizes as an "engineer-oriented" company, as distinct from those that are "customer-oriented" and "finance-oriented". Engineer-oriented companies are distinguished by their "general agreement that quality is the primary consideration (or rather the primary consideration after safety)" (p. 133). For such organizations quality in design and construction is placed centrally with profit-making as an enabling condition rather than a primary objective.

Since Sir Ove Arup's death in February 1988, the firm has continued its geographic and disciplinary expansion. A copy of The Key Speech is given to every new employee which, in the preamble, states that the firm is still committed to the principles outlined within it, including Total Design, and that it is required reading for anyone who wants to know what the firm is "all about" (Arup Ltd in Arup 1970b, p. 1). With more than 10 000 staff in 37 different countries, it now includes engineering and related professionals working on all elements of building and infrastructure design, including; planning, economics, architecture, and project and management consultants, as well as a raft of technical specialists. The firm has contributed engineering design services for structures that include the Sydney Opera House, the Oresund link joining Denmark and Sweden, the Channel Tunnel Rail Link project connecting London to the Channel Tunnel which links England and France and, more recently in China, the 'Birdnest' stadium and 'Watercube' aquatics centre for the 2008 Beijing Olympics.

## 5.8 Conclusions

The firm that Sir Ove Arup established in 1946 has become a significant international consultancy providing a range of engineering design and related services. The extent to which this success can be attributed to Sir Ove's philosophy of design and his engagement with social and moral issues is a matter for further debate and exploration. The figure of Sir Ove, his 'Key Speech' and his theory of Total Design remain prominent in Arup's offices and are well known by Arup staff, but the degree to which his values and ideals are translated into everyday practice in the global context of the firm deserves further investigation and this work has been taken up elsewhere (see Chilvers 2013 and Chilvers and Bell 2013). These conclusions inevitably follow from our analysis of Sir Ove's writings, but do not detract from our primary aim, which has been to explore the work of a practice-based engineer-philosopher in light of fundamental categories of analysis in recent philosophy of technology.

Our purpose has been to analyze the particular issues that Sir Ove engaged with as a practice-based engineer-philosopher. The analysis shows some of the contextual influences on his thinking and provides insight into the organizational issues which underpin the practice of design for the built environment. A key part of this has been the utilisation of categories available for the consideration of moral and philosophical positions in order to foreground specific views against their broader alternatives.

Mitcham (1994) notes that the field of Philosophy of Technology, from which his categories of EPT and HPT emerge, is not well-defined, rather it engages with almost the full scope of heterogeneous problems traditionally of concern to philosophy, often with sharply contrasting aims and methods. To seek confluences in the ideas of one individual with those wholly positioned within one or other of these categories would be difficult and most likely unhelpful. This is especially so when dealing with practice-based thinkers whose positions are often not formally developed.

We have, however, described Arup's specific moral position and instrumentalist view on the nature and purpose of science and technological design in relation to social and political aims. Ultimately, Mitcham's work acts mainly to highlight the limits to the formal development of Arup's philosophical position when compared to other thinkers. This touches on areas associated with both EPT and HPT, but manifests itself most strongly in an organisational undertaking. Feenberg's (2002) work shows us more specifically that Arup's instrumentalism omits the possibilities raised by alternatives, which hold that "values of a specific social system and the interests of its ruling classes are installed in the very design of rational procedures and machines even before these are assigned specific goals" (p. 15).

Conditioned by his theoretical stance Arup developed a values-led agenda which focused on mitigating social contingencies impinging on design. This shaped his leadership and organisation of his firm and, at least in part, contributed to its 'engineer-led', quality-focused character through a particular model of (total) design. Ultimately Arup shows us that engineers often bring a complex mix of moral and theoretical perspectives, usually not formally expressed, to bear on their purpose and action. These can play an important role in how they individually and collectively define and orientate themselves around the challenge of achieving their design aims for human environments within the constraints and allowances of the socio-technical contexts in which they operate.

**Acknowledgement** This research was undertaken as part of Andrew Chilvers' Engineering Doctorate studies, funded by the UK Engineering and Physical Sciences Research Council and Arup.

## References

Arup, O. (1942). *Science and world planning*. Comments delivered to the British Association for the Advancement of Science's conference on science and world planning on 27 July. Unpublished (Held at the Arup Library, London, UK).

Arup, O. (1954). Structural "honesty": A lecture delivered to the Architectural Association of London, March 25th. *Irish Architect and Contractor, 4*(9), 25–30.

Arup, O. (1956). *The importance of design*. Speech delivered in Nairobi on October 26th. Unpublished (Held at the Arup Library, London, UK).

Arup, O. (1965). The problem of producing quality in building: Talk given to the Westminster Chamber of Commerce on November 27th. Unpublished (Held at the Arup Library, London, UK).

Arup, O. (1969a). Maitland lecture: The world of the structural engineer. *The Structural Engineer, 47*(1), 3–12.

Arup, O. (1969b). Aims and means: Part 1. In P. Hoggett (Ed.), *Arup newsletter no. 37* (pp. 513–515). London: Ove Arup and Partners Consulting Engineers and Arup Associates Architects and Engineers.

Arup, O. (1970a). Architects, engineers and builders; The Alfred Bossom Lecture By Ove Arup, CBE, FICE, FIStructE, delivered to the Society on Wednesday 11th March 1970, with Lord Holford, RA, FRIBA, MTPI, in the Chair. *Journal of the Royal Society of Arts, 118*, 390–401.

Arup, O. (1970b). *The key speech*. http://www.arup.com/Publications/The_Key_Speech.aspx. Accessed 9 October 2009.

Arup, O. (1972a). The built environment. *The Arup Journal, 7*(4), 2–7.

Arup, O. (1972b). Future problems facing the designer. *The Arup Journal, 7*(1), 2–4.

Arup, O. (1972c). *Aims and means continued*. Unpublished (Held at the Arup Library, London, UK).

Arup, O. (1973). *What's our line?* Transcript of opening address delivered at The Arup Partnerships Trustee Meeting 30th October. Unpublished (Held at the Arup Library, London, UK).

Arup, O. (1981). Thinking and getting things done. In P. Hoggett (Ed.), *Arup newsletter no. 124* (pp. 1–2). London: Ove Arup Partnership.

Bucciarelli, L. (2002). Between thought and object in engineering design. *Design Studies, 23*(3), 219–231.

Chilvers, A. (2013). *Engineers and values: Ethnographic studies of the normative shaping of engineering practice*. EngD thesis, University College London, London.

Chilvers, A., & Bell, S. (2013). Professional lock-in: Structural engineers, architects and the disconnect between discourse and practice. In B. Williams, J. Figueiredo, & J. Trevelyan (Eds.), *Engineering practice in a global context: Understanding the technical and the social*. Leiden: CRC Press/Balkema.

Davis, M. (1998). *Thinking like an engineer: Studies in the ethics of a profession*. Oxford: Oxford University Press.

Feenberg, A. (2002). *Transforming technology: A critical theory revisited*. New York: Oxford University Press.

Jones, P. (2006). *Ove Arup: Masterbuilder of the twentieth century*. London: Yale University Press.

Mitcham, C. (1991). Engineering as productive activity: Philosophical remarks. In P. T. Durbin (Ed.), *Critical perspectives on nonacademic science and engineering*. London: Associated University Press.

Mitcham, C. (1994). *Thinking through technology: The path between engineering and philosophy*. London: The University of Chicago Press.

# Chapter 6
# Transferable Skills Development in Engineering Students: Analysis of Service-Learning Impact

**Donna M. Rizzo, Mandar M. Dewoolkar, and Nancy J. Hayden**

**Abstract** The practice of engineering, especially the design process, involves many aspects beyond just the technical and includes such critical components as engineering ethics, sustainability and transferable skills such as communication, leadership and mentoring. Engineering educators often struggle with how to best incorporate these nontechnical aspects within their curricula. Service learning offers an opportunity to do this. The disconnect is that students often view engineering as only the technical number crunching and these other nontechnical components as less important. We report on the assessment of student written reflections across two very different service-learning engineering design projects for the purpose of evaluating student attitudes about these service-learning experiences and to assess their awareness and appreciation of transferable-skills development. In the spirit of service-learning pedagogy, we divided the contents of the written reflections into three categories – academic enhancement, civic engagement and personal growth skills. The commonality across both courses centered on academic enhancements and the value of transferable skills (*i.e.*, leadership, teamwork, negotiation skills, mentoring, scheduling, verbal and written communication skills). Assessments show our current service-learning pedagogy improves students' understanding of the importance of written and oral presentation skills. However, as of yet, many students do not consider leadership, negotiation skills, design setbacks, scheduling and mentoring skills to be part of "real" engineering.

**Keywords** Engineering design course assessment • Student reflection assessment • Service-learning impact • Transferable skills development • Biomimicry

---

D.M. Rizzo (✉) • M.M. Dewoolkar • N.J. Hayden
School of Engineering, University of Vermont, 301 Votey Hall,
33 Colchester Ave, Burlington, VT 05405, USA
e-mail: drizzo@uvm.edu; mandar@uvm.edu; nhayden@uvm.edu

D.P. Michelfelder et al. (eds.), *Philosophy and Engineering: Reflections on Practice, Principles and Process*, Philosophy of Engineering and Technology 15,
DOI 10.1007/978-94-007-7762-0_6, 

## 6.1 Introduction

A National Science Foundation Department Level Reform grant was awarded to the civil and environmental engineering programs at the University of Vermont in 2005. The overall goal was to incorporate a systems approach (e.g., systems thinking, systems analysis, dynamic systems modeling) throughout our two ABET (Accreditation Board for Engineering and Technology)-accredited B.S. civil and B.S. environmental engineering programs. A systems approach is defined here as one that challenges the engineering profession to incorporate the long-term social, environmental, and economic factors into the context of sustainable engineering designs for the purpose of preparing students to become leaders in their chosen field who can think long-term and better anticipate the co-products or unintended consequences associated with engineered solutions.

A large component of most engineering curricula includes a stepwise reductionist approach to problem solving (i.e., only one correct answer to the problem). Although many engineering programs include open-ended (i.e. more than one solution to the problem) projects and capstone design courses, the focus is all too often on the technical aspects of the problem and its solution. To that end, we focused our reform on educating the whole engineer with special emphasis on the nontechnical areas (e.g. ethics, personal/interpersonal skills, leadership, and teamwork). Engineering educators often struggle with how to best incorporate these nontechnical aspects within their curricula. Since the civil and environmental engineering profession is largely service oriented, we opted to incorporate service-learning projects into key required courses throughout the curricula as a means of practicing civic engagement, social and sustainability awareness, and enhancing teamwork and other personal/interpersonal skills, henceforth called *transferable skills*. One of our motivations was to use these service-learning courses to instill the importance of transferrable skills as well as their practice, as we did not believe engineering students valued these skills or understood their importance in real-life engineering practice. Our previous assessments showed that after 3–4 years students appeared to understand the importance of written and oral presentation skills (Hayden et al. 2011). However, most students do not (yet) regard *other* important transferrable skills (e.g., leadership, teamwork, negotiation skills mentoring, and scheduling meetings) as important to their future engineering endeavors, and more surprisingly, part of "real" engineering.

## 6.2 Motivation

Our Department Level Reform grant was motivated, in part, by numerous reports and papers written over the past 10 years on engineering education for the twenty-first century that focus on the importance of transferable skills development (e.g. National Academy of Engineering (NAE) 2004, 2005; National Research Council 2005;

National Science Board 2007; Duderstadt 2008; American Society of Civil Engineers (ASCE) 2006, 2008) and which promote inclusion of sustainable practices, a systems approach (Forrester 1958, 1961; Wolstenholme 1990), and inquiry-based service learning in engineering curricula. As part of our effort, service-learning projects were incorporated into eight of our required and elective civil and environmental engineering courses, as a way of practicing a systems approach both for engineering problem solving and engineering practice. These ideas are well aligned with recent initiatives at the University of Vermont (*e.g.*, service learning, the university's environmental mission, and the Office of Sustainability). Reform details may be found in (Hayden et al. 2011; Dewoolkar et al. 2009a, b; Lathem et al. 2011) and on our website: www.uvm.edu/~sysedcee. For reasons discussed in Sect. 6.4, we compare and contrast student service-learning reflections for two of these eight courses: a junior level Modeling Environmental and Transportation Systems (CE 134) and the Senior Capstone Design (CE 175). Although our original hypothesis might best be stated as: "There will be similar levels of acceptance with respect to service learning across both engineering courses", the students that really disliked the CE 134 design course (*i.e.*, ~13–16 % who rated their service-learning experience and willingness to volunteer for similar mentoring experience as unsatisfactory or poor) negatively affected classroom dynamics in such a manner that we crafted an alternative hypothesis: "Students do not value all types of transferrable skills equally."

## 6.3 Background

Service learning is a teaching and learning strategy that involves connecting community partners with university students and faculty to engage in meaningful and wanted community activities, such that both groups benefit and share in a transformative learning experience. Written critical reflections are a key component of a service-learning experience (Jacoby 1996; McCarthy 1996; Moffat and Decker 2000; Ash and Clayton 2004; Collier and Williams 2005). By reflecting on their experiences, students connect the service experience to the course content and direct attention toward a personal interpretation designed to promote deeper understanding and meaning (Bringle and Hatcher 1999/2000; Bringle et al. 1996, 2001). These reflections can be open or guided and may include in-class discussions, journals, written papers/reports, and oral presentations, among others. Guided questions allow students to address specific issues involved in the project, develop solution ideas, and work through personal feelings and relationships, as well as other aspects of the project. The reflections may also be used by instructors to keep abreast of the projects and student progress and experiences. Kezar and Rhoads (2001) identify several questions related to the implementation and assessment of service-learning projects in higher education and assert that these *dynamic tensions* (*e.g.*, philosophical tensions that currently exist within institutions trying to implement service-learning programs into their cultures) are inevitable. These questions were useful in our understanding of student attitudes toward the service-learning reform and in contemplating changes in the classroom (Hayden et al. 2011).

In this paper, we report on the assessment of the student reflections to help evaluate student attitudes about these interdisciplinary service-learning experiences: what worked well and where we had difficulties, as well as students' understanding of the importance of the nontechnical aspects of the engineering profession as part of their engineering education (*i.e.*, a systems approach to engineering education including the development of transferable skills).

Over the course of this 4-year reform, we used a variety of formative and summative assessment methods to gauge student understanding and attitudes including student surveys (*e.g.*, attitude, first-year experience, senior exit, focus groups, faculty and student interviews, and assessment of student learning throughout our reform (see Hayden et al. 2011 for more detail). The vertical integration of service learning into our curricula and the development of the service-learning research projects are discussed in more detail in (Dewoolkar et al. 2009a, b). And a complete description of the mixed method longitudinal study, including initial data and analyses of student attitudes about the roles and responsibility of engineers, is presented in (Lathem et al. 2009, 2011). Qualitative data from critical reflections are sometimes misconstrued as anecdotal evidence, but in actuality can provide additional insights into student attitudes and understanding the quantitative results. The bulk of the assessment provided in this manuscript has been extracted from the students' weekly and end-of-semester critical reflections. Specifically, we assess their appreciation of and development of transferable-skills within and across two required civil and environmental engineering undergraduate courses.

### 6.3.1 Course Development

We focused on the service-learning projects in these two design courses for a number of reasons. Both service-learning projects addressed open-ended design problems and were intended to enhance students' *academic development*, *civic engagement*, and reinforce their *transferable* skills (personal/interpersonal, teamwork, leadership and mentoring skills). The small number of students in each course (n = 31 in the junior-level systems course and n = 30 in the senior capstone course) made it possible to provide students and community members with a meaningful service-learning experience, as well as monitor the weekly written reflections. Although the student reflections in these two courses were not guided, the number of students common to each course (n = 27) provided enough interesting data to monitor and assess the commonalities and differences in student attitudes toward their service-learning experiences, and specifically, the development and awareness of transferable skills as part of engineering. Of the 27 students common to both courses, 7 were women.

### *6.3.2 CE134-Engineering Design Mentoring*

In this course, teams comprised of the University of Vermont engineering students, IBM engineers, and the ECHO Lake Aquarium and Science Center in Burlington, Vermont partnered on a service-learning project to mentor 11–14 year-old home-schooled children on the engineering design process. We challenged teams, each comprising 2–3 home-schooled children, 2–3 university students, and one volunteer from IBM to design innovative solutions to mobility problems, while using the fun and inspiration of biomimicry.

Biomimicry is often defined as innovation inspired by nature (Benyus 1997; The Biomimicry Institute). In this particular service-learning project, we emphasized that biomimicry is the examination of the natural world in an attempt to find sustainable solutions to human problems. Biomimicry leverages the evolutionary process that biotic systems use to optimize complex problems. One of the most well-known examples is Velcro. Plants have the dual problem of pollination and dispersal of their seeds, and have co-evolved elaborate systems with animals to meet these challenges. Some plants have evolved burrs containing their seeds that stick to passing animal fur. Rather than have cactus-like spines, which animals would learn to avoid, the burrs are curved at the tip both to avoid hurting the animal and to provide a better grip on hair. The invention of Velcro, by Swiss engineer George de Mestral, was inspired by the observation of burdock thistles that stuck tenaciously to his hunting dog.

The student teams were challenged to use biomimicry as the inspiration to invent methods to move people, goods, foods, waste, *etc*., given common transportation-related constraints (*i.e*., amount of congestion, pollution, safety hazards), or reduce the need for transportation altogether. The service-learning project was divided into 62-h activities and was worth 25 % of the course grade. With the exception of an icebreaker/introduction on biomimicry, the sessions mirrored the five steps of the engineering design process (Table 6.1). Representative quotes from weekly student reflections are presented next to each of the design steps to help explain the process. The first session, although devoted to defining the problem, focused on the course logistics and the explanation of biomimicry.

### *6.3.3 CE 175-Senior Capstone Design*

The senior capstone design is a comprehensive design project involving two or more civil and environmental engineering sub-disciplines (e.g. structures, transportation, geotechnical, hydrology, environmental). A significant capstone design component is required for ABET accreditation, although the format is left to the individual programs. Each year, the instructor identifies service-learning projects with local towns and non-profit organizations. Students write short proposals

**Table 6.1** Representative student quotes for each of the CE 134 hands-on mentoring design sessions

| 5 steps – engineering design process | Representative quotes from student reflections |
|---|---|
| 1. Problem definition | *"Despite the kids apparent enthusiasm, they were not too psyched about engineering design until Dr. Rizzo said our inspiration would be biomimicry. This added an environmental aspect to this experience. In a world where the health and wellbeing of our environment is in constant consideration when designing new technology, it is important to teach the impacts of our decisions on the natural world."* |
| 2. Generating ideas | *"The creativity exercise focused the kids' imagination… nothing they said could be 'wrong'… they fed off each other. It was an atmosphere everyone could thrive in."* |
| 3. Design selection/prototyping | *"So far, my UVM experience has really been lacking in design, and I think this service-learning project is the perfect way to incorporate more [design] into the curriculum. Ultimately, the biomimicry aspect wasn't very important to the project,… but the emphasis on the engineering design process was definitely useful. Students paid very close attention, and we came back to it again and again in our meetings throughout the semester."* |
| 4. Testing and refining the design | *"I was worried about the design prototype meeting, but it was uncharacteristically positive because of the hands-on work….it was a turning point for some kids…this made each child feel accomplished and important."* |
| 5. Presenting the results | *"…Earth Week's presentation was the most valuable for the children….after watching the kids' presentations improve each week….their courage showed me that I should not be afraid of speaking in front of a large audience."* |
| | *"…part of an engineer's job is to translate very technical language and ideas to an audience that may not have technical background…I learned that if I cannot communicate my ideas, I'll be of no value."* |

stating their project preferences and what qualifications they bring to the project. The instructor then develops teams based on interest, along with technical and non-technical skills.

Some examples of the Spring 2009 service-learning projects conducted included designing stormwater management systems for two towns and one school; a green roof for a historic structure; a parking lot; and mitigation alternatives for two landslides. Surveying important site features, collecting and testing soil samples, and collecting hydraulic information were required for most projects. Students analyzed site conditions, designed new systems or strategies for retrofitting/mitigating

**Table 6.2** Representative student quotes from CE 175

| |
|---|
| *"Along with improving communication skills, this project allowed us to become familiar with state regulations regarding stormwater discharge, and the permitting process for projects in general. The issue of meeting state regulations, which may or may not have a scientific background, became very important.....The ability to work with a community partner provided insight on how to deal with clients and convey information to persons who may not have an engineering background....Although completing the project required use of engineering principles and knowledge, the ability to communicate what was done and how it was done effectively was a major component and became as important as the technical skills .... and this course forced me to learn these skills."* |
| *"Overall I am proud of the final product, I believe it properly reflects our efforts, and is truly professional looking...... Also, a lot of time went into the report being like a story, where transitions are flawless and word choice is exceptional."* |
| *"My personal experiences and growth throughout this project have mostly been in communication and working in a group of other engineers. Working with any number of people on a report like this is a big challenge"* |
| *"The most important thing that I learned throughout this semester is time management."* |
| *"I particularly enjoyed working on this project because it involved both structural engineering and environmental engineering.... All of my expectations for the senior design project were met. My group created an interesting professional product that was successfully presented twice in front of a panel of engineers.... Working in a group with different personality types was certainly a learning experience for me.... I recently interviewed for an engineering position, and they asked me how well I work on group projects, about my writing skills and.... those questions were extremely easy to answer after working on this project."* |

existing problems, and developed cost estimates for each alternative. All students were expected to research relevant regulations and in some instances, helped prepare documents for necessary permits. The design, report, and final presentation accounted for most of the course grade, with a small percentage (5 %) dedicated to the written reflections. Emphasis was placed on both the technical and nontechnical project aspects that students might experience in a professional engineering setting. Economic analyses were performed. Multiple oral presentations and draft reports were required, as well as research related to the short-term and long-term environmental and social impacts. Various activities promoting teamwork, ethics and professional conduct were also emphasized and discussed. Table 6.2 presents representative quotes from the capstone design course reflections illustrating the importance of transferrable skills recognized by the students, specifically communication and teamwork.

## 6.4 Methodology

All statistical methods were implemented in JMP 8.0. In addition, the HyperRESEARCH™ 2.8.3 software allowed us to quantify the written qualitative data (e.g., students written reflections in rich text format). The software allows for the flexible coding of text (e.g., assigning a code such as *academic enhancement* or *personal growth* to text of any length: word, phrase, sentence, etc.) and then the retrieval of similarly coded material to perform simple frequency analyses or other code statistics.

For this analysis, each sentence in the students' end-of-semester reflections was identified as belonging to one (or more) of 248 codes in HyperRESEARCH™. The frequency of coded phrases was tallied and codes that occurred more than 15 times across either of the two service-learning courses were considered to have "popped." Unfortunately, the code frequencies were weighted slightly in favor of the CE 134 design-mentoring project due to the fact that the end-of-semester reflections specified a minimum 2.5 page length, while the CE 175 reflections varied in length from half a page to 2.5 pages because the page length was not specified. To account for this bias, we normalized the raw code frequencies by the total number of words written by each student prior to performing any statistical analyses.

## 6.5 Results

In Spring 2008 (CE 134) and Spring 2009 (CE 175), students were asked to submit weekly writing assignments (reflections), in part to monitor the thematic topic presented during the service-learning activities. Course assessments and written reflections reveal that mentoring home-schooled children in the engineering design process using the concepts of biomimicry was an effective means of teaching engineering design. Students frequently commented that the biomimicry aspect kept the home-schooled students engaged and several noted that when you have to teach something to someone, you really learn it (referring to the design process). Although the anecdotal quotes from students' reflections (Table 6.1) indicate positive support for the CE 134 engineering design mentoring project on a weekly basis, the end-of-semester student reflections and evaluations (summarized in Table 6.3) show a slightly different picture.

Overall, the majority of students (26 in CE 134 and 24 in CE 175) responded favorably (4 = good or 5 = excellent) in both courses. However, there were no low scores in the CE 175 course for the first question (Table 6.3), while four students in the CE 134 course rated their service-learning experience as 2 or below (unsatisfactory or poor). And when students were asked to rate their willingness to volunteer for the CE 134 design-mentoring project again, five (out of 31) CE 134 students indicated they would not be willing to volunteer in the future.

In addition, this same minority (~13–16 %) often dominated and affected the classroom dynamics during in-class service-learning discussions. These end-of-semester ratings raised a red flag, which focused our student reflection assessment toward understanding the disapproval generated by this select group. It is important to keep in mind that the weekly reflections, a few of which are highlighted in Table 6.1, revealed little as to why students disliked the CE 134 service-learning component.

To mirror the intent of our service-learning pedagogy, we divided the contents of all end-of-semester student reflections into three categories – academic enhancement,

**Table 6.3** End-of-semester student evaluations rating their overall learning experience and willingness to volunteer for the service-learning project again

| | CE 134 engineering design mentoring | CE 175 senior capstone design |
|---|---|---|
| *1: poor – 5: excellent* | *1 2 3 4 5* | *1 2 3 4 5* |
| Rate overall learning experience | (# of responses)<br>**3 1** 1 18 8 | (# of responses)<br>0 0 0 9 13<br>*(8 students did not respond)* |
| Rank your willingness to volunteer to do this project again? | (# of responses)<br>**3 2** 3 15 8 | Because it is the final course in their program, this question was not asked. |

1: poor, 2: unsatisfactory, 3: satisfactory, 4: good, 5: excellent

**Table 6.4** Percentage of student written reflections classified into the three service-learning categories and further subdivided into positive and negative sentiments

| | CE 134 engineering design mentoring | | | CE 175 senior capstone design | | |
|---|---|---|---|---|---|---|
| *Key phrase* | Total (%) | Positive (%) | Negative (%) | Total (%) | Positive (%) | Negative (%) |
| Academic enhancement | **40** | 88 | 12 | **33** | 89 | 11 |
| Civic engagement | **15** | 71 | 29 | **9** | 52 | 48 |
| Personal growth | **45** | 77 | 23 | **58** | 84 | 16 |

civic engagement and personal growth skills (Table 6.4). Students wrote the most about personal growth, then about academic enhancements, and lastly about civic engagement. The division of the student reflections into these three categories alone provides little information other than what the students chose to write about. Our initial thought was that civic engagement (15 % for the design-mentoring vs. 9 % for the capstone design projects) was the key factor in identifying the cause of the student unrest. However, when these categories were further subdivided into positive and negative reflections, we found that, on average, the students had more positive things to say about civic engagement with respect to the CE 134 design-mentoring project (71 % positive vs. 29 % negative) when compared to the senior capstone design (52 % positive vs. 48 % negative). Academic enhancement showed almost no difference across the two courses. The greatest difference in positive vs. negative reflections across the two courses occurred in the area of personal growth reflection.

HyperRESEARCH™-coded phrases from the students' end-of-semester reflections occurring more than 15 times across either of the two service-learning courses are shown in the first column of Table 6.5. Note that the majority of these phrases fall into the category of *transferable* skills. The code frequencies (prior to being normalized by total word count) for each course are shown in columns two and three.

**Table 6.5** Phrases coded in HyperREASEARCH™ that occurred more than 15 times in either of the two service-learning courses and results of two-sample t-test

| Codes phrases | CE 134 engineering design mentoring | CE 175 senior capstone design | Prob > ltl |
|---|---|---|---|
| **Comfort zone** | **60** | **0** | **0.0008** |
| Communication skills | 89 | 40 | 0.0173 |
| Communication – lack of | 15 | 19 | 0.5566 |
| Community partnership | 30 | 24 | 0.7892 |
| Confidence | 29 | 13 | 0.6617 |
| **Engineering design** | **55** | **29** | **0.0087** |
| **Group dynamics** | **54** | **32** | **0.0097** |
| **Innovative ideas** | **85** | **3** | **0.0007** |
| **Leadership** | **71** | **10** | **0.0012** |
| Learning experience | 22 | 24 | 0.8243 |
| **Mentoring** | **107** | **1** | **0.0048** |
| **Service learning** | **97** | **9** | **0.0007** |
| **Teamwork** | **66** | **56** | **0.0014** |

Eight of the coded phrases elicit statistically significant differences across the two service-learning projects using a paired t-test or Wilcoxon signed rank test. The codes that are statistically different at the 99 % confidence interval are identified in bold in Table 6.5.

## 6.6 Discussion

In this section, we compare and contrast student service-learning reflections to test whether students value all types of transferrable skills equally. The results of Table 6.4 show that commonalities across the two courses focus mostly in the area of academic enhancement. Students recognized that this experience enhanced their academic learning. Greater differences were observed primarily in civic engagement and personal growth categories. While students reflected more negatively on the civic engagement aspects with respect to the CE 175 senior capstone design projects, the CE 134 engineering design mentoring reflections concentrated more negatively on the personal growth aspects (e.g. leadership skills, teamwork, mentoring, organizational skills, communication to nontechnical audiences).

It is interesting to note that of the 13 coded phrases to "pop" across the students' critical reflections, two of the five phrases that were not statistically different were the written and oral communication skills. The reflections across both projects indicate that students considered report writing and oral presentations an important part of being engineers. This is encouraging, since our reflections and assessments earlier in the reform process indicated that students did not value written and oral presentation skills. These annual evaluations and data analyses not only informed

**Table 6.6** Additional student quotes from CE 134

| |
|---|
| "...*I learned to deal with* design setbacks... *how to* reach consensus, cooperation ... *the biomimicry and* engineering design *I offered to the mentees helped develop their* problem solving skills...*Overall, this project taught me* negotiation skills, mentoring *and* communication skills *necessary for public interactions* but did little to help with engineering." |
| "... *I learned that I enjoy being part of something that helps me be a part of 'the change I want to see in the world'. I seem to take the role as* leader...*I'm not sure it had much to do with engineering.*" |
| "...*I learned a number of things...most important* – teamwork *and* leadership. I understand that this was not an engineering project, *but I know for sure that I got a very valuable experience out of it...I benefited the most from being* the liaison and leader ... It made me feel good about myself." |
| "*I am walking away from this project with greater appreciation for educational differences.... The need for keeping to schedules, importance of email....*nothing related to engineering." |

Underlined words illustrate that some students did not consider these transferable skills to be part of engineering

our teaching, but provided indicators that ABET requires for continuous improvement. As a result, we made revisions and concerted effort to explicitly identify and outline our goals, ABET outcomes, and the importance of transferrable skills during class. Communication skills are now practiced more regularly throughout our 4-year engineering curricula; and their importance is now emphasized in almost every course. The student reflections across both courses (representative quotes of Tables 6.1 and 6.2) show students understand (or can at least repeat back) the importance of written and oral presentation skills to their future endeavors.

Of the eight codes that elicit statistically significant differences between the two service-learning projects (shown in bold in Table 6.5), service learning, innovative ideas and engineering design are perhaps the easiest to explain. Despite our best in-class attempts to emphasize that service learning differs from community service, we believe many students viewed the mentoring of home-school children in the engineering design process as more "service" than the service component of the senior capstone design course. In addition, because of the focus on biomimicry as inspiration for the design in CE 134, the word innovation seemed to "pop" more frequently in the student reflections when compared to the CE 175 reflections. An understanding of the remaining statistical differences lies in reviewing the coded phrases – comfort zone, leadership, group dynamics, teamwork, and mentoring. A more qualitative assessment of individual student reflections shows a common theme (Table 6.6).

Although significant discussion of personal growth appears in the CE 134 engineering design mentoring reflections, the students do not necessarily recognize certain transferable skills (i.e., leadership, mentoring, negotiation skills, design setbacks, scheduling and others underlined) in Table 6.6 as "real" engineering. This may stem from their rejection of these ideas as a result of their underlying attitudes and perceptions about what engineering should be. At a minimum, this identifies an area for improvement in our future service-learning endeavors. If service-learning instructors value transferable skills such as leadership, teamwork, mentoring and scheduling, then we need to find ways to better articulate these as part of our learning goals and provide incentive for students who take on these leadership qualities during the service-learning implementation. Some students, however, were able to connect the dots and saw that this is all part of engineering, as reflected in the following quote:

> I have learned things that I never would have gotten from a class, and for the first time I really have a good idea of what it is like to be working as a professional engineer. Since finishing the project, I now have a greater understanding of service learning and what it's about. Once you have experience with something, it becomes a lot easier to explain. I will never forget this experience; it really changed me and I give it a 5 out of 5.

Despite the well-crafted language in our mission statement and ABET criteria outlining the necessary outcomes engineering graduates must have to prepare them for engineering careers and the many challenges of the future (many of which relate to transferable skills), no one prepares the engineering faculty responsible for meeting these lofty goals for this task. In addition, more attention is needed in understanding and articulating the importance of transferrable skills within civil and environmental engineering. National Science Foundation funding for the Department Level Reform grant allowed us the resources for such preparation and incentive for ongoing evaluation. What we learned as educators from this experience was invaluable and has helped further modify our curricula and pedagogy. The theme of what is engineering has been integrated much earlier into our programs. It is now included in multiple required courses to make students mindful that engineering is much more than applied math and science, and that their education is much more than performing mathematical calculations. Some of the design, creative elements and biomimicry aspects from the junior-level CE 134 course are now incorporated in first and second year courses. We also recognize the need to be more explicit with students about the reasons for learning and doing various assignments, and the benefits of such experiences especially in regard to projects and hands-on learning opportunities. Our syllabi for example, have evolved to be more than a list of topics for the course. They have now become roadmaps to the expected learning objectives and outcomes of the course, often with reasoning from ABET or other educational documents included.

## 6.7 Conclusions

Results showed that students across both service-learning courses regarded two of the transferable skills (i.e., verbal and written communication) as important skills for engineers. However, the disconnect in the student evaluations made it apparent that many did not regard leadership, teamwork, negotiation skills, design setbacks, scheduling and mentoring skills as "real" engineering. And most likely, these engineering students are not as interested in these issues compared to other aspects of engineering. Although, we as engineering educators still struggle with how best to incorporate these nontechnical aspects within their curricula, understanding the students' viewpoints is the first step toward making further adjustments to our curricula and teaching.

One of the major benefits of the CE 134 service-learning project, and experiences like it, is the opportunity for students to engage with other students and engineers outside of traditional settings and roles. While some students relished this experience, many were pushed to the edge of their "personal growth" comfort zones. However, this is what "higher" education should do; take students *and* faculty out of their comfort zones, because that is where real growth occurs.

## References

American Society of Civil Engineers (ASCE). (2006). *The vision for civil engineering in 2025.* Reston: American Society of Civil Engineers (Prepared by the ASCE Steering Committee to Plan a Summit on the Future of the Civil Engineering Profession in 2025).

ASCE. (2008). *Civil engineering body of knowledge for the 21st century preparing the civil engineer for the future* (2nd ed.). Reston: American Society of Civil Engineers (Prepared by the Body of Knowledge Committee of the Committee on Academic Prerequisites for Professional Practice).

Ash, S. L., & Clayton, P. H. (2004). The articulated learning: An approach to guided reflection and assessment. *Innovative Higher Education, 29*(2), 137–154.

Bringle, R. G., & Hatcher, J. A. (1996). Implementing service hearing in Higher Education. *Journal of Higher Education, 67*(2), 221–239.

Benyus, J. M. (1997). *Biomimicry: Innovation inspired by nature.* New York: Harper Collins.

Bringle, R., & Hatcher, J. (2000). Reflection in service learning: Making meaning of experience. In *Introduction to service-learning toolkit.* Providence: Campus Compact. (Original work published 1999)

Bringle, R. G., Hatcher, J. A., Hamitton, S., & Young, P. (2001). Planning and assessing campus/community engagement. *Metropolitan Universities, 12*(3), 89–99.

Collier, P. J., & Williams, D. R. (2005). Reflection in action, the learning-doing relationship. In Cress, C. M., Collier, P. J., Reitenauer, V. L. & Associates (Eds.), *Learning through serving, a student guidebook for service-learning across the disciplines* (pp. 83–97). Sterling: Stylus Publishing LLC.

Dewoolkar, M. M., George, L. A., Hayden, N. J., & Neumann, M. (2009a). Hands-on undergraduate geotechnical engineering modules in the context of effective learning pedagogies, ABET outcomes, and curricular reform. *Journal of Professional Issues in Engineering Education and Practice, 135*(4), 161–175.

Dewoolkar, M. M., George, L. A., Hayden, N. J., & Rizzo, D. M. (2009b). Vertical integration of service-learning into civil and environmental engineering curricula. *International Journal of Engineering Education, 25*(6), 1257–1269.

Duderstadt, J. J. (2008). *Engineering for a changing world, a roadmap to the future of engineering practice, research, and education, The Millennium Project.* Ann Arbor: The University of Michigan.

Forrester, J. W. (1958). Industrial dynamics: A major breakthrough for decision makers. *Harvard Business Review, 38*(4), 37–66.

Forrester, J. W. (1961). *Industrial dynamics.* Waltham: Pegasus Communications.

Hayden, N. J., Rizzo, D. M., Dewoolkar, M. M., Neumann, M. D., Lathem, S., & Sadek, A. (2011). Incorporating a systems approach into civil and environmental engineering curricula: The effect on student work, and student and faculty attitudes. *Advances in Engineering Education, 2*(4), Available at: http://advances.asee.org/vol02/issue04/04.cfm (Accepted, in revision).

Jacoby, B. (1996). *Service-learning in higher education*. San Francisco: Jossey-Bass.
Kezar, A., & Rhoads, R. A. (2001). The dynamic tensions of service learning in higher education: A philosophical perspective. *The Journal of Higher Education, 72*(2), 148–171 (Special issue: The social role of higher education).
Lathem, S., Neumann, M. D., & Hayden, N. (2009). *The socially conscious engineer: Fostering student awareness in a global society*. Paper presented at the American Education Research Association Annual Meeting, San Diego, CA.
Lathem, S., Neumann, M., & Hayden, N. (2011). The socially responsible engineer: Assessing student attitudes of roles and responsibilities. *Journal of Engineering Education, 100*(3), 444–474 (Accepted with revision).
McCarthy, M. D. (1996). One-time and short-term service-learning experiences. In Jacoby, B., & Associates (Ed.), *Service-learning in higher education* (pp. 113–134). San Francisco: Jossey-Bass.
Moffat, J., & Decker, R. (2000). Service-learning reflection for engineering: A faculty guide. In E. Tsang (Ed.), *Projects that matter, concepts and models for service-learning in engineering* (pp. 31–39). Washington, DC: American Association for Higher Education.
National Academy of Engineering (NAE). (2004). *The engineer of 2020: Vision of engineering in the new century* (Parts I and II). Washington, DC: National Academies Press.
NAE. (2005). *Educating the engineer of 2020: Adapting engineering education to the new century*. Washington, DC: National Academies Press.
National Research Council. (2005). *Rising above the gathering storm: Energizing and employing America for a brighter economic future*. Washington, DC: National Academies Press.
National Science Board. (2007). *Moving forward to improve engineering education*. NSB-07-122. Arlington: National Science Foundation.
The Biomimicry Institute. http://www.biomimicryinstitute.org/
Wolstenholme, E. F. (1990). *System enquiry: A system dynamics approach*. New York: Wiley.

# Chapter 7
# Future Reflective Practitioners: The Contributions of Philosophy

**Viola Schiaffonati**

**Abstract** Reflection on engineering practice and the essentiality of reflection for the development of engineering cannot leave aside an analysis of engineering education, its potentialities and its current limitations. This paper addresses the possible role of philosophy in engineering education, with particular attention to how philosophy may impact the creation of future practitioners able to be reflective in their professional practice. We hold that to educate future *responsible* professionals not just ethics, but also other fields of philosophy—such as the critical history of scientific ideas, philosophy of mind, philosophy of science, philosophy of technology, and philosophy of engineering—play an important role. We mean "responsibility" quite broadly: from responsibility concerning specific design choices to responsibility regarding moral attitudes. We claim that a key factor in achieving such responsibility is to teach future engineering professionals to critically reflect on their tools, methods, and results. The analysis presented is based on personal experience in teaching a philosophy course to computer engineering students at Politecnico di Milano, Italy.

**Keywords** Philosophy and engineering education • Critical thinking and history of ideas • Reflective practitioners, responsibility

V. Schiaffonati (✉)
Dipartimento di Elettronica, Informazione e Bioingegneria, Politecnico di Milano,
Piazza Leonardo da Vinci 32, 20133 Milan, Italy
e-mail: schiaffo@elet.polimi.it

D.P. Michelfelder et al. (eds.), *Philosophy and Engineering: Reflections on Practice, Principles and Process*, Philosophy of Engineering and Technology 15,
DOI 10.1007/978-94-007-7762-0_7, 

## 7.1 Introduction

Despite a growing interest in the relationship between philosophy and engineering (van de Poel and Goldberg 2010), save for some important exceptions, the role of philosophy in engineering education has not received much attention (Goldberg 2008, 2010).

Reflection on engineering practice and the essentiality of reflection for the development of engineering cannot leave aside an analysis of engineering education, its potentialities and its current limitations.

This paper addresses the possible role of philosophy in engineering education, with particular attention to how philosophy may impact the creation of future practitioners able to be reflective in their professional practice. We hold that to educate future *responsible* professionals not just ethics, but also other fields of philosophy—such as the critical history of scientific ideas, philosophy of mind, philosophy of science, philosophy of technology, and philosophy of engineering—play an important role. We mean "responsibility" quite broadly: from responsibility concerning specific design choices to responsibility regarding moral attitudes. We claim that a key factor in achieving such responsibility is to teach future engineering professionals to critically reflect on their tools, methods, and results.

Philosophy, especially some areas of philosophy, can contribute to the kind of critical analysis that aims at stimulating students to become more reflective and conscious about some of the issues encountered during their formation. Critical analysis can take different forms: it can be the analysis of foundations, it can be the inquiry into meaning and truth conditions of questions and statements, it can present problems from a historical point of view to show their evolution, or it can be a reflection on general unifying themes. In the following paper, we will try to concretely spell out all of these aspects of critical analysis with specific examples.

The analysis presented in this chapter is based on personal experience in teaching a philosophy course to computer engineering students at Politecnico di Milano, Italy. Although this analysis is based on a very limited experiment, we believe that it is significant in revealing how the overall approach to engineering education is slightly changing. Moreover, even this fairly narrow experiment provides a starting point for the discussion of some educational issues from a concrete point of view. While we are careful not to inappropriately generalize on the basis of our limited results, we are at the same time hopeful that some of these results can inspire further reflections on how at least some parts of philosophy and some ways of doing philosophy could be incorporated in engineering curricula.

We start by presenting the context and the history of this pilot project, as we believe that this information is useful to understand the rationale behind it (Sect. 7.2). From there, we go on to describe the general aims and goals of the course 'Philosophical Topics in Computer Engineering'. We provide examples of the topics addressed in the different parts of the course, the reasons for including these topics in the course, as well as the results obtained in teaching them (Sect. 7.3). We then present some critical points, related in particular to the quantitative evaluations of the results, along with some open questions to be further investigated (Sect. 7.4).

## 7.2 Introducing Philosophy at Politecnico di Milano

To explain how philosophy can contribute to engineering education we discuss the case of Politecnico di Milano, a leading Italian technical university that, just in the last few years, has introduced some philosophy courses for the students in the master degree programs of Computer Engineering, Systems Engineering, and Mechanical Engineering.

Politecnico di Milano was established in 1863 by a group of scholars and entrepreneurs belonging to prominent Milanese families. It is now ranked as one of the outstanding universities in Engineering, Architecture, and Industrial Design in Europe. It is organized into 16 departments and a network of 9 schools spread over 7 campuses. The number of students enrolled in all campuses is approximately 40,000, which makes Politecnico di Milano the largest institution for engineering, architecture, and industrial design in Italy.

The engineering curriculum at Politecnico di Milano reflects some traits common to all the other engineering schools in Italy and shows the origin of the institution. Education is primarily based on mathematics and applied science in order to guarantee a strong scientific and technical preparation. The choice of elective courses is limited and basically concentrated only in the last year of the master degrees. The array of course offerings other than science and engineering courses is very limited.

Despite being in existence over 100 years, it was just few years ago that Politecnico di Milano introduced a small number of philosophy classes as engineering elective courses and only in the curricula of the Schools of Information Engineering, Systems Engineering, and Mechanical Engineering. This delay, when compared to other similar institutions in Europe, along with the skepticism of a consistent part of the faculty involved, points to some issues that need to be examined.

When about 10 years ago we started with the project of offering philosophy classes in engineering curricula, the first concern was to avoid simply importing standard philosophy classes, without the effort to rethink them for the education of future engineers. Although everybody recognized the cultural centrality of philosophy, most of the people involved considered this of secondary importance and concentrated more on the useful impact of philosophy for engineers' education. Moreover, standard philosophy courses would have been, in most cases, too far removed from engineering students' background. In concrete terms, this has meant that these philosophy courses had to be designed not for philosophy students but for engineering ones.

We detected some gaps in the critical abilities of engineering students. Their intensive training in learning and manipulating concepts from a scientific and technical point of view made them strong in these areas, but extremely weak when it came to the ability to critically analyze the same scientific and technical notions. This discovery was not the result of a rigorous survey designed to test students about these critical capabilities, but rather emerged from extensive conversations with leading instructors at our institution.

Much effort was devoted to analyzing how philosophy could be taught to engineering students and how it could help in enhancing their capabilities if philosophical concepts were integrated with scientific and technological ones. During this process the central questions were: What can philosophy introduce that is different from the traditional topics already taught in an engineering curriculum? In which ways could this introduction enhance the students' capabilities?

With regards to the aim of offering some conceptual tools useful for an informed reflective education, the motivation is both specific and general. It is specific because the goal of these philosophy courses is to increase the capability of engineering students for reflecting on concepts used throughout the entire course of their formation, but which are seldom critically analyzed from a foundational point of view. It is general because teaching of philosophy is considered a way to learn some conceptual tools useful not only for present engineering students but also for future professional engineers. What characterizes a reflective practitioner in comparison to a non-reflective one is the ability to evaluate the consequences of some design choices from a wider perspective than a purely technical one. The idea behind this is that being able to better articulate the foundations of engineering disciplines can improve conceptual clarity, as well as help in diagnosing errors and in considering the future consequences of some choices.

For this reason, it was decided that in the pedagogy for these courses scientific and technological notions would need to be integrated with philosophical ones; for instance, in the study of the philosophical history of ideas particular emphasis would be placed on the birth of scientific concepts and their functioning. We firmly believe that considering how current notions and concepts have been developed can have a deep impact on understanding these notions and concepts and, thus, on how to put them into practice. It is worth noting, once again, that the objective of these philosophy courses is not to teach philosophy and its history, but to teach how to apply philosophical analysis to engineering problems.

## 7.3 Philosophical Topics in Computer Engineering

In order to spell out more concretely the reflections presented in the last section, let us now turn to analyze one of these philosophy classes. This section discusses the author's personal experience in teaching the course 'Philosophical Topics in Computer Engineering' offered in the last year of the Computer Engineering Master Degree program at Politecnico di Milano.

The aims and goals of this course, inspired by Rapaport (2005), are to increase computer engineering students' awareness of some central concepts of their curriculum, to improve their critical thinking skills, and to encourage reflection on metaphysical, epistemological, and ethical issues of computer science.

The course is organized as follows. In the first part of the course, scientific and technological issues are introduced from a philosophical perspective. Typical topics are: what is science and how it was born during the Scientific Revolution; the experimental scientific method from Galileo Galilei to current science; and the philosophical and practical issues concerning theories and models in science and engineering. The second part presents a critical analysis of fundamental concepts and topics in computer science and engineering. Typical topics are: what is the philosophy of computer science and its method; what is computer science; the role of simulations and their experimental capabilities; the debate on intelligence and machine intelligence; the mind-body problem; computational models of consciousness; good experimental methodologies in robotics; information and computer ethics; epistemological and methodological issues of biorobotics. The third part is devoted to supervising students' critical essays. During class hours, while the students are working on their critical essays, the instructor is on hand to answer questions, discuss problems, and provide advice.

A central feature of the course is that students do not learn about specific philosophical problems *per se*. Instead they learn, by means of these problems, the *modus operandi* of philosophical analysis. Again we do not want to diminish the cultural impact of philosophy in general, but to remember that the aim of teaching philosophy in this context is different from that of teaching philosophy to students in the humanities.

The main advantages we identified in teaching philosophy to engineering students were that engineering students would:

- learn not to take concepts for granted, and develop better critical abilities;
- learn to see problems from perspectives (historical, conceptual, …) other than the usual technical one, in order to develop a pluralistic attitude toward problem solving;
- get more used to qualitative rigor even in absence of numbers and formulas.

The philosophical problems presented and discussed in the course 'Philosophical Topics in Computer Engineering' are representative of the opportunity offered by teaching philosophy to computer engineering students in such a way that the topics and methods used to teach them are specifically tailored to meet their needs. This specific tailoring can be seen in the organization of this course, which is centered on the following activities.

- The critical analysis of the fundamental concepts of computer science and engineering (computation, machine, information).
- The investigation of the meaning and truth conditions of some recurrent questions ('Is the brain a computer?').
- The presentation of problems from an evolutionary, historical point of view.
- The reflection on topics differently declined in different areas of computer science (i.e., the notion of experiment).

In what follows we present a more detailed analysis of some of the topics taught in the course, and some thoughts about their importance for the education of future engineers. As will be clear from the discussion, this course incorporates different sub-areas of philosophy: from the critical history of scientific ideas to the philosophy of mind, the philosophy of science, the philosophy of technology, the philosophy of engineering, and ethics. We believe that all these different parts of philosophy can, in different ways, contribute to enrich and complement the education of future engineers.

### 7.3.1 *Critical History of Scientific Ideas*

This part of the course is devoted to introducing the basic vocabulary of science and scientific concepts both from a historical and a conceptual point of view. To this purpose, we present the birth of the modern conception of science during the Scientific Revolution of the seventeenth century (Kuhn 1957), the development of the modern conception of science, and the more recent philosophical debate about its nature (Godfrey-Smith 2003). Beside these general themes, we discuss more specific ones as well, by focusing in particular on the transition from specific astronomic issues to general scientific ones promoted by Galileo Galilei's work.

Why is this part of the course important for the education of engineering students? First of all, we believe that showing the evolution of concepts can help in widening the reference frame of every discussion. Without a historical perspective, students could have the mistaken idea that scientific concepts have not evolved, and thus are untouchable and cannot be criticized. Moreover, the analysis of the evolution of ideas opens up a way of bringing in the influence of social, political, cultural, and other factors that engineering students are not used to consider. Finally, the detailed study of a part of the history of science (in this case the so called Copernican Revolution) is a good way to show that details are important in the humanities as well as in engineering.

### 7.3.2 *Philosophy of Mind*

One of the topics addressed in this part is the mind/body problem, presented from the perspective of the results achieved by Artificial Intelligence (AI). We start by considering the relationship between mind and body, asking questions such as: Does the mind emerge from the brain? Are they basically the same? Are they different and why? These questions trigger some other important ones, such as: What is the brain? Is the brain a digital computer? Is the brain a physical symbol system? Do mental properties emerge by the ability of the brain in processing symbols? (Newell and Simon 1976; Searle 1980).

In considering the mind-body problem, we have observed that several computer engineering students trained in AI take for granted the analogy between a human brain and a computer. Thus, our aim is to reflect on the presuppositions of this analogy by considering the meaning and truth conditions of this analogy. The question 'Is the brain a computer?' requires a careful analysis of its *meaning conditions* and, in case, of its *truth conditions*. The analysis of the truth conditions of such a question can also help us in understanding whether considering the brain in terms of a computer can shed new light on the traditional philosophical problem of the mind and the body. Moreover, the answers to all these questions require a reflection on what a computer is, given for granted the shared and intuitive definition of the brain as an organ. This triggers other interesting questions, such as 'What is a machine?', that require further analysis.

Why is this part important for the education of engineering students? In our opinion, presenting these debates is useful in teaching students how to avoid starting from bad questions or ill-posed problems. This is also connected to learning how conceptual clarity is the first step for meeting challenges emerging in future professional practice. This analysis, moreover, shows how every question mentioned above needs to be reformulated to be addressed in a computer engineering context. Therefore, an apparently simple question, such as 'Is the brain a computer?' becomes translated into: 'In what sense is it correct to identify the brain and a computer and in which sense is it not?' This can help students to concretely realize how the conclusions of a line of thought strongly depend on how the concepts involved have been defined.

### *7.3.3 Philosophy of Science*

Among the topics addressed in this part of the course are the concept of simulation, computer simulation in particular, and the question of whether computer simulations can be considered as kinds of experiments. After a careful analysis of a possible definition of simulation (Humphreys 2004), students are introduced to the problem of considering under which conditions computer simulations can be used as experiments, starting from the acknowledgment that computer simulations are essential tools of current scientific activity. This requires introducing the notion of scientific experiment (Radder 2003) and the conditions under which this parallelism can be accepted. Moreover, the epistemological limits of the explorative use of simulations in doing science are presented, as well the reasons for trusting simulation results. The concept of reliability is also introduced, together with the set of strategies that can be adopted to validate simulation results.

Why is this part important for the education of engineering students? While this part of the course further emphasizes the importance of clear definitions of concepts it also, and just as importantly, introduces a fallibilist perspective in science and engineering (Hacking 1983). In our opinion this may have a deep impact in the education of engineering students, as it shows how traditional scientific concepts, such as validity and truth, need to be revised, and in some cases substituted for by weaker ones, such as reliability. Moreover, it presents a fallibilist perspective not as a renunciation of more solid concepts, in particular for engineering students so used to looking for 'objective' results, but as a better articulation for some problems.

If we consider simulation results, for example, not only does there exist no single solution to the validation problems, but these results are always fallible, even in cases when we have strong reasons to believe in them.

### 7.3.4 Philosophy of Technology

This part of the course focuses on the nature of technological artefacts. We consider, for example, the nature and role of computational ontologies in computer science, intended as explicit and formal specifications of conceptualizations. More precisely, an ontology is a specific artefact expressing the intended meaning of a vocabulary in terms of the nature and structure of the entities which it references. Computational ontologies are a means to formally model the structure of a system, i.e., the relevant entities and relations that emerge from its observation, and which are useful to our purpose. An example of such a system can be a company with all its employees and their interrelationships. The ontology engineer analyzes relevant entities and organizes them into *concepts* and *relations*, being represented, respectively, by unary and binary predicates. The backbone of an ontology consists of a generalization/specialization hierarchy of concepts, i.e., a taxonomy.

Besides the issues traditionally dealt with in the field, such as the different levels of precision or the problem of accuracy, this part of the course concentrates in particular on how conceptual analysis is a necessary prerequisite for the creation of sound ontologies (Guarino et al. 2009). Even if, in the common practice of computer science, the representation of knowledge does not usually take into account a prior conceptualization, we show by means of several examples how this process is fundamental to have robust, well founded, and reusable tools. In other words, we try to demonstrate how it is impossible to design good computational ontologies without an adequate ontological analysis.

Why is this part important for the education of engineering students? Ontological analysis focuses on the study and articulation of content *per se*, independently of the way in which this content is represented and of the tools used to represent it. In other words it is an explicit specification of a conceptualization. Let us consider a company as the system that has to be represented and suppose we are interested in aspects related to human resources. *Person*, *Manager*, and *Researcher* might be relevant concepts, where the first is a superconcept of the latter two. *Cooperates-with* can be considered a relevant relation holding between persons. A concrete person working in a company would then be an instance of the corresponding concept. This is a relatively new task for computer scientists and a new dimension for the users of computer systems, whereas philosophy has a long and strong tradition in this kind of analysis. We believe, therefore, that philosophy here can play a central role with important collateral effects for engineers' education. To design a conceptual model to be transferable within a computer system is a good way of understanding what objects are at the basis of the context to which the computer system is applied. Philosophical analysis can be useful here in raising some foundational questions that are usually taken for granted during the design and realization of a computer system. To learn this foundational approach is a good way, in our opinion, to learn how to realize not only more robust systems, but also more easily reusable ones.

### *7.3.5 Philosophy of Engineering*

This part of the course deals with topics that require expanding the traditional boundaries of philosophy of science to include topics within the novel field of the philosophy of engineering. We present the problem of good experimental methodologies in one area of robotics, namely mobile autonomous robotics, a field in which a lively debate on the issue of adopting more rigorous experimental practices has recently started up. We reflect on the possibility of importing traditional principles of experimental science, such as *comparison*, *repeatability*, *reproducibility*, *justification*, and *explanation* in this field of autonomous robotics (Amigoni et al. 2009). Although we do not claim that they are the only principles that should be adopted in defining experimental methodologies for autonomous robotics, we hold that they are at the very foundation of any experimental activity and, hence, cannot be ignored.

The analysis of experimentation in autonomous mobile robotics reflects the peculiar position of this discipline at the intersection of engineering and science. On the one hand, robotic systems are man-made artefacts, which seem to bring the discipline closer to engineering than to science, which is focused instead on natural phenomena. Accordingly, experiments in autonomous mobile robotics have the goal of demonstrating that a given artefact is working in some way, or that it is better than another. On the other hand, the most advanced autonomous robotic systems are so complex that their behavior is hardly predictable, even by their own designers, especially when considering their interaction with the physical world. From this perspective, experiments in autonomous mobile robotics are somehow similar to experiments in natural sciences since, broadly speaking, both have the goal to understand how complex systems work.

Why is this part important for the education of engineering students? Several things stand out here. First, this section makes the concept of experiment problematic by looking at the way it has historically been considered, beginning from the Scientific Revolution of the seventeenth century. Experiment is a multi-faceted concept, and we value that to be exposed to these different facets can have a positive impact on engineering students. Moreover, this analysis can help to get students to consider the design of experiments not as a mere list of clear-cut steps where the results are always guaranteed, but a complex process requiring specific solutions where a certain degree of fallibilism is not eliminable. Finally, when reflecting on experiments at the intersection between science and engineering, the necessity of moving from the traditional philosophy of science categories to novel ones calls for the development of the new discipline of philosophy of engineering (or philosophy and engineering, depending on one's emphasis).

### 7.3.6 Ethics

This part deals with the discussion of ethical issues, particularly those associated with computer and information contexts. This choice is due to the fact that the course is offered to computer engineering students and, hence, we try to exploit what they have already learnt during their education and work on these notions. We introduce first a brief history of computer ethics, its problems, and tools (Johnson 1985). Then, we focus on specific problems whose discussion requires a conceptual analysis of their presuppositions. For example, in presenting the ethical issues related to intellectual property, we present not only the pros and cons of different moral positions, but also the concept of software and the different aspects of software that can be owned. Here again the idea is to present the complexity of each problem and to show how a single choice can have deep consequences, both from an intellectual point of view and from a practical one as well.

Why is this part important for the education of engineering students? We consider fundamental to teach how technical problems are always connected to more general problems, both conceptual and moral in nature. This is one of the great challenges in engineering education for the development of future reflective practitioners: helping them learn that single, specific problems are associated with other ones that are more complex, and that a single choice has a profound impact at various levels.

## 7.4 Conclusions

This chapter has addressed the issue of how philosophy can contribute to the education of engineering students, through the discussion of a personal experience of teaching philosophy to computer engineering students at Politecnico di Milano. We argue that the teaching of philosophy can have a deep impact on forming future generations of reflective engineers who have the capacity for enlarging their scenarios of analysis, but that to achieve this goal philosophy and engineering need to be integrated along two different dimensions. The first one is a *historical dimension*. It needs to be shown that current concepts and ideas of engineering have not always been the same, but have evolved along different directions. This promotes a more pluralistic view of science, technology, and engineering disciplines. The second one is a *pragmatic dimension*. Philosophy has to be directly connected to the needs of engineering students; it needs to be shown that conceptual clarity is essential in practice and can be achieved in qualitative as well as in quantitative terms.

The experience reported is very limited: just one course of about 60 students offered once a year over a 8 year span. However, it is significant as expressing a change of mind when reflecting on engineering education. Unfortunately, we do not possess concrete assessment data, as quantitative evaluations were not taken. This is due both to the lack of proper methodologies and to the specific nature of the object that would be evaluated. It would be very difficult to evaluate whether and how taking such a course could have a positive impact on the students' curricula.

How to 'measure' the effect of this philosophy course is an open issue. One of the reasons is due to the designation of the course as an elective, which means it is usually taken by highly motivated students who generally do better than others. Nevertheless, we do possess some qualitative data that show a very positive impact of this course on students that have taken it. These students were asked to fill the 'official' course evaluation form provided by Politecnico; on these forms, overall grade of the course was very high. Moreover, students were asked by the instructor to answer more specific questions about the course, its organization, its goals, and method. The answers to these questions revealed that students strongly appreciated not only the topics taught and the way in which they were taught, but also that they were conscious of the potentiality to learn critical abilities by means of philosophical analysis. The results of the final projects done by the students provided further evidence for this last point. Because students were asked to write a paper on one of the topics discussed in the course, they had to critically analyze a problem, a topic, or an issue encountered during their formation. Most of the papers revealed that students learned to apply philosophical analysis to engineering problems. In the future we aim at improving our evaluation approach and at tracking and comparing the careers of students who have taken philosophy courses to the ones that have not taken them. A possible solution to overcome the 'elective bias' mentioned above could be to introduce this course as mandatory for some tracks of the Computer Engineering Master Degree program. In this way the careers of the students belonging to the track with the philosophy course and those belonging to the track without it could be more objectively compared.

Moreover, we plan to integrate this so called *ex-post* methodology in teaching philosophy with a type of *ex-ante* one. As already stated, the few philosophy courses at Politecnico di Milano are offered the last year of the master degree program, at the end of the education process. The idea has been that in these philosophy courses students could explore what they have already learned, as objects of critical analysis by means of philosophical tools. We believe that this methodology could be complemented with a different one introducing philosophical elements of critical thinking the first year of the bachelor degree program. In such a case, our aim would not be to offer a whole course of philosophy, but rather to insert philosophical issues and methods within engineering courses themselves. This would give students the opportunity to get acquainted from the beginning with critical skills typical of philosophy, which can contribute to the education of future reflective practitioners.

## References

Amigoni, F., Reggiani, M., & Schiaffonati, V. (2009). An insightful comparison between experiments in mobile robotics and in science. *Autonomous Robots, 27*(4), 313–325.

Godfrey-Smith, P. (2003). *Theory and reality: An introduction to the philosophy of science*. Chicago: The University of Chicago Press.

Goldberg, D. E. (2008). What engineers don't learn and why they don't learn it: And how philosophy might be able to help. In *Abstract Booklet: 2008 Workshop on Philosophy and Engineering (WPE)* (pp. 85–86). London: The Royal Academy of Engineering.

Goldberg, D. E. (2010). Why philosophy? Why now? Engineering responds to the crisis of a creative era. In I. van de Poel & D. E. Goldberg (Eds.), *Philosophy and engineering: An emerging agenda* (pp. 255–263). Dordrecht: Springer.

Guarino, N., Oberle, D., & Staab, S. (2009). What is an ontology? In S. Staab & R. Studer (Eds.), *Handbook on ontologies* (pp. 1–17). Berlin: Springer.

Hacking, I. (1983). *Representing and intervening*. Cambridge: Cambridge University Press.

Humphreys, P. (2004). *Extending ourselves: Computational science, empiricism, and scientific method*. Oxford: Oxford University Press.

Johnson, D. (1985). *Computer ethics* (1st ed.). Upper Saddle River: Prentice-Hall.

Kuhn, T. S. (1957). *The Copernican revolution: Planetary astronomy in the development of western thought*. Cambridge, MA: Harvard University Press.

Newell, A., & Simon, H. (1976). Computer science as empirical inquiry: Symbols and search. *Communications of the ACM, 19*, 113–126.

Radder, H. (Ed.). (2003). *The philosophy of scientific experimentation*. Pittsburgh: University of Pittsburgh Press.

Rapaport, W. J. (2005). Philosophy of computer science: An introductory course. *Teaching Philosophy, 28*(4), 319–341.

Searle, J. R. (1980). Minds, brains, and programs. *Behavioral and Brain Sciences, 3*, 417–424.

van de Poel, I., & Goldberg, D. E. (Eds.). (2010). *Philosophy and engineering: An emerging agenda*. Dordrecht: Springer.

# Chapter 8
# Fitting Engineering into Philosophy

**Joseph C. Pitt**

**Abstract** It is argued that old ways of thinking and philosophical hubris have ossified categories in such a way as to keep engineering out of the philosophical discussion. By showing how common sense marks the fundamental method of reasoning across all disciplines, and that engineering epitomizes this form of reasoning, it is shown that a consideration of engineering concerns should be at the heart of the new philosophy.

**Keywords** Common sense • Engineering • Reasoning • Categories

## 8.1 Introduction

According to the American philosopher Wilfrid Sellars, "the aim of philosophy is to see how things, in the broadest possible sense, hang together in the broadest possible sense" (Sellars 1963). And by "things" he quite literally meant everything from death to 'cabbages and kings'. I not only like Sellars' injunction here, I think that if we attempt to do as he says it will result in a significant revision in the nature of philosophical thinking, one that will bring philosophy back into our daily lives and bring engineering into the philosophical enterprise playing the central role it deserves.

The problem with fitting engineering into the philosophical dialogue stems from the philosophy side of things. Philosophers are fond of drawing distinctions and creating categories they then carve into stone. More often than not, arguments abound over whether these are the right distinctions and categories, but sometimes things slip into place and just stay there. In this paper I will be looking at some of the

J.C. Pitt (✉)
Department of Philosophy, Virginia Tech, Blacksburg, VA, USA
e-mail: jcpitt@vt.edu

D.P. Michelfelder et al. (eds.), *Philosophy and Engineering: Reflections on Practice, Principles and Process*, Philosophy of Engineering and Technology 15,
DOI 10.1007/978-94-007-7762-0_8, 

things that have been in place too long and that need to be reevaluated. The primary distinction at fault here is the one between the life of the mind and the life of action. It a bogus distinction but it is responsible for many missteps, including the increasing irrelevance of philosophy.

As freshmen everywhere learn in the first day of their Introduction to Philosophy class, "Philosophy" means love of wisdom. While philosophers are not necessarily wise, if they really love wisdom they should do something to show it by seeking it out. Wisdom consists in knowing how to live well and how to improve the lives of those around you. Needless to say, wisdom is rare and clearly hard to achieve. So it seems prudent to start with lowering our sights slightly. Instead of demanding that we only seek wisdom, how about trying to achieve understanding – specifically understanding of how we think about things. For coming to understand why we think the way we do may lead us, if not to wisdom, at least to some clarity about why we make mistakes and knowing that may help improve things.

Like Sellars, I too am a Peircean pragmatist, meaning by that, among other things, I see an understanding of inquiry as the key to much else of what we seek to understand. Thus, understanding how we go about inquiry, rather than merely analyzing the specific results of inquiry, is crucial to understanding the meaning, i.e., the consequences, of those results. How you arrived at whatever conclusions you have is key to understanding their reliability and hence what they bode for future inquiry. But the primary consideration here is this: the mark of knowledge, understanding, and wisdom is action. Talk about it all you want, write volumes and fill up the libraries, but if what you say does not lead to action, successful action, then you are just blowing smoke. Thus spoke the pragmatist.

## 8.2 Origins of the Topic

I came to my current topic, as I do to most of my research, after teaching several undergraduate and graduate courses in the philosophy of technology. That resulted in the publication in 1999 of a little book entitled *Thinking about Technology* in which I asked, among other things, the following question: If science is what scientists do, then isn't technology what technologists do? That seems obvious, but it led to a problem: who are the technologists? Working at a land grant university with a strong engineering program, I thought engineers were as likely candidates for my technologists as anyone. Since then I have spent a lot of time learning about what engineers do and how they do it, which is not to say that I am an expert by any means. It is also the case that I have come to appreciate that engineers do not exhaust the category of "technologists" – they simply provide a convenient starting point for identifying the larger group. A more comprehensive account would include information technologists, farmers, beekeepers, teachers, scientists, etc.

As noted, I tend to take my research problems from my students' concerns. In the latest incarnation of my philosophy of technology course, I had my undergraduates read parts of my 1999 book. As we all are aware, today's undergraduates are the first generation raised in what increasingly appears to be a totally digital world. They love their technologies and the life styles they make possible. But, much to my surprise, they really objected to my identification of a technologist with an engineer. The class was composed of about 50 % engineering students – they objected too. But, as typical undergraduates, they found it difficult to articulate the reason behind their objections. After much pulling and prying it finally came out: to the rest of the students engineers are perceived as dull, boring, single-minded and unimaginative. Worst of all they turned everything into an equation. How could all these neat and innovative technologies have come from them? Good question. The engineering students objected as well, arguing that they had richer lives than the label "technologist" suggested. Mostly they offered up their hobbies as examples, many of which were associated with the world of music.

## 8.3 Common Sense and Feed-Back Loops

However, there are some misguided assumptions behind it. First, don't confuse young, immature but very intense engineering students still growing out of their teenage years with adults. Second, don't assume that just because someone is an engineer that they aren't like you in a large number of ways. For example: they marry, they have families, they go to their kid's ballet recitals and ball games, they go to the opera and old time country music concerts. Most importantly, they think like everyone else does. That's right. The difference between the way non-engineers approach a problem and how an engineer approaches an engineering design problem, for example, may be that engineers are sometimes dealing with materials that lend themselves to quantification. However, the fundamental thinking process is the same. To understand this is to take a major step toward dissolving the distinction between the life of the mind and the active life.

To solve a problem, any problem, we all begin by making decisions. In making decisions, we first lay out the various alternative courses of action among which we have to choose. We also all come equipped to deal with that decision-situation with basically the same categories: some knowledge, some values and some goals. The particulars may vary, but the three categories will remain the same. Thinking about what we want to achieve, we choose the option that seems to have the best chance of getting us to our objective. If we fail, as we often do, we go back and look at what we started with and try to figure out where we went wrong. We reexamine what we thought we knew, the set of beliefs that gave us the confidence to choose that option as the best course of action to take to get us to our desired result. Likewise, we examine our values and our goals. But we look at values and goals after we adjust our knowledge base, implement a new course of action and see what happens. If we fail a second time, then we start looking elsewhere for the culprit, i.e., that which we had assumed to be acceptable, but which in fact lead us to the wrong result. While not necessary, there does seem to be an order in which this probing of our knowledge base, values and goals takes place.

We are toughest on what we have easiest access to – our assumed knowledge base – probing it for flaws and false assumptions. It is the easiest to access because we think we know how to check our facts, that is something we are taught how to do. Our next target of assessment is usually our goals, with our values being the last thing we challenge. This is more difficult because we are not taught how to check to see if we have right values, whatever that may mean, and reasonable goals, whatever that may mean. Thus, if, after having selected an option and acted on it we do not get the result we want, we go back and first examine our knowledge. Then we try again. If we fail again, we go back and look at our goals. Adjusting our goals is difficult, but it can be done. We just have to "get realistic." If we fail again, it is time to look at our values. Challenging values is also difficult but not impossible. It is easiest when we challenge cognitive values such as truth.[1] We may decide that requiring that all our evidence be true is too strong a requirement and settle instead for high probability. Challenging non-epistemic or non-cognitive values is a lot more difficult. What are the kinds of reasons you can give, for example, for adopting the conclusion that this is not the time to insist on an elegant solution, when elegance is seen as an aesthetic value? In essence, the entire process of evaluating a failed course of action by reviewing what we thought we knew, our goals, and our values is to be understood as a feedback loop and we employ such loops in every aspect of our lives.

When a feedback loop is not involved in how we think about things, we usually find ourselves in a mess. I have argued that the failure to use a feedback loop in making decisions is irrational and that rationality should be defined as learning from experience, which inevitably requires using a feedback loop (Pitt 1999/2006). Thus, consider the 2008 financial crisis. The failure to rethink key assumptions about how the market works led to collapse. The assumption that not only is there an invisible hand that guides the market toward equilibrium, but that all players in the market share a common value system has rarely been examined.[2] But the present financial system is falling apart because one of those assumptions, that we share a common moral system, is manifestly not the case. Thus, while most of us will resist spending our neighbors' money on a risky venture, current events have shown that Wall Street investors think nothing of risking their investors' trusted funds.

I want to call the form of thinking I am describing here *common sense*. It's that basic notion that ends up with "humph, well that didn't work, back to the drawing board." This is how we approach raising a child, or plowing a field for maximum drainage. It is how scientists think – in other words, if there is a scientific method, this is it. Sometimes, because science is sometimes described as self-correcting, it is assumed that common sense and self-correcting science are the same, but that is really too simple. However, it is how engineers think. Let's start with how scientists think.

[1] In a much neglected paper, Richard Rudner (1953) speaks to the issue of how scientists necessarily employ cognitive values.

[2] Adam Smith's (1776) two conditions for the viability of a capitalist economic system.

Scientists are taught certain theories about how the phenomena in the domain in which they are interested behave and they develop experiments based on the assumption that those theories are correct. That is, the doing of science is constrained by the theories in vogue at the time. Then one day someone brings a new device into the lab and they start fooling around and discover things they never had discovered before. Think here of Galileo training his telescope on the heavens – he saw things using his telescope that he had been taught shouldn't be there. It was time to rethink some assumptions. His values were still the same: truth, rigor, etc. His goals were still the same: to explain the way the world works. But it turned out that what he thought he knew he didn't. It turns out, for example, not to be the case that there is one center in the universe, i.e., the earth, around which all heavenly bodies rotate. The Medicean planets revolve around Jupiter! Time to rethink. It also turns out that it is not the case that the earth is the only corruptible and imperfect body in the heavens, the surface of the moon is not perfectly smooth – it has mountains and valleys on it. That there is only one center around which objects in the universe rotate and that the heavens are perfect were key assumptions whose reconsideration forced a major overhaul in what Galileo and his contemporaries thought they knew. In particular it set off a search for another theory that could capture what was now being discovered. The difference between the self-correcting nature of science and common sense is that at some point in time the little self-corrections are not enough and we have to throw out the entire theory. We never throw out all of common sense.

Further, the reevaluation of the knowledge base had a major effect on the value system that guided most people at the time. For most of the Christian West, that marvelous metaphysical and astronomical system, the Aristotelian/Ptolemaic universe, with the earth located at the center, confirmed and supported the theological position of the Catholic Church that man was at the center of God's creation. Moving the earth and man out of the center of the universe undermined the Christian value system in ways that had major ramifications for western culture.

Now look at the Tacoma Bay Bridge incident. Based on what we thought we knew about the effect of crosswinds on a suspension bridge we built a bridge that collapsed in high winds due to the oscillation the winds created. Back to the drawing boards. It was time to rethink what we knew about suspension bridges and weather and geology. In this case, however, it is not quite so clear what the ramifications for goals and values were, if any. So it may be the case that we don't always have to rethink everything – it is just that some revisions of knowledge are more far-reaching than others and some less.

This fundamental template of how we tend to approach problems is, I want to argue, universal. When I ask my freshmen engineering students what distinguishes engineering from other disciplines, they tell me "engineers are problem solvers." Well, here is a news flash: we are all problem solvers. From the owner of a dog kennel trying to figure out how to reduce the barking when she shows up at feeding time to an engineer designing a widget, we start with certain assumptions and, this is important, certain constraints. No decision is made in a vacuum. We are always restrained by economics, by materials, by aesthetics, etc. These constraints help define the options among which we have to choose. But in acknowledging how those constraints condition our decision-making, we are acknowledging the basic common sense approach to problem solving. Feedback loops are the common factor here. View the flawed assumption that our knowledge is reliable as a constraint.

## 8.4 Philosophical Issues

Now how does that relate to putting engineering into philosophy? Well, since common sense problem solving involves questions of prior knowledge, values and goals, and engineering involves the utilization of those factors in the creation of artefacts of various complexities, it seems we are smack dap in the middle of the major areas of philosophy: epistemology, value theory and metaphysics. Here are some questions that immediately spring to mind, and I am sure there are others.

- What is the nature of the knowledge that engineers employ when solving engineering problems? (Is it different from ordinary knowledge and scientific knowledge?)
- What are the values at play in engineering decision-making (and are they different from ordinary values and the values at work in science)?
- What is the ontological status of the artefacts engineers create?

The same questions could be asked of biologists, artists, architects, teachers and househusbands about their respective domains of problem solving.

- Is the knowledge a biologist requires for his research different from the knowledge a househusband needs to run a household successfully?
- Does an architect employ the same values a teacher does?
- Do all artistic creations exist in the same way? That is, does the performance of a ballet exist in the same way as a painting does? What is the ontological status of a clean house?

So here is the punch line. There is nothing in the domain of human inquiry that does not admit of philosophical analysis. If we distinguish between the activities, doing biological research, designing a new can-opener, creating a symphony, on the one hand, and the people doing those activities, on the other, we will see that the decision-making processes employed by people in these various domains are fundamentally the same despite the differences in the content of the knowledge, the goals and the values. The people are different – now that's real news! We are all attracted to different areas of inquiry (and I will include the arts as an area of inquiry, exploring relationships of sounds, colors, shapes, and motion). And it may be that certain personality types are drawn to certain fields. But don't judge the nature of the thinking taking place in solving the problems that emerge in those different areas by some characteristics displayed by some of the people attracted to working in those fields.[3]

[3] It is important to note here that I am talking about ordinary day-to-day problem solving, not the big "Ah, Ha!" insight into the nature of material being that oh so rarely comes along. There is no question about the fact that we do not know what gives rise to those insight and bursts of creativity that take us to the next stage in our development.

Once we understand this, we can proceed to examine the contents of the three categories in order to achieve some degree of understanding of how the decisions that were made came about.

Let's assume for the moment that we are justified in calling some engineering students "geeks" because they are socially inept and are constantly calculating something or other. We are probably just as justified in calling budding artists and architect students "weird" because they talk of expressing emotions or envisioning spaces. Philosophy students and mature philosophers (oxymoron?) are surely "strange" because who knows what they think about! Furthermore, we can't say that all engineers are alike any more than we can say all artists or all philosophers are alike. Some of us, and by "us" I mean, engineers, artists, scientists and whomever else, are urbane, articulate and sensitive, others are clumsy, inarticulate and bumbling – you tell me who is which *necessarily*. People are people – they are interesting because of their quirks and their differences. But you can't judge a field by one or two practitioners. Furthermore, it is rare that you can predict what someone employed in one field will be doing tomorrow. In fact, some of the most interesting and creative people are marked by just that spontaneity that makes their next move so unpredictable.

One of the best examples of this kind of unpredictability is Wayne Clough, a civil engineer by training, former chair of civil engineering at Virginia Tech, former dean of engineering at Georgia Tech, former provost and president of the University of Washington and now, oh my goodness, Secretary of the Smithsonian Institution – how did that happen? A civil engineer is running a bunch of museums? But to say that is to commit the kind of fallacious reasoning that should be avoided. Dr. Clough is not a civil engineer any more than he is a dean. He is a person who employs different kinds of knowledge, values, and goals to make decisions, depending on what problem he is facing at the moment. The job does not define the person.

## 8.5 Some Speculations on How Engineering Got Left Out of Philosophy and the Possible Death of Philosophy

It is not clear to me how we came to think that engineering was different from other human endeavors and outside of philosophical interest. Maybe it was because engineering got a bad rap when it was cursed by the quantification label. But engineering was not always so closely aligned with quantification. In the Middle Ages it was part of the *media scientia,* the middle sciences whose domain of inquiry lay between pure physics and theology. It was that set of techniques for building things. Both Leonardo da Vinci and Galileo worked in the *media scientia,* Leonardo built the defenses of Milan, Galileo built military compasses and telescopes. Here there was no stigma attached to the thinking or the creativity of the man who made things.

But it cannot be simply the case that being coated with the quantification brush resulted in removing engineering from the philosophical discussion. Other areas of human inquiry involve intense quantification and still merit philosophical attention, such as economics, the social and natural sciences, mathematics, etc. Oh, by the way logic is a field in philosophy. My guess is that stereotyping engineers as interested only in equations and hence as otherwise uninteresting may have been a minor source of the problem but certainly not the whole story.

For there is a deeper issue lurking here – it is not exactly a reiteration of C.P. Snow's (1959) "Two Cultures," but it seems to arise out of some of the same sources. Since the mid-nineteenth century there has been a bifurcation in western culture between what for lack of a better vocabulary I will call the quantifiable world and the expressive world. And the interesting thing is that as this split developed Philosophy got caught in the middle. The deeper problem, the problem that led to our failure to include engineering as part of the larger human enterprise is that Philosophy lost its way. Let me explain. I will be drawing a picture here using a broad brush, but I think the basic outline is secure.

Given the explanatory successes of the sciences (biology, physics, chemistry and geology) and the emerging professionalism of engineering in the nineteenth century, quantification was increasingly seen as a mark of accomplishment. The distinction between moral and natural philosophy began to splinter as many of the human sciences sought to become "real" sciences by endorsing quantitative methods. Political Economy broke apart into Political Science, on the one hand, and Economics on the other, both seeking means to quantify their domains. Psychology split from Philosophy. And they all headed towards the numbers. Everyone was rushing to become quantified and hence raise their profile in the pantheon of intellectual merit. At roughly the same time there was a movement in Philosophy to also become scientific, chronicled in Hans Reichenbach's (1951) *The Rise of Scientific Philosophy*. What could this possibly mean? It has as much to do with misperception as anything else.

At the end of the nineteenth and beginning of the twentieth centuries, one perception of the way science proceeded was that scientists solved their problems by breaking big problems into little ones. Usually the big problems were unmanageable, but by breaking them into more manageable pieces headway could be made. But this way of proceeding was not new. Aristotle knew about this as well. He saw inquiry as a two-part process: analysis and synthesis. However, at the turn of the last century there was no insistence that the scientist solving the smaller problems go back and put it all together. Every once in a while some genius could do that. (I am not arguing that this was in fact the case, only that there was a popular perception that this is how science proceeded, and perception is a very powerful club.)

Philosophy, on the other hand, was still stuck in the old game of system-building. Even into the early twentieth century, philosophers like Whitehead (1929/1978) were still building wildly imaginative metaphysical schemes that relied on strange rubrics such as "process" to create a universe into which to put people and have it make sense. And, as such, what philosophers were doing was increasingly seen as unconnected to the dynamic scientific way of progress. The popular imagination also included engineering in the scientific model of progress because engineers solved problems in much the same way that scientists did: by breaking them into smaller and smaller parts – and while this much remains true, I would rephrase it as saying that scientists solve problems in much the same way as engineers do, by breaking them into smaller and smaller parts (Vincenti 1979). For the model for common sense problem solving isn't some special scientific method, it is epitomized by the engineering feed-back model.

During the twentieth century, in an attempt to remain current, Philosophy attempted to cash in on the model of science and gave up system building in favor of solving smaller and smaller problems. It wasn't the medievalist who counted the number of angels on the head of a pin (as Marjorie Grene so often reminded me) it was the Anglo-American analytic philosophers of the twentieth century. The irony is that in so doing Philosophy became increasingly irrelevant, for its small problems were increasingly perceived by the outside world as having nothing to do with anything at all.

At this time it is also the case that Philosophy simultaneously became preoccupied with the revolutionary programme of explaining away philosophical problems. This is the famous Linguistic Turn coined by Richard Rorty (1963). It is a very complicated story, but here is a simplified version. The insight, perhaps Frege's, perhaps Russell's, was that what were deemed to be eternal philosophical problems were really confusions engendered by the misuse, or vagueness of language. The goal of the new programme was to eliminate philosophical problems by either constructing languages in such a way as to prohibit the possibility of their formulation or to explain away these problems by appeal to the misuse of language. The result of the two trends, scientific philosophy and the Linguistic Turn, has been the disappearance of philosophy from the world of common discourse.

Two towering figures in the world of analytic philosophy Nicholas Rescher and Wilfrid Sellars, waged a backdoor battle against the minute problem solving approach, paradoxically by boldly championing the building of systematic explanatory schemes that tried to make sense of the world as we know it and live it. They did this by using the tools of analytic philosophy and by reinserting human beings into the picture. Philosophy is not all about how language works, but how people use it to communicate better and solve problems affecting their immediate lives. It is not all about logic, but also about how people reason informally and why that is, in the long run, more important than producing a consistency theorem. It is not just about the structural flaws in various ethical theories, but about what kind of a world we should be building in order truly to live the good life. In short, it is about people. It is about how all of us relate to each other and to the world. And if it is about all of us, then it necessarily includes engineers.

Now the interesting result of all this is the glimmer of a resurgence of Philosophy. If we start by concentrating on people and what we do and then try to relate the rest of the universe in which we do it to us, rather than trying to put Man into harmony with some apriori universe, we not only have a different starting point, we have hope for a Philosophy that will actually be of some use. Philosophy is seriously in danger of disappearing from American intellectual life because it has forgotten it is about wisdom and wisdom is about people doing things for the better.

There is one more brick in the wall. Whatever the reason, there is a serious streak of anti-intellectualism in American culture. It may come from our pioneer heritage, where surviving was the problem and anything that could not be immediately seen to contribute to that goal was irrelevant. It could stem from the successor to the pioneer, the builders of nineteenth century America, i.e., the steel and railroad barons. Anything that got in the way of the creation of these empires was seen as irrelevant.[4] It could come from a misplaced super patriotism that saw any arguments against American use of power to advance its political and economic world agenda as threatening and irrelevant. These and other factors formed the background to post WWII developments resulting in driving American intellectual life, here in the very bastion of free speech, out of daily life behind the walls of academe. The McCarthy era and then the damage done to American cultural unity by the Vietnam War conspired to keep American intellectuals from speaking out. They retreated to the universities and insisted on keeping politics out of academic life – we were only to deal with real matters of intellectual interest. Interestingly, it fell to the humanities to set up the battle lines. These are the people who refused to walk the picket lines and participate in demonstrations. Instead, they wrote books and they complained about the state of American life, but they *did* nothing. And of course, if you aren't seen as committed to the life of the mind but, rather, are interested in, for example, making things, then you were clearly anti-intellectual. Scientists were cut some slack because it was claimed that they were interested in "knowledge for its own sake."[5] But not all university professors were seen as intellectually curious in that arrogant and evasive manner favored by the humanities. Chief among the perceived enemy were "the engineers." So in an odd reversal, curious and creative designers and tinkerers were written out of the story. Clearly if engineers are anti-intellectual, then what they do cannot be of intellectual, i.e., philosophical, interest.

## 8.6 Conclusion

However, engineers are people like you and me. They think like you and I do. If what was suggested earlier about feedback loops and common sense is anywhere near the mark, then engineering as a human process in which engineers strive to improve the world is an integral part of the world in which we live. Studying what engineers know, how they know it, what they do and how they do it and why, what they design and build and how it affects us, is as central to the philosophical enterprise of seeing how it all hangs together as anything can be.

---

[4] Of course there is a problem here with the philanthropy of giants such as Andrew Carnegie. But we can perhaps explain away there later concerns with the common good by appeal to a bad conscience.

[5] This is not the time or place to talk about the buying out of the American scientific enterprise by the United States' government in the form of research grants or the developing alliances between universities and industry.

## References

Pitt, J. C. (2006). *Thinking about technology: Foundations of the philosophy of technology.* New York: Seven Bridges Press. (Original work published 1999). http://www.phil.vt.edu/Pitt/jpitt.html

Reichenbach, H. (1951). *The rise of scientific philosophy.* Berkeley: University of California Press.

Rorty, R. (1963). *The linguistic turn.* Chicago: The University of Chicago Press.

Rudner, R. (1953). The scientist qua scientist makes value judgments. *Philosophy of Science, 20*(1), 1–6.

Sellars, W. (1963). Philosophy and the scientific image of man. In *Science, perception and reality.* London: Routledge, p. 1.

Smith, A. (1776). *An inquiry into the nature and causes of the wealth of nations.* London: W. Strahan and D. Cadell.

Snow, C. P. (1959). *The two cultures and the scientific revolution.* New York: Cambridge University Press.

Vincenti, W. (1979). *What engineers know and how they know it.* Berkeley: University of California Press.

Whitehead, A. N. (1978). *Process and reality: An essay in cosmology.* New York: Free Press. (Original work published 1929)

# Chapter 9
# Engineering as Willing

**Jon Alan Schmidt**

**Abstract** Science is widely perceived as an especially systematic approach to knowing; engineering could be conceived as an especially systematic approach to willing. The transcendental precepts of Bernard Lonergan may be adapted to provide the backdrop for this assessment, which is manifest when the scientific and engineering methods are compared. In science, although the will is implicitly involved, the intellect is primary, because the goal is ideal—additional "objective" knowledge. In engineering, although the intellect is implicitly involved, the will is primary, because the goal is pragmatic—some "subjective" outcome, which is often selected by a manager or client, rather than the engineer. Furthermore, engineering problems are rarely well-defined; uncertainties and resource constraints dictate that they be conceptualized and solved heuristically. As a result, different engineers will follow different design procedures and develop different models, none of which is uniquely "correct." Because tradeoffs are always necessary, engineering decision-making—and human behavior in general—is more intentional than rational. Recognizing this can help today's society to overcome its traditional bias in favor of knowing over willing and to engage engineers more explicitly in addressing the many challenges that it faces, technological and otherwise.

**Keywords** Heuristics • Intentionality • Models • Social captivity • Tradeoffs

---

J.A. Schmidt (✉)
Aviation & Federal Group, Burns & McDonnell, Kansas City, MO, USA
e-mail: JonAlanSchmidt@gmail.com

D.P. Michelfelder et al. (eds.), *Philosophy and Engineering: Reflections on Practice, Principles and Process*, Philosophy of Engineering and Technology 15,
DOI 10.1007/978-94-007-7762-0_9, 

## 9.1 Introduction

Philosophy is not so much about finding answers—it is really all about asking questions. For example, what is engineering? At first glance, it appears that engineers would simply be those who operate engines. However, in many languages other than English—including French, German, and Spanish—the word for "engineer" starts with the letter I, not the letter E. In fact, what it really means to be an engineer is to be a person who exercises *ingenuity*. Unfortunately, common views of engineering tend to omit this essential aspect of it.

The classic definition of engineering is attributed to Thomas Tredgold and dates to the 1828 Royal Charter of the Institution of Civil Engineers in Great Britain. It says, "Engineering is the art of directing the great sources of power in nature for the use and convenience of man." This certainly sounds important and impressive, but the problem with this definition is that it does not really capture what engineers do on a daily basis. Most would be a bit reluctant to claim that their job is "directing the great sources of power in nature."

There is another definition that has been attributed to various individuals. It is so pithy, and has been repeated so often, that no one really knows for sure who originally uttered it. The earliest documented instance was in a structural analysis textbook (Brown 1967): "Engineering is the art of moulding materials we do not really understand into shapes we cannot really analyze, so as to withstand forces we cannot really assess, in such a way that the public does not really suspect." This seems much closer to the mark.

One of the interesting things about these definitions of engineering is that, contrary to popular usage, neither refers to it as a *science*. Instead, they both refer to it as an *art*. What is the difference? The dictionary (*Merriam-Webster's* 1993) provides three relevant definitions of science:

- "a department of systematized knowledge as an object of study";
- "a system of knowledge covering general truths or the operation of general laws"; and
- "principles and procedures for the systematic pursuit of knowledge."

Notice that one word is common to all three: knowledge. Apparently science is all about knowledge. Is engineering all about knowledge? In light of these definitions, is engineering a science? Should engineers be calling themselves scientists?

What about art? Once again, the dictionary (*Merriam-Webster's* 1993) provides three definitions that are potentially relevant:

- "skill acquired by experience, study, or observation";
- "an occupation requiring knowledge or skill"; and (my personal favorite),
- "the conscious use of skill and creative imagination."

Once again, notice that one word is common to all three: skill. Apparently art is all about skill. Is engineering all about skill? In light of these definitions, is engineering an art? Should engineers be calling themselves artists? Do engineers perceive what they do for a living as "the conscious use of skill and creative imagination"? Is that how non-engineers—including philosophers—typically perceive what engineers do for a living?

Another way of highlighting this distinction is to say that science has to do with *knowing*, while art has to do with *willing*—that is, making decisions that may or may not have a rational justification. This chapter will explore the notion that, just as science is widely perceived as an especially systematic approach to knowing, so engineering could be conceived as an especially systematic approach to willing.

## 9.2 Knowing and Willing

Presenting this thesis poses an immediate challenge, because—as Mitcham (1994) has noted—"willing, although clearly a theme, is itself so poorly articulated by philosophy." This discussion of willing and its relationship to knowing is grounded in the "transcendental precepts" of Bernard Lonergan (1957). These are not commandments or even guidelines, but rather distinct levels of awareness and function that are inherent, to some degree, in every person. Lonergan argued that humans innately seek, legitimately gain, and properly apply *knowledge* by means of these operations. His formulation may be adapted as follows.

- The first level is *experience*, and the precept is to be *attentive* in examining the data presented.
- The second level is *understanding*, and the precept is to be *intelligent* in envisaging possible explanations.
- The third level is *judgment*, and the precept is to be *reasonable* in evaluating which account is most likely.
- The fourth level—an insertion into Lonergan's scheme—is *deliberation*, and the precept is to be *considerate* in exploring potential courses of action.
- The fifth and final level is *decision*, and the precept is to be *responsible* in electing to proceed accordingly.

Attentive experience, intelligent understanding, and reasonable judgment lead people to adopt *beliefs* about how the world *was* in the past and *is* now; considerate deliberation and responsible decision lead people to make *choices* about how the world *will be* in the future.

In fact, although Lonergan was mainly concerned with developing a cognitional theory to explain the process of human *knowing*, the transcendental precepts also call for *willing*. It takes an act of the *will* to be attentive, to be intelligent, to be reasonable, to be considerate, and to be responsible. These are non-compulsory inner demands, especially the last two. Considerate deliberation and responsible decision require not only apprehending an obligation, but also striving to fulfill it—setting priorities and selecting the best way forward from among multiple options. Assistance is provided by a tender and well-informed conscience, disciplined through habitual exercise of the transcendental precepts, which will consistently evoke attraction to the good or the better, and repulsion from the bad or the worse.

## 9.3 Science and Engineering

Lonergan's framework greatly clarifies the interactions between knowing and willing, as well as their distinctions. With this in mind, consider these contrasting concepts:

- intellect vs. volition,
- adopting beliefs vs. making choices,
- having reasons vs. having motives,
- exercising judgment vs. reaching a decision, and
- being reasonable vs. being responsible.

Which ones are commonly associated with science, and which with engineering? Again, the difference between science and engineering is analogous to the difference between knowing and willing.

This becomes even more evident when simply comparing the scientific and engineering methods. Scientists observe natural phenomena, propose hypotheses in an effort to explain them, and conduct careful experiments to test their theories. Although the will is implicitly involved, the *intellect* is primary, because the goal is ideal: additional knowledge that is supposed to be objective.

By contrast, as Koen (2003, 2010) has long been arguing, engineers use "state-of-the-art heuristics to create the best change in an uncertain situation within the available resources." Koen has recently revised his wording here—note that this definition of the engineering method invokes the "state of the *art*," not the "state of the *science*," and explicitly acknowledges that engineering is a *creative* activity. Although the intellect is implicitly involved, the *will* is primary, because the goal is pragmatic: some outcome that is usually subjective. Knowledge serves mainly as a necessary but insufficient means to that contingent end—"the best change"—which is usually not something that is up to the engineer to determine.

## 9.4 Social Captivity

In fact, as Goldman (1991) has observed, engineering is a captive enterprise. For one thing, it is widely—and wrongly—perceived as nothing more than *applied* science. In addition, the practice of engineering has been limited in terms of what problems engineers are allowed to address and what solutions are considered acceptable. The autonomy of the profession is restricted by the need for someone to retain an engineer before he or she can undertake a specific project. Engineers rarely have the opportunity to influence the process that leads a manager or client to decide that a certain product or facility is necessary or desirable. Furthermore, a variety of constraints are imposed on engineering designs by others as a result of aesthetic, functional, or other considerations, which are often non-technical.

These two types of captivity—one intellectual, the other social—are not separate, but interdependent; the first anchors the second, which in turn reinforces the legitimacy of the first. Goldman's thesis was that technology and innovation are generally dominated by market-driven value assessments, rather than by technical knowledge. Even when managers or clients are engineers by training, the decisions that they make inevitably reflect the agendas and priorities of the organizations that they serve—not necessarily the capabilities and limitations of the engineers whom they supervise or retain.

As a result, engineering tends to be *instrumental* in nature; it is exploited by non-engineers to achieve their own objectives, which may be quite arbitrary. In other words, the willfulness of engineering is both enabled and restricted by the willfulness of the institutions that appropriate it.

## 9.5 Heuristics and Design Procedures

According to Koen (2003), heuristics are central to engineering—and everything else, for that matter—and a heuristic is "anything that provides a plausible aid or direction in the solution of a problem but is in the final analysis unjustified, incapable of justification, and potentially fallible." This formulation reflects how engineering is intrinsically at odds with the dominant tradition in Western culture going all the way back to Plato's triumph over the Sophists.

Goldman (1984, 1990, 2004) has pointed out that science—along with philosophy, as well—is supposed to be concerned with necessity, certainty, universality, abstractness, and theory. It seeks objective knowledge of timeless truth that is based on reality, for the purpose of intellectual contemplation and understanding. By contrast, engineering is characterized by contingency, probability, particularity, concreteness, and practice. Engineers rely on subjective know-how and historical opinions that are derived from experience, with the goal of willful action and use. Heuristics cannot be "proven" in the absolute sense, but their utilization is legitimately warranted, frequently on the grounds of successful past implementation.

In fact, each individual engineer has a unique collection of relevant heuristics at his or her disposal, along with "meta-heuristics" for selecting which heuristics are most appropriate in a given set of circumstances. When these are combined to facilitate converting a client's technical and non-technical requirements into a viable solution that adequately accounts for the unknowns and satisfies all applicable constraints, they constitute what Addis (1990, p. 46) calls a design procedure. This is analogous to a hypothesis in the scientific method; however, a design procedure does not lead inevitably to a particular outcome. In fact, Addis notes "that it is possible to produce very similar structural designs using different design procedures and that similar design procedures can lead to significantly different structures—there is no logical connection between the two."

Most design procedures include the development of mathematical *models* that are supposed to capture the important aspects of reality. The engineer's challenge is to ascertain what those features are and what assumptions and simplifications can safely be incorporated in order to keep everything manageable, while still yielding a meaningful assessment of likely performance. As de Vries (2010) has observed, two common strategies are abstraction and idealization. Abstraction involves neglecting certain aspects of reality in order to gain a better understanding of the remaining aspects. Idealization involves replacing a complicated aspect of reality with a simplified version.

Fundamentally, models are approximate representations that serve as epistemic tools (Boon and Knuuttila 2009)—they help engineers learn about the parameters of a design problem and evaluate possible solutions. Although analysis of a model is usually straightforward, conforming to fundamental principles derived from *science*, its initial construction and subsequent adjustment require "the conscious use of skill and creative imagination"—which, again, is one of the dictionary definitions of *art* (*Merriam-Webster's* 1993).

## 9.6 Engineering Intentionality

The bottom line here is that engineering is not deterministic; it routinely requires setting priorities and selecting the best way forward from among multiple options when there is no one "right" answer (Addis 1997). Consequently, attempts to apply a theory of *rationality* to engineering are probably misguided (e.g., Kroes et al. 2009); *intentionality* seems like a more appropriate concept. Take the example of a bridge. Is there a single optimal span for a particular location? The one across the Golden Gate might be a candidate, but the reality is that there is always a staggering array of variables that influence what is ultimately constructed. For example:

- What kind of traffic will the bridge carry—cars, trucks, or trains?
- What type of bridge will it be—suspension, cable-stayed, or box girder?
- What primary material will be used—wood, steel, or concrete?

Tradeoffs are inevitable, and not just for technical reasons—there are often budget constraints, legal restrictions, or political considerations that come into play. As Goldman (2010) has discussed, these kinds of tradeoffs—both technical and non-technical—are at the heart of the design process, and therefore constitute the essence of engineering intentionality.

In the end, how does a bridge designer settle on the final form of the structure? There is no rigid and inerrant formula that will provide the "proper" outcome. This is the challenge that engineers face—routinely having to make and justify seemingly arbitrary decisions with an understanding of the situation that is incomplete, at best.

The contrast here is really between deduction and induction. *Deduction* conclusions that are absolutely certain, because they contain no new inf… that was not already present in the premises. A rational person who accepts the premises must also accept the conclusion. *Induction* reaches conclusions that are *not* certain, because they *do* contain new information; the argument has to add something to the premises.

With this in mind, which approach do engineers most commonly employ? Put another way: Is every design a foregone conclusion, derived deductively from the specified criteria and other project requirements? Of course not! Engineering is clearly an *inductive* activity—it requires the addition of information along the pathway from thought to thing, or from concept to completion. In other words, engineers are *information creators*.

There is no law of nature or mathematics that dictates a particular design. There are constraints, to be sure. There are decisions made by others that limit the engineer's options. In the end, though, the engineer makes certain choices; and another, equally competent engineer might well go in a different direction. Each designer has a singular point of view that informs how he or she "sees" a problem and its (potential) solution. Certain concepts, especially within a particular discipline, are familiar to all; but each individual also develops and refines a certain amount of technical intuition through professional experiences—"tacit" knowledge that is difficult to capture and communicate to others.

As a consequence of all this, two models of the same system can both be "correct," yet yield different results. The advantage that an engineer has over a scientist is the ability not only to adjust the model to represent reality better, but also to adjust *reality* to suit the model better—yet another example of engineering practice being more of an art than a science. This is especially true for unconventional projects that go beyond standard designs and details, requiring the engineer to develop *custom* designs and details using first principles. In such cases, it is often the model that dictates reality, rather than the other way around.

In this way, engineering practice involves reinterpreting an inductive situation to facilitate deductive analysis. Once the model is set up, the results follow inexorably; but setting up the model is the real work of the engineer. In other words, engineering includes problem *recognition* and *definition*, not just problem solution. Engineers have to convert all of the relevant design criteria—which are often dictated by clients, codes, and other authorities—into the "language," so to speak, of engineering.

As in the case of natural language translation, the formulation of engineering problems and their solutions is inherently indeterminate. They tend to be ill-structured (Simon 1973) or even "wicked" (Rittel and Webber 1969). Consequently, design—in fact, all human behavior—is not ultimately governed by *reasons*, but by *motives*. Although common usage treats these two terms as virtually synonymous, the prevalence of the first one in both ordinary and philosophical discourse reflects an ancient prejudice that subordinates action to contemplation; that is, willing to knowing.

## 9.7 Conclusion

As members of the profession that exemplifies willing, engineers should strive to resist and reverse this tendency. For example, although it is common in many fields—including engineering—to perceive the existence of a "gap between theory and practice," Addis (1990) advocates abandoning this widespread notion in favor of an alternative classification: engineering *science* vs. engineering *design*. The essential difference between these two activities is not in the types of *knowledge* that they employ, but rather in their distinct *purposes*: further understanding and explaining the natural world vs. efficiently producing useful artefacts in a context of uncertainty.

The fact of the matter is that the outcomes of engineering are rarely black and white, right or wrong. Managers and clients give engineers their problems and expect them to be solved, even though there are no *objective* solutions: there are simply too many parameters and too many criteria. Those who hire engineers depend on them to exercise good *judgment*, grounded in formal education and honed by subsequent experience. As Vick (2002) wrote, "The novice begins with data and ends with a number; the expert begins with knowledge and ends with understanding."

There is a widespread perception that engineers are little more than number crunchers. This is not only inaccurate, it is *dangerous*. Data and numbers are meaningless, unless and until they are *properly* interpreted by someone who *knows* where they came from and *understands* what they mean.

With this in mind, consider the suggestion made by Samuel Florman (1996): "I propose that we take the time to think about who we are, as citizens living in a 'technological society,' and—for engineers—as the profession most essential to the well-being of that society." After all: Is there any situation in human existence that is *not* subject to uncertainty and resource constraints? Or that does *not* require the use of successful heuristics? Or that does *not* demand willing, as well as knowing? It is precisely when there is more than one path to follow that it is possible and desirable to exercise *wisdom*—and is that not what *philosophy* is supposed to be all about?

This being the case, because of their training and temperament, engineers are uniquely suited to help society wrestle with the many challenges that it faces—not only in the technological realm, but across all aspects of human existence.

## References

Addis, W. (1990). *Structural engineering: The nature of theory and design*. New York: Ellis Horwood.

Addis, W. (1997). Free will and determinism in the conception of structures. *Journal of the International Association for Shell and Spatial Structures, 38*(2), 83–89.

Boon, M., & Knuuttila, T. (2009). Models as epistemic tools in engineering sciences. In A. Meijers (Ed.), *Philosophy of technology and engineering sciences* (pp. 693–726). Amsterdam: Elsevier.

Brown, E. H. (1967). *Structural analysis volume I*. New York: Wiley.

de Vries, M. J. (2010). Engineering science as a "discipline of the particular"? Types of generalization in engineering sciences. In I. van de Poel & D. E. Goldberg (Eds.), *Philosophy and engineering: An emerging agenda* (pp. 83–93). Dordrecht: Springer.

Florman, S. C. (1996). *The introspective engineer*. New York: St. Martin's Griffin.

Goldman, S. L. (1984). The *techné* of philosophy and the philosophy of technology. *Research in Philosophy & Technology, 7*, 115–144.

Goldman, S. L. (1990). Philosophy, engineering, and western culture. In P. T. Durbin (Ed.), *Broad and narrow interpretations of philosophy of technology* (pp. 125–152). Dordrecht: Kluwer Academic.

Goldman, S. L. (1991). The social captivity of engineering. In P. T. Durbin (Ed.), *Critical perspectives on nonacademic science and engineering* (pp. 121–145). Bethlehem, PA: Lehigh University Press.

Goldman, S. L. (2004). Why we need a philosophy of engineering: A work in progress. *Interdisciplinary Science Reviews, 29*(2), 163–176.

Goldman, S. L. (2010). Beyond satisficing: Design, trade offs and the rationality of engineering. *2010 Forum on Philosophy, Engineering & Technology.*

Koen, B. V. (2003). *Discussion of the method: Conducting the engineer's approach to problem solving*. New York: Oxford University Press.

Koen, B. V. (2010). Debunking contemporary myths about engineering. *2010 Forum on Philosophy, Engineering & Technology.*

Kroes, P., Franssen, M., & Bucciarelli, L. (2009). Rationality in design. In A. Meijers (Ed.), *Philosophy of technology and engineering sciences* (pp. 564–600). Amsterdam: Elsevier.

Lonergan, B. J. F. (1957). *Insight: A study of human understanding*. New York: Philosophical Library.

*Merriam-Webster's collegiate dictionary* (10th ed.). (1993). Springfield: Merriam-Webster.

Mitcham, C. (1994). *Thinking through technology: The path between engineering and philosophy*. Chicago: The University of Chicago Press.

Rittel, H. W. J., & Webber, M. M. (1969). Dilemmas in a general theory of planning. *Panel on Policy Sciences American Association for the Advancement of Science, 4*, 155–169.

Simon, H. (1973). The structure of ill-structured problems. *Artificial intelligence, 4*, 181–201.

Vick, S. G. (2002). *Degrees of belief: Subjective probability and engineering judgment*. Reston: American Society of Civil Engineers.

# Part II
# Reflections on Principles

# Chapter 10
# Debunking Contemporary Myths Concerning Engineering

**Billy Vaughn Koen**

**Abstract** Efforts to understand the human activity we call engineering and to develop a Philosophy of Engineering are hampered by a number of myths. One oft-heard and over-used example will demonstrate this point. Repeatedly we read in the newspaper or hear on television that "engineering is applied science" in spite of the demonstrable fact that this could not possibly be true. The objective of this paper is to debunk some of the most egregious of the contemporary myths concerning engineering. Rather than rely on conjecture and personal opinion, the strategy employed is to use an inordinate number of direct quotations from classical texts and living experts. This is supplemented by extensive quotations, images, and commentary from documentaries produced by the most reputable sources such as the *History Channel*, the *National Geographic Channel*, the *Discovery Channel*, and the *Smithsonian Encyclopedia* where there would be credible fact checking by content specialists if it was to exist anywhere. At the conclusion of our investigations, we will consider the archetypical engineering project and meet the earliest engineer in history whose name is known and stare directly into his face.

**Keywords** Engineering myths • Engineering artefacts • Engineering method • Engineering trial and error • Ancient engineering

B.V. Koen (✉)
Professor Emeritus, Department of Mechanical Engineering, The University of Texas at Austin, Austin, TX, USA
e-mail: koen@uts.cc.utexas.edu

D.P. Michelfelder et al. (eds.), *Philosophy and Engineering: Reflections on Practice, Principles and Process*, Philosophy of Engineering and Technology 15,
DOI 10.1007/978-94-007-7762-0_10, 

## 10.1 Introduction

Recently the number of conferences and articles seeking to develop a Philosophy of Engineering has increased significantly. These conferences bring together philosophers and engineers who, although recognized experts in their respective fields, have very different, often contrasting, world-views. As a result, efforts to understand the human activity called engineering are hampered by a number of myths about engineering put forth on both sides of the divide.

This paper brings together some of these misconceptions collected from international conferences that appear naïve from the point-of-view of the engineer, with the sincere hope that someday a philosopher will return the favor and correct the naïve views of the engineers about philosophy to the mutual benefit of the two groups of scholars and the benefit of the development of a Philosophy of Engineering.

This paper is divided into two parts: first, a very brief discussion of a definition of engineering method that has appeared frequently in the literature is given and then an analysis of a series of contemporary myths concerning engineering is proposed.

During these investigations, we will have occasion to meet the earliest engineer who has ever lived whose name is known as an example of an engineer using the engineering method, examine his engineering work, and then—finally—stare directly into his face.

## 10.2 Definition of Engineering Method

Engineering is most appropriately understood and recognized in terms of behavior: it is an activity, it is something an individual does, it is a creative undertaking. If we look in on an individual and see that he or she is doing certain specific, identifiable things, we can infer that he or she is an engineer actively engaged in engineering work. Therefore, engineering should be understood in terms of *method* instead of in terms of one of the multitude of common, arbitrary, egocentric *definitions* often put forth. The simple fact that engineering is behavior is confirmed by a quotation from one of England's most noted nineteenth century engineers, Sir William Fairbain (Burke 1919):

> The term *engineer* comes more directly from an old French word in the form of the verb *s'ingénieur*... and thus we arrive at the interesting and certainly little known fact, that an engineer is anyone who seeks in his mind, who sets his mental powers in action, in order to discover or devise some means of succeeding in a difficult task he may have to perform.

An accurate understanding of what engineering is depends on an understanding of what an individual must be doing to be called an engineer.

As a result we began our investigations with—but not belabor—a slightly revised and improved definition of *engineering method* that has frequently appeared in the literature as a starting point for our considerations, to wit:

> The engineering method is the use of state-of-the-art heuristics to create the best change in an uncertain situation within the available resources.

**Fig. 10.1** State of the art

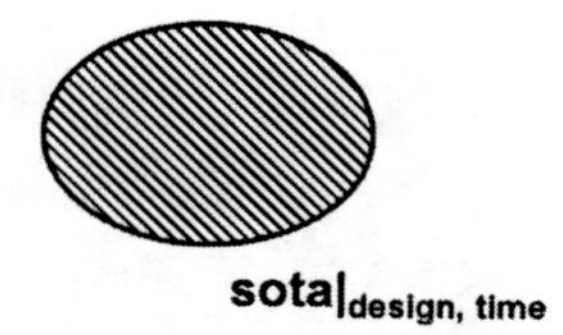


This definition uses many important concepts in highly technical senses. But space considerations are such that focus in this paper is limited to the two terms *state of the art* and *resources.* Those interested in investigating this definition in more detail are referred to the seminal book that forms the basis of this article, *Discussion of the Method: conducting the engineer's approach to problem solving,* published by Oxford University Press in 2003 (Koen 2003). It will be referred to by the acronym *DOM* in what follows.

### *10.2.1 State of the Art*

The noun *state of the art* or the adjective *state-of-the-art* in the definition just given is one of the most important concepts in engineering. In the literature, it has frequently been discussed in conjunction with. but apart from, an analysis of engineering method. But for present purposes it has been absorbed into the definition itself.

Seldom does an engineering project require only one heuristic. This introduces the concept of a collection or set of heuristics that we will call the state of the art or to use an acronym *sota.* Figure 10.1 shows pictorially a set of heuristics or sota. It must have a label and time stamp and can be written as $\text{sota}|_{design,\ time}$ to mean the set of heuristics used in a specific design at a specific time.

The notion of a set of heuristics or sota evaluated at a specific time is a very powerful concept. The sota of an individual both confines and restricts the range of the possible in engineering design for him or her.[1] It can refer not only to the individual, but also to a group of individuals—even to countries. It is reasonable to speak of the sota of French engineers, Japanese engineers, and American engineers and to compare them. It is reasonable to compare the sotas of a developed and an developing country or to talk of technological transfer as a strategy for transferring the appropriate heuristics from one nation to another. It is, also, reasonable to consider engineering education as converting the entry sota of a freshman engineering student to that of a competent, practicing engineer.

[1] See an article in a previous volume of this series by Springer for more detail (Koen 2010).

**Fig. 10.2** Effect of different resources. (**a**) Karnak, (**b**) Parthenon, (**c**) Pantheon

## *10.2.2 Limitation by Resources*

The second concept in the given definition of engineering to be considered is the important notion of the resources and the constraints of those resources on the typical engineering problem. From DOM, page 15:

> An engineering problem is defined and limited by its resources, but the true resources must be considered. Because we tend to think only in terms of depletable resources, because we confuse nominal and actual resources, and because we neglect the efficiency of allocating resources and the probability of exchanging one kind for another, often the true resources are hard to determine.

A recent documentary on the Science Channel (2008) gives an interesting and unexpected example of the importance of resources that appear in the definition of engineering method. The host, Ian Steward, is a Scottish geologist who is interested in the impact of geology on civilizations. He argues that the rocks found in an area influence its art, buildings, etc. This parallels the point that the resources, in this case the rocks, impact the local engineering design. The example Dr. Steward uses is given in Fig. 10.2.

At issue is the design and construction of *enclosed space.* Since the beginning of civilizations, humans have designed houses, meeting rooms, kivas, temples, and so forth from the available materials. Dr. Steward compares the enclosed space in large rooms in Egypt, Greece, and Rome and concludes that the local rocks dictated and constrained their design. Figure 10.2a is a picture of a portion of the hypostyle hall in the Karnak temple in Egypt which consists of 134 huge columns seven stories tall. Only the sedimentary rock, sandstone, was available for its construction. As is well known to engineers, building material such as sandstone is relative strong in compression, but weak in tension. As a result, for Karnak, massive, closely spaced columns were needed to support the architrave or beam that rests on the capitals of the columns. Note the tiny man in the figure to give a sense of proportion. As a result the enclosed space is crowded and has a claustrophobic feeling. Figure 10.2b is a picture of the Parthenon in Greece. Here the stronger metamorphic rock, marble, was available and the columns are slenderer and the enclosed space has a more airy feeling. Finally, Fig. 10.2c is a picture of the Pantheon in Rome. Here the much stronger igneous rock was available to make very strong concrete. As a result of this building material in conjunction with the innovation of the arch, the load of the roof is transferred down the walls to the ground and a truly spacious room with no visible means of support could be designed. The point is that the resources, in this case the indigenous rocks, available to the engineer affect the sota and ultimately the final design.

## 10.3 Debunking Contemporary Myths

Attention now turns to the second objective of this paper, debunking specific myths concerning engineering. To be examined are the claims that (1) the definition of engineering method previously discussed is vacuous, (2) engineering is a relatively recent human activity, (3) engineering is applied science, (4) engineering is trial and error, and (5) engineering artefacts must be concrete objects that persist over time.

### *10.3.1 Myth: The Definition of Engineering Method in Terms of Heuristics Is Vacuous*

A vague feeling that the definition of engineering method just discussed is vacuous is a concern that was raised concerning the present paper at a recent fPET conference.[2]

The complainant poses an extremely important question that is very subtle, although somewhat outside of the scope of this article. The response has already been extensively developed in a variety of forums, most notably in the philosophical journal, *The Monist* (Koen 2009) and in DOM (Koen 2003). To quote the example given:

> A person placing a wager on the daily double at the nearest race track may also be using state-of-the-art heuristics to create the best change in an uncertain situation within the available resources.

and as a result the proposed definition is vacuous.

This myth results from misinterpreting the nature of the term *state of the art* as can be understood by considering Fig. 10.3. This figure shows a large, grey, irregularly shaped sota labeled $\text{sotal}_{Overall,\ t}$ inside of which are two overlapping sotas labeled $\text{sotal}_{engineering,\ t}$ and the solid black one, $\text{sotal}_{Daily\ Double,\ t}$. While it may be true, that the latter two may share some heuristics in common as indicated by the extent of the overlap between them, it is certainly not true that they are identical.

[2] The 2010 Forum on Philosophy, Engineering and Technology (fPET-2010) held on 9–10 May 2010 at the Colorado School of Mines in Golden, CO.

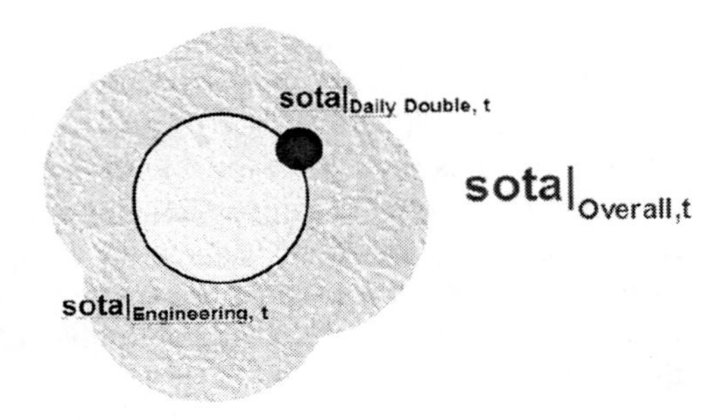


**Fig. 10.3** Comparison of sotas of daily double winner and engineer

The interesting subtlety is in defining the heuristics that properly defines each. This is done heuristically.

> Or to quote the cited article in *The Monist*, to define each sota,
> Use the heuristics heuristically thought to be appropriate for that domain.

We can anticipate that there could be other sotas representing the method of the novelist, artist, flautist, and so on. Each would appear within the overall sota with varying amounts of overlap.

The astuteness of the critique highlights the similarity of all of these definitions of method that ultimately leads to a definition of universal method given in the cited references.[3]

Properly understood, the definition of engineering given above withstands the criticism and is hardly vacuous.

### *10.3.2 Myth: Engineering Is a Relative New Human Activity*

The feeling that engineering is a relatively new human invention is a notion that once made a curious appearance at a recent international conference on the Philosophy of Engineering.[4]

In a room with several engineers and philosophers who were very well-known in their respective specialties present, the question was asked "Do you believe there were engineers in ancient Egypt?" The philosophers immediately responded "of course not"; the engineers responded "but of course. Why do you ask?"

---

[3] For a consideration of this definition and its relationship to other methods, specifically, to universal method from a more philosophical view, an article that appeared in the journal *The Monist* (Koen 2009) might prove useful. Finally, two oral histories, one entitled "The Search for Universal Method" can be found at the persistent URL, http://www.me.utexas.edu/~koen/etc-lecture/ (Accessed Nov. 1, 2011) and the other a keynote address for the Workshop for Engineering and Philosophy at The Royal Academy of Engineering, London, England entitled "Towards A Philosophy of Engineering"(Koen et al. 2008), are available.

[4] Workshop on Philosophy and Engineering, Royal Academy of Engineering, London, England, November, 2008.

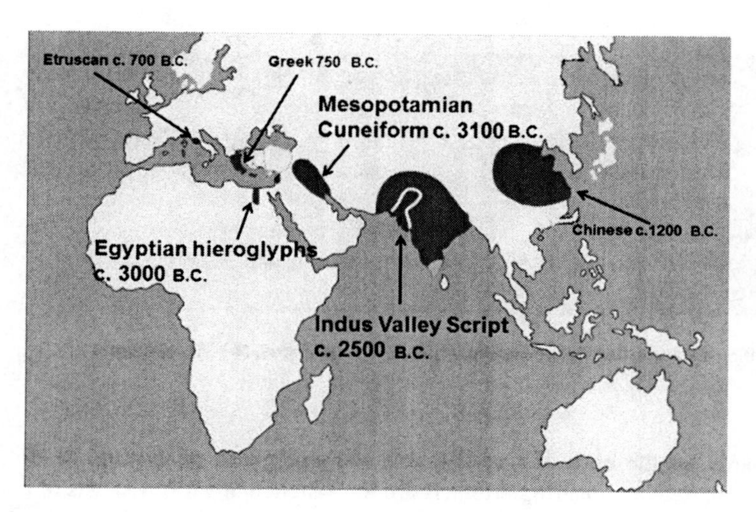


**Fig. 10.4** First appearance of writing

The erroneous feeling among some that engineering is a relatively new human activity may derive from the fact that the English word *engineer* first entered the language in the early fourteenth century as "constructor of military engines" (Harper) with the understandable implication that that is when the behavior we associate with engineering first appeared.

Earlier in this paper the notion that *engineering* necessarily has anything to do with engines was challenged by citing Sir Fairbain. A quotation from another etymological dictionary will substantiate Sir Fairbain's view (Spiritus-temporis).

> It is a myth that engineer originated to describe those who built engines. In fact, the words engine and engineer (as well as ingenious) developed in parallel from the Latin root ingenious, meaning "skilled". An engineer is thus a clever, practical, problem solver.

Once again we must insist that we should base the notions, *engineer* and *engineering*, on behavior—on engineering method—and then ask when the behavior we associate with engineering first appeared.

Let's listen in rapid succession to the testimony of a large group of credible witnesses.

We begin with a quotation from the classic book *The Ancient Engineers* by L. Sprague de Camp (1963):

> The story of civilization is, in a sense, the story of engineering—that long and arduous struggle to make the forces of nature work for man's good.

To see that what Spargue de Camp says is true, consider Fig. 10.4. This is a redrawing and simplification of a published map that preserves the essential dates when *writing first appeared* in various countries (Robinson 2009).

Although some of the precise data may be in dispute, Egypt, Mesopotamia, and Indus are arguable the oldest civilizations we know based on one of the common

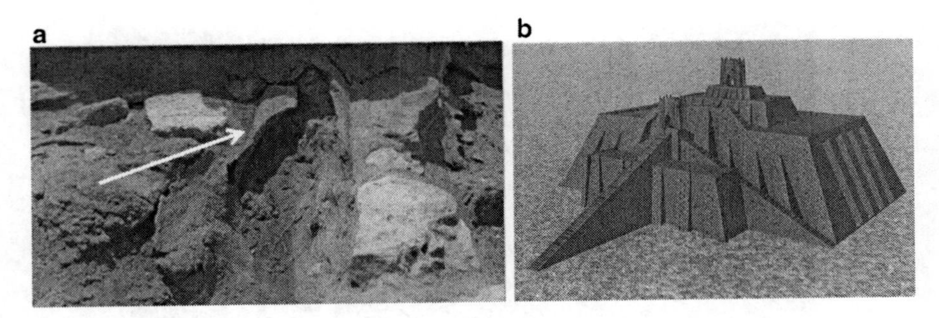


**Fig. 10.5** Extant examples of ancient engineering. (**a**) Indus, (**b**) Mesopotamia

standards for the birth of a civilization, the emergence of writing. In all of these three, concrete engineering artefacts are in existence in various stages of disrepair.

Figure 10.5a is a screen capture from the documentary, *What the Ancients Did For Us: The Indians,* produced by the BBC TWO for The Open University (Hart-Davis ). The white arrow is pointing to the broken end of a clay pipe in ancient Indus[5]. The moderator reaches down and picks up a broken piece of the pipe and says in a truly astounded voice:

> This is a 4500 year old sewerage bath water collection pot chard.

and then somewhat later in a voice over

> …showing the extraordinary skills in *engineering* and planning.

On this evidence alone, surely we should admit that there were engineers in ancient India.

In Fig. 10.5b we see a modern depiction of the Sumerian Ziggurat at Ur[6] in ancient Mesopotamia. The rubble is still there and efforts are underway to build a reconstruction. If that is not sufficient to win an argument that there were engineers in Mesopotamia, we can turn to another in the series of documentaries produced by the BBC—this time to the one entitled *What the Ancients Did for Us: The Mesopotamians* for added evidence. And finally, an exhaustive and definitive treatment of Mesopotamia, the Ziggurat, and the state of the art of science, mathematics, and engineering works is to be found in a Britannica guide to ancient civilizations in reference (Kuiper 2011). Surely we should admit that there were engineers in ancient Mesopotamia.

Finally, ancient Egypt seals the case that there were engineers in the ancient world. The number of colossal monuments, temples, fortifications, and buildings that have been very well preserved in the dry climate and buried under the sands

[5] This is in Dholavira in the Western area of present day India.

[6] Located in southern Iraq, the Ziggurat was part of a massive temple complex where the moon god Nanna lived.

**Fig. 10.6** Examples of Egyptian engineering

should leave no doubt. See Fig. 10.6. This figure shows us the huge monuments at Abu Simbel, the Temple at Karnak, and the sphinx and pyramids as just a few samples of the fruits of the engineers labor in ancient Egypt.

In fact, an entire documentary produced for the History Channel aptly entitled *Engineering an Empire: Egypt* (Cassel 2006) is completely dedicated to these ancient achievements.

Dr. Kent Weeks, American University of Cairo claims that:

> Twenty-five hundred years before the reign of Julius Caesar, the ancient Egyptians were deftly harnessing the power of engineering on an unprecedented scale. Egyptian temples, fortresses, pyramids and palaces forever redefined the limits of architectural possibility.

and from the same documentary, Dr. Zahi Hawass, Secretary General, The Supreme Council of Antiquities says:

> The Egyptians put the foundation of engineering—they were the people who invented engineering.

This importance of engineering in Egypt is a sentiment Dr. Hawass has repeated in numerous documentaries produced by a wide variety of organizations.From another documentary called *Secrets of Egypt: The Valley of the Kings* produced by Five TV (Halliley), an archeologist and practicing engineer, Steve Macklin who is a professional tunneling engineer with Tunnelling & Geology, Arup appearing *in situ* in the documentary and shows us:

> [how]he recognized the technique [being used] because it is one the engineers still use today.

Surely we should admit that there were engineers in ancient Egypt.

Considering the evidence in Indus, Mesopotamia, and Egypt, it is hard to dispute the claim in Wikipedia (Wikipedia: Civil Engineering) that

> Engineering has been an aspect of life since the beginnings of human existence.

Or as was succinctly stated in DOM, page 7:

> **To be human is to be an engineer.**

Based on these comments by professional engineers, comments from credible documentaries, extant engineering artefacts in the earliest civilizations (and we have only scratched the surface), we must conclude that there were engineers in ancient times—and the claim that there were no engineers in ancient Egypt is a myth.

### *10.3.3 Myth: Engineering Is Applied Science*

"Engineering is applied science." This is undoubtedly the most common definition of engineering. It appears frequently in the newspaper; on the television; and from the lips of the sophisticated, of the uneducated, and even, unfortunately, on occasion, of the engineer. Some credence, or more appropriately blame, for this myth should be given to the definition of engineering of the Engineering Council for Professional Development (ECPD) around 1932 (with emphasis added) (Wikipedia: Engineering):

> [Engineering is] the creative application of *scientific principles* to design or develop structures, machines, apparatus, or manufacturing *processes*, or works utilizing them singly or in combination; ...

The problem is that the definition of engineering in terms of science is not true—in fact, it cannot possibly be true. To see this, we need only return to Fig. 10.4 and this time focus attention on the appearance of Greece as a civilization based on the appearance of writing in that country. This figure shows, in a smaller font, that writing in Greece appeared about 750 BC.

Although there are minor disputes in the literature, science made its appearance somewhat later in about the sixth century BC with the Ionian Philosophers—Thales, Anaximander, and Anaximenes. See DOM (Koen 2003).

There is a rich literature in the History of Science concerning the birth of science, several quotations are representative (Burnet 1930):

> ...it is an adequate description of science to say that it is thinking of the world in the Greek way. That is why science has never existed except among people who came under the influence of Greece.

and from another scholar

> The Greeks were the first scientists and all science goes back to them.

For comparison as to age, what is one of the earliest examples of engineering on a significant scale in the literature? A likely candidate would have to be the city of Memphis, the capital of Egypt during the Old Kingdom, founded by the pharaoh

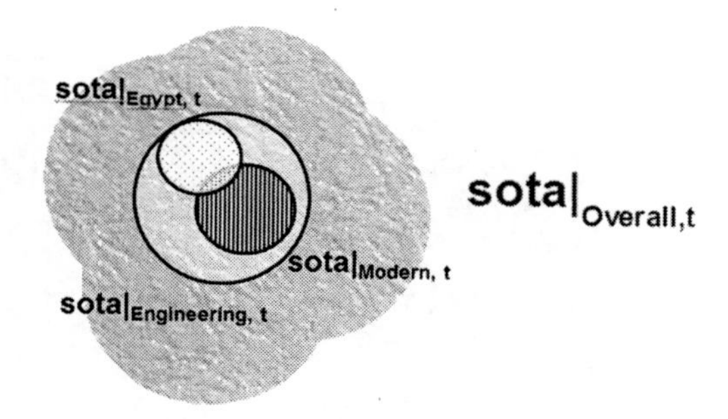


**Fig. 10.7** Comparison of Egyptian and modern sotas

Menes around 3000 BC. The ruins of Memphis are 20 km (12 miles) south of Cairo, on the west bank of the Nile.

Memphis has had several names during its history of almost four millennia. Its Ancient Egyptian name was Inebou-Hedjou, and later, Ineb-Hedj (translated as "the white walls"), because of its majestic fortifications and crenellations (battlements). These historical fortifications were certainly the work of engineers (Wikipedia: Memphis).

Clearly, engineering predated science—by millennia. Science cannot logically be used as a definitive definition of engineering as it existed throughout history.

At best, we might try to argue that engineering and its relationship to science is as given in Fig. 10.7.

This figure is interpreted as follows. Modern engineering, represented by the crosshatched circle indicated by $\text{sota}|_{Modern,\ t}$ contains science, the use of modern tools, and contemporary design techniques. The small plain circle indicated by $\text{sota}|_{Egypt,\ t}$ would contain the skills for working with copper tools and other appropriate heuristics used by Egyptian engineers but long forgotten in the present day. On this basis the overall definition of engineering would be represented by the large circle surrounding both of them, but it is somewhat larger to account for other engineering traditions.

One disclaimer concerning this view of engineering is worthy of note. As a matter of fact, even *modern* engineers do not *always* use science as we see in the design of the Mars rover, the deep space probe, and more recently a deep water oil exploration where the exact scientific conditions are impossible to know and what science that does exist is used more as heuristics.

If we forsake science as the *sine qua non* of engineering and try mathematics instead as some have tried to do, we again run into trouble.

The earliest extant treatise on mathematics showing mathematical calculations is the celebrated Rhind papyrus and to a lesser extent, the Moscow papyrus, shown in Fig. 10.8. One can just make out a triangle on the former and a truncated pyramid on the latter. Actually, there are three other minor papyri that could be vaguely relevant here (Darling). But the mathematics depicted in all of these is very imperfect

**Fig. 10.8** Rhind and Moscow papyri

and used more as heuristics than the more certain mathematics we think aids the engineers of today.

In any event, the persistent claim that engineering is applied science rests on an irresponsible anachronism.

## *10.3.4 Myth: Engineering Artefacts Must Be Concrete Objects*

Whether or not an engineering *artefact* must be a concrete object as opposed to the claim that a *process* would also qualify as the result of engineering design caused discussion at a recent conference[7]. Some philosophers present insisted that an artefact must be a physical object—something a person could touch. This view is not consistent with engineering practice.

First, refer back to the definition of engineering by the ECPD given in the previous section on page 10. The engineers who developed that definition specifically give a *process* as one of the specific ends of engineering design.

Second, the etymology of the word *artefact* makes it clear that an artefact is "anything made by human art" which would, of course, included a process and then

[7] Norms, Knowledge and Reasoning in Technology Conference at Boxmeer, the Netherlands, sponsored by University of Technology, Eindhoven, 2005 (Koen 2005).

**Fig. 10.9** Egyptian wall painting

specifically singles out the archaeological connotation of the word as having entered the language at a more recent time, certainly millennia after the Egyptian engineers lived. To quote the etymological dictionary (Harper):

> (artefact) "anything made by human art," from It. artefatto, from L. arte "by skill" (ablative of ars "art;" see art (n.))+factum "thing made," from facere "to make, do".
> Archaeological application dates from 1890.

Third, there is a branch of engineering called Operations Research that specifically deals with the best way to carry out operations to achieve a goal. It includes such topics as the design of assembly lines, supply chain management, queuing theory, and the best way to attack the traveling salesman problem.

Consider the *assembly line* as one example to make the point that the creation of processes are an important part of engineering. We are all familiar with the assembly line and usually attribute its invention to the Ford Motor Company in the manufacture of the automobile in 1908 AD. What is less well known is that the Egyptians used an assembly-line technique as sophisticated as the ones we use today in the creation of the famous wall paintings in their tombs and tunnels. Figure 10.9 shows a sample of one passage way in the tomb of Horemheb known as KV5l dated to about 1319 BC and Fig. 10.10 shows a detail from another place in that tomb (Wikipedia 2008).

The second figure is a screen capture from a television documentary that has a voice over by Dr. Kent Weeks, whom we met earlier saying (Halliley):

We have examples of almost every stage in the process of smoothing the walls, outlining the decoration, covering the decoration, modeling the details of the relief, and painting the relief. Almost every single step is shown.

**Fig. 10.10** Egyptian wall painting (detail)

The color version of the figure given clearly shows the first preliminary sketch of figures in black, and finally the corrections by the master artist in red made by teams that moved along the wall one after the other.

In an extremely relevant and interesting documentary entitled *Engineering an Empire: Egypt* (Cassel 2006) that describes the assembly line in detail, we find the voice over statement by the narrator, Michael Carroll and then the comment by Salima Ikram of the American University of Cairo:

> …the work took on the efficiency of an assembly line…. Some people would specialize in hands, some would do faces…

It is clear that the design of strategies to achieve specific purposes has been a part of engineering for a very long time.

These process strategies are as important to our understanding of how an engineering design was achieved as the concrete object itself. They are also passed on from generation to generation. As one well-documented example, consider the construction of the Empire state building in New York and the construction of the Great Pyramid in Egypt. At the time of its construction, each was the tallest man-made structure.

We have almost complete knowledge of how the Empire state building was built. It is a 1,453-ft, 103-story structure built in just over 13 months. Time had to be scheduled down to the minute. Workers would swing the girders into place and have them riveted as quickly as 80 h after coming out of the furnace and off the roller. The frame of the skyscraper rose at the rate of four and a half stories per week, or more than a story a day (Grabianowski 2001; Tauranac 1995).

On the other hand, almost nothing exists that preserves the state of the art or set of heuristics used in the construction of the Great Pyramid apart from the concrete engineering structure itself. To quote Robert Partridge, chairman of Manchester Ancient Society (History Channel 2004):

> There are no representations whatsoever of building the pyramids.

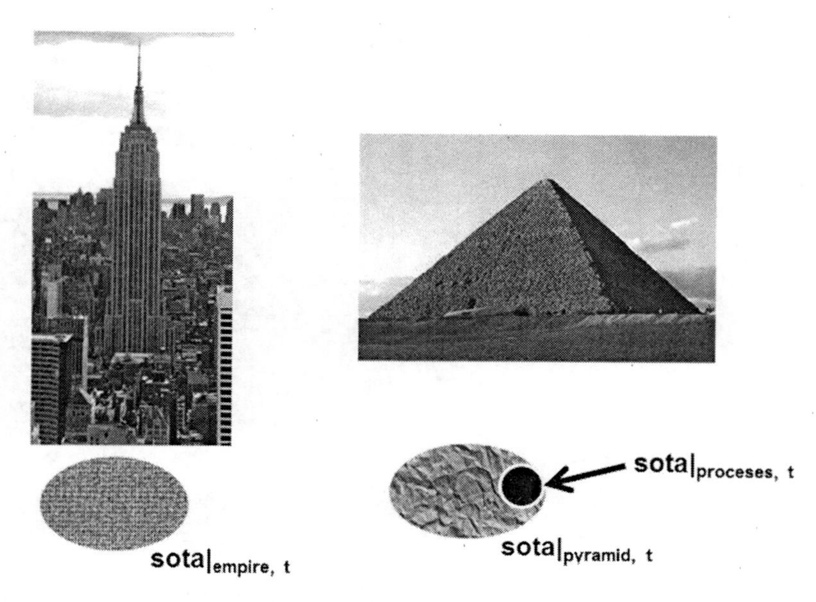


**Fig. 10.11** Comparison of the sotas of the Empire State Building and Great Pyramid

Modern engineers are convinced that they could not reconstruct the Great Pyramid using the tools of ancient Egypt and knowledge gleaned from the completed design as it stands. On the other hand, they are confident they could exactly duplicate the Empire State Building in the same period of time required in the original using the same tools based on examination of the building and the heuristics used in its construction. This situation is shown in Fig. 10.11. The complete $\text{sotal}_{empire,\ t}$ is known in one case; the $\text{sotal}_{pyramid,\ t}$ or set of heuristics needed for the construction of the Great Pyramid are not all known in the other. What is missing is the set of heuristics represented by the small black circle, the $\text{sotal}_{process,\ t}$. The process by which an engineering object is made is certain something "made by human art" and, hence, qualifies as an artefact in the true meaning of the word. It is not, however, a "concrete" object.

### 10.3.5 Myth: Engineering Is Trial and Error

It is undeniable that on occasion engineers make errors, sometimes even very dramatic ones. One of the most celebrated failures from the past was the Tacoma Narrows Bridge failure in 1940 shown in Fig. 10.12. It is also undeniable that engineers will not build an exact duplicate of the Tacoma narrows bridge in the future. But the issue here is whether or not *trial and error* is a legitimate definition or valid characterization of engineering. That it is not is evident for a variety of reasons.

Fig. 10.12 Tacoma narrows bridge failure

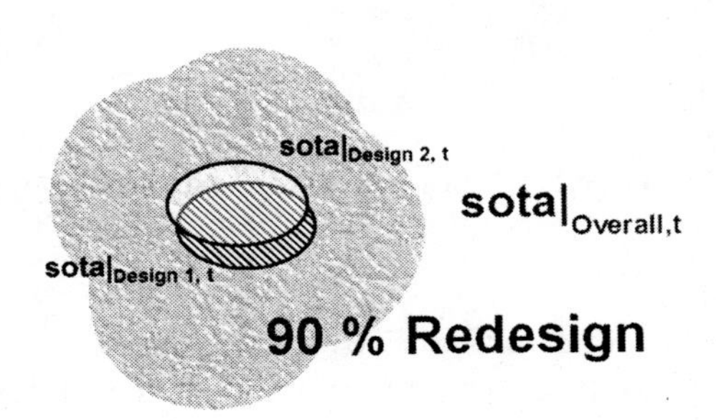


Fig. 10.13 Redesign

First, there are a very large number of engineers in the world. The exact number is hard to determine and depends on who is doing the counting and whether one is counting engineers in general, professional engineers, or only practicing engineers, etc. One Internet search engine reports that the Bureau of Labor Statistics data for the number of engineers in the U.S. as 1,512,000 in 2006.[8] Daily these individuals are making decisions, solving mathematical problems, sizing equipment, testing designs, and marketing the product, etc. It is hard to believe that a significant percentage of the truly huge number of engineering decisions made world-wide every day are errors.

Second, limiting ourselves to the overall design of a finished product, credible engineers have estimated that 90 % of all engineering designs are redesigns (Otto and Wood 2000). Figure 10.13 illustrates this point. The set of heuristics of a later design, $sota|_{design1,\ t}$, is based on or just a small tweak of a previous set of heuristics 90 % of the time.

[8] Reported by Semerich, a computer engineer, from Google on 12/4/2010 in answer to the query, "What is the number of engineers worldwide?" A defensible number for the engineers worldwide appears difficult to obtain.

Third, by its nature engineering is a risk taking activity. As stated in DOM:

> To qualify as design, a problem must carry the nuance of creativity, of stepping precariously from the known into the unknown, but without completely losing touch with the established state of the art. This step requires the heuristic, the rule of thumb, the best guess.

And, finally, since human life is often involved in engineering design and creativity is the essence of engineering, some risk of tragic error is unavoidable. $Sota1_{Modern,\ t}$ shown in Fig. 10.7 on page 11 contains very powerful heuristics developed over at least seven millennia to reduce risk to an acceptable level. A small, but representative, sample of risk avoidance engineering heuristics includes:

- Make small changes in sota
- Give yourself a chance to retreat
- Develop a project by successive approximations
- Allocate resources to the weak link
- A project usually squeaks before it fails
- Do a feasibility and pilot study

For all of these reasons, we are compelled to conclude that it is inadequate to characterize modern engineering as trial and error and to do so grossly misrepresents the true state of affairs.

## 10.4 Conclusions

The preceding sections have considered the following contemporary claims concerning engineering: (1) a popular definition of engineering method is vacuous, (2) engineering is a relatively recent human activity, (3) engineering is applied science, (4) engineering is trial and error, and (5) engineering artefacts must be concrete objects that persist over time and have given reasons why they should be considered as myths.

By way of conclusion, let's look at a positive unifying characterization of engineering, instead of lingering on these negative myths. Now the archetypical engineering project, the construction and evolution of the Egyptian Pyramids over four centuries, will be examined in some detail to demonstrate what engineering is really all about and the folly of the contemporary myths just considered.

As this is being written, 138 pyramids have been found with almost certainty that another one has been located. Others undoubtedly await discovery and still others have surely degraded and vanished from the earth forever. Out of the 138 only 6 of the most characteristic and well-known will be described. Refer to Fig. 10.14 for pictures of this selection and to Table 10.1 on page 20 for the specific design criteria of each.

Then a brief discussion of the implications of this review of Egyptian engineering design will be given, and, finally, we will meet—face to face—so to speak the image of the very *first* engineer whose appearance, name, works, and reputation is positively known in the historical record.

**Fig. 10.14** Evolution of Egyptian Pyramids

**Table 10.1** Egyptian Pyramid design data

| Name | Date | Height (m) | Slope |
|---|---|---|---|
| Mastaba | c. 2649 BC | – | – |
| Step Pyramid | c. 2630 BC | 62 | – |
| Meidum Pyramid | c. 2630 BC | 92 | – |
| Bent Pyramid | c. 2600 BC | 104 | Started 60°; then shallower angle of 55°; finally, slope reduced to 43° |
| Squat (Red, North) Pyramid | c. 2600 BC | 104 | Slope of 43°22′ |
| Great Pyramid | c. 2250 BC | 141 | Slope of 51°52′ |

Nothing in Egyptology is beyond dispute because of the age of the ancient Egyptian civilization. The following outline drawn from highly credible sources is sufficiently accurate for present purposes and we will leave the often contentious squabbles to others. Except as noted, information comes from The National Geographic (2008), the Encyclopedia Smithsonian (2008), or the MSN Encarta (Nolan 2008).

An abbreviated history of pyramid construction is as follows:

**Mounds of Sand** During the 1st dynasty which began in 2920 BC and the 2nd dynasty, the Egyptian Pharaohs were buried in graves topped with piles of clean sand inside low-lying walls.

**Mastaba** In the 3rd dynasty, the Pharaohs were buried under *mastabas* See Fig. 10.14;

**Step Pyramid** The *Step Pyramid* is considered Egypt's first pyramid. The Step Pyramid and later pyramids of the 3rd dynasty were constructed of small, almost brick-sized stones that were laid in vertical courses and inward-leaning to create the sloped sides; Patterned after the Step Pyramid are other smaller step pyramids, for example: the Seila Pyramid, the Zawiyet el-Meiytin Pyramid, the Sinki Pyramid, the Naqada Pyramid, the Kula Pyramid, the Edfu Pyramid, and the Elephantine Pyramid.

**Meidum Pyramid** The *Meidum* Pyramid was influenced by the step pyramid and is considered the first "true" pyramid. A step pyramid was built, the steps filled in with stones, and a smooth casing was added. It was a straight-sided pyramid whose inward-leaning walls ultimately collapsed;

**Bent Pyramid** The bottom of the *Bent Pyramid* looked like a mastaba, but the middle and upper portions resembled a true pyramid. This came about from purely engineering considerations.

> The architects had designed it with an angle of 60° (to the ground), but as the pyramid rose, it started to sink because of the weight and angle of the stones. To solve this problem, the builders put up an outer supporting wall, giving the half-finished pyramid a shallower angle of 55°. After this, the architects finished the upper portion of the pyramid off with a slope of only 43°. This shift in angle from 55° to 43° gives this pyramid its name—the Bent Pyramid. (Nolan 2008)

Another engineering innovation was made during the construction of the Bent Pyramid's upper portion. Instead of leaning the stones inward, they were laid down in horizontal layers with each level slightly smaller than the one it lay upon;

**Squat (Red, North) Pyramid** The stones of the *Squat Pyramid* were again laid down in horizontal layers suggesting that the ancient engineers followed the state of the art of the upper portion of the Bent Pyramid design. This gave the pyramid an unpleasing squat look;

**Great Pyramid** The *Great Pyramid* and all of the pyramids built during the 4th dynasty were built based on the heuristics previously used. It is the largest pyramid ever built and incorporates about 2.3 million stone blocks, weighing an average of 2.5–15 tons each. The workers would have had to set a block every two to two and a half minutes for 20 years according to both James Allen from the Metropolitan Museum of Arts (Allen 2008) and National Geographic (2008). Some recent estimates of the number of workers are as low as 10,000 individuals. Carefully placed shafts pierce The Great Pyramid and are thought to have been situated to aid the dead pharaohs journey into the afterlife.

**Later Pyramids** By the 5th dynasty (Nolan 2008),

> The quality of royal pyramid construction declined. The cores were made of smaller blocks of stone, laid more irregularly and by 2134 BC, the pyramids had a core of shoddy masonry and debris covered with a veneer of fine limestone.

This decline is thought to be from changing economic conditions and the tendency of the pyramids to become less secure as a resting place for the Pharaohs.

This abbreviated chronology of the evolution of the Egyptian pyramids reveals many of the interesting characteristics of the engineering method that have already been discussed and the importance of the engineering concept of the state of the art. Consider the following by way of review.

- No science was involved in the construction of the pyramids, yet engineering problems were solved. With science nonexistent, we might ask: during the years of pyramid design, what changed? What changed was the set of heuristics of pyramid design—that is, the sota.
- The sota of one pyramid was clearly a function of the sota of previous ones. The angle of the top of the Bent Pyramid is the same as the next "squat" one.
- The designs were defined and limited by the resources of money, talent, pharaoh's pride, and organization, not by some external, true norm.
- Engineering failures do happen when the engineer exceeds the range of applicability of the current heuristics, but he quickly retreats to a solidified information base and strikes out again.
- Trade-offs clearly existed between the aesthetic and the technical heuristics. The squat pyramid was surely a function of the earlier bent pyramid that failed. When technology improved i.e. stones were no longer laid at an angle, but put in courses, the angle increased again.
- The importance of the sota is hard to overstate. With no extant records of means of construction, even today we have no idea how the huge number of engineers was organized to build the pyramids. In technical terms, we do not know what the engineering artefact called *supply chain management* in modern terminology was like.
- Reality as conceived today had nothing to do with the placement of the shafts that pierced the walls of the later pyramids. They were, however, clearly important to the Egyptian civilization of the time. Constructing a pyramid is a complicated and difficult task, but doing so when the design of each level is constantly changing so that a straight shaft will pierce the completed structure at an angle is almost unbelievable. The engineer designs, not for the truth about the afterlife as we think we know it in the *twentieth century*, but as it was understood at the time the design was made.

Far from just building engines, the Egyptian engineers were certainly "clever, practical, problem solver[s]".

Even this abbreviated example of the evolution of the Egyptian pyramids shows that the definition of engineering:

> *The engineering method is the use of state-of-the-art heuristics to cause the best change in an uncertain situation within the available resources.*

is valid.

As mentioned in the introduction, this paper concludes with an introduction to the earliest engineer in history whom we know by name and to an example of his most famous engineering achievement. We can even look into his eyes. His name is

**Fig. 10.15** Imhotep with name in hieroglyphs

Imhotep and Fig. 10.15 is a small statue in the Louvre, Paris, France. A large number of similar statues have been found throughout Egypt.

According to Wikipedia (emphasis added) (Wikipedia: Imhotep):

> Imhotep (2655–2600 BC)... was an Egyptian polymath, who served under the Third Dynasty king, Djoser, as chancellor to the pharaoh and high priest of the sun god Ra at Heliopolis. He is considered to be the first architect and engineer and physician in early history.

The full list of his titles translated into English from the hieroglyp (probably by way of French) is (with important emphasis added):

> *Chancellor of the King of Egypt, Doctor, First in line after the King of Upper Egypt, Administrator of the Great Palace, Hereditary nobleman, High Priest of Heliopolis,* ***Builder****, Chief Carpenter, Chief Sculptor, and Maker of Vases in Chief.*

Certification that he was indeed an engineer is undoubtedly derived from his title as *builder* emphasized in the quotation of his titles above since, of course, the word *engineer* did not exist in the twenty-seventh century BC.

And the greatest achievement of the first engineer in history known by name? He dreamed, designed, created, and built the very first pyramid in Egypt—the Step Pyramid. Note the size of the people beside it in Fig. 10.16 to establish the scale.

The example of the evolution of the Egyptian pyramids from the Step Pyramid to the Great Pyramid is one of the greatest sustained examples of the practice of

**Fig. 10.16** Imhotep's step Pyramid

engineering over a long period in history. It should aid philosophers in avoiding the myths in the literature as they collaborate with engineers to develop a cogent Philosophy of Engineering. We can only hope that the future will bring a philosopher willing to return the favor and aid engineers in achieving their side of the bargain by debunking the myths concerning contemporary philosophy that engineers surely believe.

## References

Allen, J. (2008). *Seven wonders: Egyptian wonders*. England: Travel Channel.

Burke, J. G. (1919). *Technology and change*. New York: Van Nostrand Reinhold.

Burnet, J. (1930). *Early Greek philosophy* (4th ed.). London: A. and C. Black.

Cassel, C. (2006). *Engineering an empire: Egypt*. History Channel. http://www.amazon.com/History-Channel-Presents-Egypt-Engineering/dp/B000H5U5T4. Accessed 21 Dec 2010. ASIN: B000H5U5T4.

Darling, D. *Rhind papyrus*. http://www.daviddarling.info/encyclopedia/R/Rhind_papyrus.html. Accessed 13 Dec 2010.

de Camp, L. S. (1963). *The ancient engineers* (Paperback ed.). Garden City: Doubleday and Company.

Encyclopedia Smithsonian. (2008). *The Egyptian Pyramid*. http://www.si.edu/Encyclopedia_SI/nmnh/pyramid.htm. Accessed 21 May 2008.

Grabianowski, E. (2001). *Empire State Building completed*. http://science.howstuffworks.com/engineering/structural/empire-state-building.htm. Accessed19 Dec 2010.

Halliley, M. Secrets of Egypt: Valley of the Kings [TV documentary]. Five TV secrets of Egypt Valley of the Kings. *Egypt Unwrapped*, Season 1, Episode 6, Release Date: 13 Jan 2009, in the UK, Director David Lee.

Harper, D. Engineer. In *Online etymology dictionary*. http://www.etymonline.com/index.php?term=engineer. Accessed 10 Dec 2010.
Harper, D. Artefact. In *Online etymology dictionary*. http://www.etymonline.com/index.php?search=artefact&searchmode=none. Accessed 10 Dec 2010.
Hart-Davis, A. *What the ancients did for us: The Indians, number 5 of 9 programs* [TV documentary]. British Broadcasting Corporation TWO, The Open University.
History Channel. (2004). *Flying Pyramids: Soaring stones*. History Channel, UPC:133961222814.
Koen, B. V. (2003). *Discussion of the method: Conducting the engineer's approach to problem solving*. New York: Oxford University Press.
Koen, B. V. (2005, June 3–4). An engineer's quest for universal method. In *Norms, knowledge and reasoning in technology conference*, Boxmeer, the Netherlands.
Koen, B. V. (2008) Towards a philosophy of engineering. In *Workshop for engineering and philosophy* (Keynote address). London: Royal Academy of Engineering.
Koen, B. V. (2009). The engineering method and its implications for scientific, philosophical, and universal methods. *The Monist, 92*(3), 351–386.
Koen, B. V. (2010). Quo Vadis, humans? Engineering the survival of the human species (chap. 27). In I. van de Poel & D. E. Goldberg (Eds.), *Philosophy and engineering: An emerging agenda, Philosophy of Engineering and Technology* (Vol. 2) (pp. 321–341). Dordrecht: Springer.
Kuiper, K. (Ed.). (2011). Mesopotamia: the world's earliest civilization. In *The Britannica guide to ancient civilizations*. New York: Britannica Educational.
National Geographic. (2008). *Egypt secrets of an ancient world: Explore the Pyramids*. http://www.nationalgeographic.com/pyramids/pyramids.html. Accessed 27 May 2008.
Nolan, J. (2008). *Pyramids Egypt*. http://encarta.msn.com/encyclopedia_761555128_2/Pyramids_(Egypt).html. Accessed 29 May 2008.
Otto, K., & Wood, K. (2000). *Product design: Techniques in reverse engineering and new product development*. Upper Saddle River: Prentice Hall.
Robinson, A. (2009). *Writing and script: A very short introduction* (Paperback ed.). Oxford/New York: Oxford University Press.
Science Channel. (2008). *Hot Rocks: Geology of civilization: Architecture* (Ian Steward is host). http://science.discovery.com/convergence/hotrocks/tunein.html. Accessed 9 Dec 2010.
Spiritus-temporis. *Engineer. Spiritus-temporis online etymology dictionary*. http://www.spiritus-temporis.com/engineering/etymology.html. Accessed 10 Dec 2010.
Tauranac, J. (1995). *The Empire State Building: The making of a landmark*. New York: Scribner. (Richmond Shreve Quote p. 204).
Wikipedia. (2008). *Kv57*. http://en.wikipedia.org/wiki/Main_Page. Accessed 15 Dec 2010.
Wikipedia. *Civil engineering*. http://en.wikipedia.org/wiki/Civil_engineering. Accessed 16 Dec 2010.
Wikipedia. *Engineering. Web site*. http://en.wikipedia.org/wiki/Engineering. Accessed 14 Dec 2010.
Wikipedia. *Imhotep*. http://en.wikipedia.org/wiki/Imhotep. Accessed 16 Dec 2010.
Wikipedia. *Memphis*. http://en.wikipedia.org/wiki/Memphis_(Egypt). Accessed 13 Dec 2010.

# Chapter 11
# The Engineer's Identity Crisis: *Homo Faber* or *Homo Sapiens*?

**Priyan Dias**

**Abstract** Engineers have an identity crisis arising from questions regarding their influence, role and knowledge. These questions relate to ethics, ontology and epistemology respectively, and demonstrate that philosophy is indeed relevant to engineering. The tensions, differences and similarities between philosophy and technology, engineering and science, and theory and practice are explored, in order to shed light on some of the above crises. It is argued that engineers should remain proud of their contributions to society, but work at developing an acute awareness of technology's ill effects. They should see themselves as holistic managers dealing with real world complexity, but possessing a kernel of scientific knowledge. They must recognize that the knowledge required for engineering is mostly practical, but should work at formalizing practice at both the conceptual and technical levels. It is concluded that while an engineer is primarily a *homo faber*, there is enough justification for him/her to be placed high on the scale of *homo sapiens* too.

**Keywords** Ethics, ontology and epistemology • Philosophy and technology • Engineering and science • Theory and practice • Practice-based knowledge

## 11.1 Do Engineers Have an Identity Crisis?

While every engineer may not have looked into a mirror and asked the question "Who am I?", there are at least three reasons why engineers could suffer from an identity crisis. First, there is a crisis regarding the engineer's *influence*. Although there was a time when engineering was synonymous with the progress and uplifting of humanity, the technological society and environmental crisis have raised the

P. Dias (✉)
Department of Civil Engineering, University of Moratuwa, Moratuwa, Sri Lanka
e-mail: priyan@uom.lk

D.P. Michelfelder et al. (eds.), *Philosophy and Engineering: Reflections on Practice, Principles and Process*, Philosophy of Engineering and Technology 15,
DOI 10.1007/978-94-007-7762-0_11, 

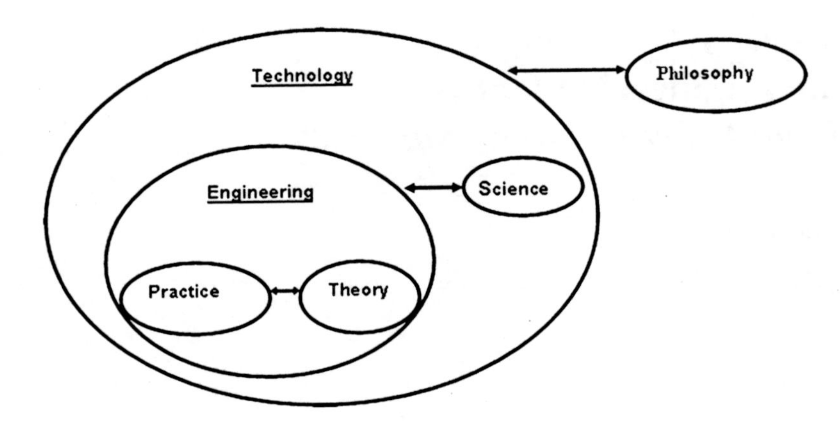


**Fig. 11.1** Broad framework for discussion (with focus on technology)

question as to whether *engineers are doing more harm than good*. The study of such actions, their motivations, and impacts is that branch of philosophy called *ethics*.

Next, there is a crisis regarding the engineer's *role*. Most students who enroll in engineering undergraduate programs have a strong background and interest in science. They are good at analysis. Practicing engineers on the other hand have to produce something or make something happen. That involves integrating products, processes and people. In other words, they must be good at management and synthesis. The question then arises as to whether *engineers are scientists or managers*. Genuine scientists and capable managers are both valued in most societies, but engineers run the risk of becoming neither in trying to be both. The study of roles within the wider study of *being* is that branch of philosophy called *ontology*.

Finally, there is a crisis regarding engineering *knowledge* (which overlaps the crisis regarding role). Most university programs in engineering are filled with theoretical subjects that are largely 'mathematics in disguise'. Engineering practice on the other hand is predominantly practical in nature, and great reliance is placed on established procedures (or 'rules of thumb'), specified guidelines (or 'codes of practice'), and that indefinable element called 'engineering judgment'. Therefore, we can ask whether *engineering knowledge is theoretical or practical*. In some situations, engineers have difficulty in explaining how their knowledge differs from that of a technician or even craftsman, because of this reliance on rules of thumb. The study of knowledge is that branch of philosophy called *epistemology*.

The above questions are valid for engineers in most if not all societies. It is the duality posed in the questions that creates the *angst*. It is the undermining of self-worth or social value inherent in the questions that constitutes the crisis.

Some of the answers to the above questions can be found by identifying the tensions and clarifying the issues in the wider framework encompassing engineering and philosophy. This framework is represented in two ways by Figs. 11.1 and 11.2. The double-headed arrows indicate debates or tensions. Figure 11.1 conveys the idea that the debates at each successive lower level take place only within a single

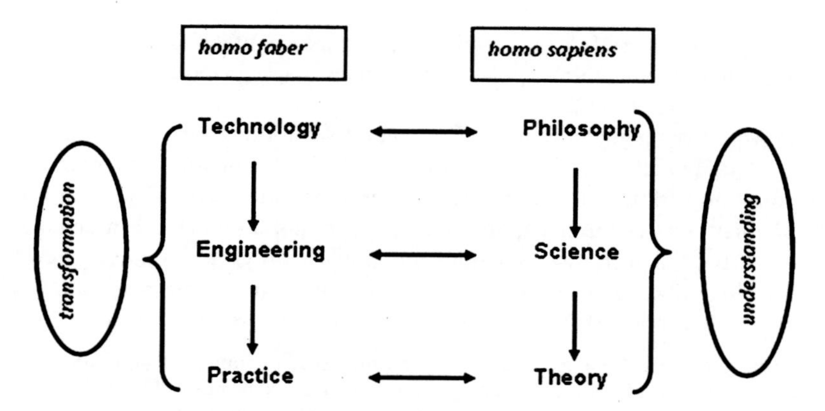


**Fig. 11.2** Broad framework for discussion (transformation vs. understanding)

component of the upper level, with the focus on technology. However, there are also vertical links between entities, as shown in Fig. 11.2. The rationale for this latter representation is that the entities on the right hand side are concerned with *understanding* (which is the goal of *homo sapiens* or 'wise man'), while those on the left hand side with *transformation* (which is the goal of *homo faber* or 'man the maker'). Recall Karl Marx's comment, i.e. "philosophers have tried to understand the world; the point however, is to change it." All three identity crisis questions raised above can be related to the question of whether an engineer is a *homo faber* or a *homo sapiens*.

## 11.2 The Engineer's Influence: More Harm than Good?

We answer this question in the context of the tension between philosophy and technology, given that engineers are the main purveyors of technology (Fig. 11.2). Some twentieth century philosophers (Ellul 1948; Heidegger 1977) have charged technology with being a pernicious influence, quite in contrast to the humanizing influence of philosophy and other liberal arts. The ill effects of technology can be categorized into at least four aspects (Dias 2003). The most obvious is the hazardous nature of some technologies, the prime example being nuclear technology. In addition, technology can promote injustice, for instance through infrastructure projects where social costs are borne by the poorer segments of a country while the benefits are reaped by the wealthier ones. Technology can have adverse sociological impacts too – consider the way in which visual screens (whether televisions or computers) tend to destroy family conversation and interaction. Finally, and most subtly, it can have undesirable psychological impacts. Has technology created a society where 'technique' is all important, as opposed to understanding (of phenomena) or even genuineness (in relationships), reflected in the growing number of 'how to' books? Michael Shallis (1984) argues that the invention of the clock resulted in persons being judged for their efficiency, while that of the computer resulted in them being judged for their logical thinking. Heidegger states that man has been 'enframed' by his own technology (1977).

The American engineer Samuel Florman (1994) refutes these charges, and also points to the benefits bestowed upon the world by technology, in areas such as transportation and health, and by the general improvement of the standard of living. In other words, 'humanization' can be seen, not so much as an engagement with the arts (which in most societies are enjoyed only by a relatively few), but rather as the liberation from 'slavery' brought about through technology, as described below by Karl Popper (1999), himself a philosopher of science.

> Perhaps even more important, morally, was the great liberation of domestic slaves (also known as maids), which became possible largely through household mechanization. This tremendous revolution, and the emancipation that all but the very richest women experienced at that time, is today remarkably little remembered, even though it was a liberation from heart-rending slavery. Who today has any idea what it meant when all water had to be fetched and carried, when coal had to be brought in for heating, when all washing had to be done by hand, and when there were still oil lamps with wicks? (p. 104)

Florman (1994) also contends that the engineer's activity of making things and engaging in work is a way of experiencing his humanness, relating to the earth and producing existential joy. He does admit, however, that the work of engineers may lead to inadvertent negative consequences, but lauds them for trying to improve the world. Florman asks engineers to take courage from Sisyphus, the character from Greek mythology who was condemned to keep rolling a stone up a hill, only to have it falling back as he approached the summit. Florman sees Sisyphus as heroic – someone who refuses to give up even though his work is undone from time to time. In this context, it is interesting to note that modern movements against some of the ill-effects of technology want to use technology itself to cure those ills (Feenberg 1999). For example, underground sequestering of carbon dioxide is being considered for reducing the consequences of burning fossil fuels. Again, it is the internet that is used for getting greater access to knowledge and to communicate, by people who feel they are marginalized and alienated by technology.

The anti-technology attitude does not arise just because of technology's rather recent negative effects. For many centuries university education was seen primarily as a humanizing process, through the dissemination and discovery of knowledge that was largely non-utilitarian. In Ancient Greece for instance, 'pure speculation' was considered to be a loftier pursuit than utilitarian pursuits. Consider the following description of Archimedes given by Plutarch (Blockley 1981):

> Yet Archimedes possessed so high a spirit, so profound a soul, and such treasures of scientific knowledge, that though these inventions had now obtained him the renown of more than human sagacity, he would not deign to leave behind him any commentary or writing on such subjects; but, repudiating as sordid and ignoble the whole trade of engineering, and every sort of art that lends itself to mere use and profit, he placed his whole affection and ambition in those purer speculations where there can be no reference to the vulgar needs of life.

Florman (1994) says that this mind-set, together with the Biblical New Testament emphasis on the spiritual as opposed to the material, has given technology a bad image or low status in Western culture. In Eastern cultures too, the role of the sage has been exalted over that of the worker, with the strict caste system in India, for example, perpetuating social barriers for generations. Blockley (1981) says that "We must break the chains of Ancient Greece"; but how? Florman (1994) suggests, as least for Western culture, that engineers dig deeper into their heritage for evidence that 'making' is indeed a noble pursuit, e.g., into the Old Testament, where the ability to perform various skilled crafts is ascribed to the indwelling of the Spirit of God; also into the pre-Socratic era, where craftsmanship was held in high esteem by Homer, who gives great technical detail regarding the making of Odysseus' raft and Achilles' shield, covering both tools and materials.

Meanwhile, the past century has seen a vast expansion of university education and the mobilization of the academy for wealth creation and problem solving through research. Science and engineering faculties were well funded while humanities faculties, other than in the most prestigious universities, experienced inexorable decline in both funding levels and student numbers. Technologists now use the term 'soft' in a derogatory manner to describe what goes on in arts faculties, while those in the humanities bemoan the swamping of human values by the technological juggernaut. At the end of the day however, in many societies an 'educated' person (or 'intellectual') is considered to be one who has knowledge of literature and culture, rather than one who can describe an internal combustion engine or an integrated circuit.

Despite his critique of technology, Heidegger is probably a good 'patron' philosopher for engineers. On the one hand, his view of the 'human way of being' was an instrumentalist one – we 'are' and 'do' before we 'think' (Dias 2006). On the other hand he advocated a suspicion of technology where it destroyed diversity through reductionism (Dias 2003).

So, do engineers do more harm than good? Whatever accusations are made against engineers, those who level such charges would probably not want to live in a world without technology and engineering influence. Where the capacity for humanization is concerned, technology has credentials that can rival those of philosophy, as Popper has articulated. Furthermore, in the tradition of Sisyphus (as seen by Florman), engineers can be proud of their being men and women of action – being *homo faber* in other words – rather than merely engaging in 'pure speculation'. Engineers may feel inferior about their intellectual status however, i.e. their place on the scale of *homo sapiens*, and we deal with this at the next level of the framework (Fig. 11.2), which considers engineering (very much a part of technology) and science (a development of philosophy).

## 11.3 The Engineer's Role: Scientist or Manager?

In order to illuminate the role of an engineer, we consider engineering design, which is a good reflection of engineering practice as a whole. We could view (engineering) science as the core or kernel of engineering design knowledge; a core however that

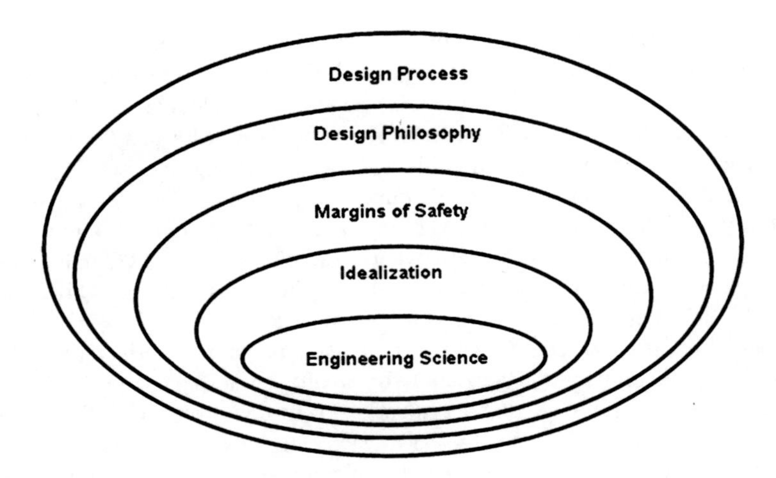


**Fig. 11.3** Engineering design knowledge (after Dias 1994)

is encapsulated by 'rules of thumb' (also called 'heuristics') such as engineering idealizations, margins of safety, design philosophy and the design process – see Fig. 11.3 (Dias 1994). Before employing engineering science theories, we have to adopt a particular design philosophy, decide on margins of safety and idealize the real world into a model to which scientific or mathematical theories can be applied; and all this has to be done within a design process that may involve collaboration and communication, not least with those who will fabricate, maintain and utilize the designed artefact.

Let us use the beam supported between two columns in Fig. 11.4 as a simple example. It is the idealized beam that is analyzed using engineering science (to find, say, the design bending moment). Before that, however, we have to decide that the beam is best idealized as a simply supported beam and apply appropriate safety factors to the load that is assumed. We also have to adopt some design philosophy to make allowance for the restraint moments at the column supports, where the fixity is not known precisely. Blockley (1980) uses the term 'calculation procedure model' to describe this entire process; it is not confined to calculations alone, but incorporates all of the other decision making procedures.

The message of Fig. 11.3 is that *engineering is broader and richer that science.* This breadth and richness create complexity that has to be managed for practical problem solving. It should be noted that the term 'complex' is used to denote richness in structure, whereas the term 'complicated' denotes abundance of detail. It is this richness in structure that constitutes the intellectual challenge of engineering. Engineering complexity arises from many things. One of them is the layered nature of the structure, as described above, and in Figs. 11.3 and 11.4. Another is the uncertainty associated with engineering; this uncertainty has been classified (Blockley and Godfrey 2000) as Fuzziness, Incompleteness and Randomness (FIR). Fuzziness relates to the imprecision in assigning states to an entity – for example,

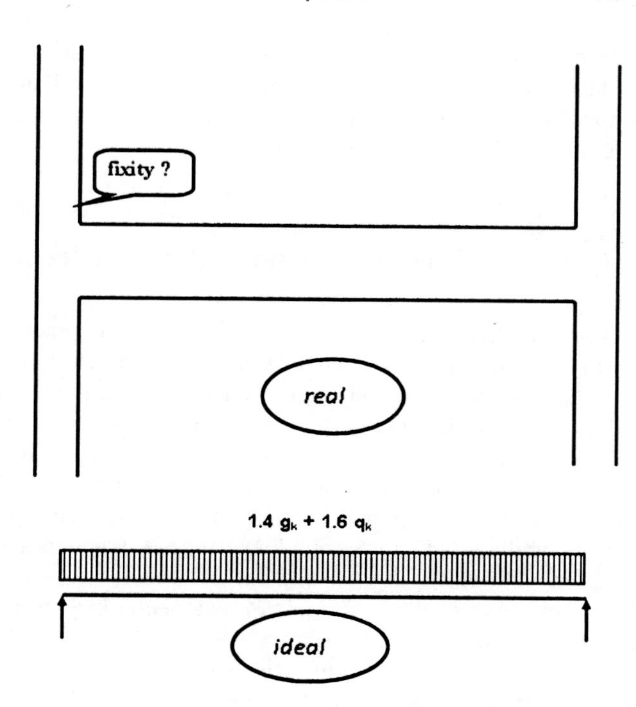


**Fig. 11.4** Idealization of a real structural element

the judgment as to whether the beam in Fig. 11.4 is fixed ended or simply supported. Randomness describes the variations to be expected in loading and material strength. Incompleteness has to do with the lack of knowledge about possible future scenarios, for instance the question as to whether the beam will be overloaded at any time during its design life, and by how much. There are mathematical approaches to deal with all these types of uncertainty, but rules of thumb too. Some rules of thumb are calibrated against mathematical simulations. An engineer can choose to use mathematics, or rules of thumb or a combination of these, depending on the situation.

Yet another type of complexity is that engineering problem solving often requires *abductive* reasoning, where a cause has to be posited, given an effect (observed or desired) and known rules (Dias 2010). The fact that there will inevitably be more than one cause that fits the effect is what constitutes the challenge. This is also called the solution to an 'inverse problem'. Apart from all of this, the 'calculation procedure model' has to make allowance for human error (or even malice) and accidents as well. So the engineering role is that of managing a process that involves people, procedures and products to deliver quality (including both safety and economy).

So, are engineers managers or scientists? From the above discussion we must conclude that engineers act more like holistic managers than like specialized scientists, although their practice is grounded in science. This emphasizes yet again that the engineer is a *homo faber* – not only in the narrow sense of 'making', but also in the broader sense of 'making it happen'. However, the complexity that (s)he has to tackle requires a particular kind of knowledge, understanding and even wisdom, thus making it appropriate for the engineer to be high on the scale of *homo sapiens* too. We are now ready to consider the tensions between practice and theory, the third level in the framework of Fig. 11.2.

## 11.4 The Engineer's Knowledge: Theoretical or Practical?

There are many dimensions to the theory versus practice debate. As stated at the start of this paper, most engineering programs are dominated by theoretical subjects, probably to ground engineering students in the 'kernel' of science (Fig. 11.3), but also to justify the existence of engineering programs in the academy, which prizes theory over practice. Engineering graduates discover, however, that 'rules of thumb', 'codes of practice' and 'engineering judgment' dominate the actual practice of engineering. This can be considered an acceptable compromise. We can say therefore that engineering knowledge is largely practical, although it has to be based on theory.

Heidegger's (1962) example of a carpenter hammering a nail is very insightful for resolving this practice-theory tension. The 'primordial' experience of the carpenter is a seamless web of activity without any deliberate rationality on his part. However, when there is a breakdown in this 'everyday' experience, say when the hammer is too heavy, the carpenter will have to resort to 'mentality' and study properties such as the weight of the hammer object; or if the head comes off the handle, once again he will have to give careful attention to solve the problem. In fact, Heidegger considered that scientific observation and reflection took place at such breakdowns. This then underlines the necessity for the theoretical training of engineers. Although they may be using only practical intelligence (e.g. 'rules of thumb') in their routine work, they will need a bedrock of theoretical knowledge to fall back on when faced with problems that intrude into their practice. Many professional engineering organizations, in the process of admitting engineers to full membership after a period of work-based training, are interested in finding out about problems encountered during the engineer's work, and how engineering 'first principles' were used to overcome them (Dias 2006).

There are some other aspects of the interplay between theory and practice that are worth looking at also. For example, most engineering academics would hold that the 'theory' components of a course should be taught before introducing students to 'practical applications'. In an overall sense, an engineering graduate would be seen as putting into practice the theory learnt at university. However, Patrick Nuttgens (1980), an architecture professor at the University of Edinburgh who became the founding director of Leeds Polytechnic in the U.K. in the early 1970s, argues that children first learn about the world by practice before they acquire a theoretical framework, and that technical education should reflect this.

Also, practice itself is now considered to be a rich source for theory, especially theories regarding the engineering design process itself; and the process of engineering design has been equated to theory building (Monarch et al. 1997). There are echoes here of 'grounded theory' (Glaser and Strauss 1967), where sociological generalizations are derived from the analysis of documents and transcripts of unstructured interviews.

A broad philosophy of practice has been actively developed (e.g. Skill 1995), with contributions from philosophers, engineers, craftsmen and actors; parallels have been drawn between actors and engineers. The attempt is to show that knowledge is very often acquired from practice (perhaps under apprenticeship), rather than from theory alone. Donald Schon (1983) wrote a very influential book called *The Reflective Practitioner*, which was subtitled 'How professionals think in action'. The main theme of the book is that 'reflective practice', i.e. reflection on one's professional practice, generates practice based knowledge that is invaluable and very different from the theoretical knowledge that is embedded in 'technical rationality'. His ideas have been applied to engineering in general (Blockley 1992) and to engineering design in particular (Dias and Blockley 1995; Dias 2002), where it is argued that a combination of reflective practice and technical rationality is required.

We conclude this section by affirming that the knowledge used by engineers during their professional careers is mostly practical in nature, once again reinforcing the *homo faber* image of the engineer. We have shown however, that there is considerable interplay between theory and practice, and that many recent initiatives promote the intellectual status of practice. Despite this new focus on practice however, the greatest 'shortcoming' in practice-based knowledge is its lack of formalism. It is its theoretical formalism that gives science its credibility and prestige in the academy and indeed even in wider society. The practical knowledge of engineers is often perceived as 'just common sense'. In fact even craftsmen and technicians are seen as having such knowledge, so that it is not valued, especially in the academy. The place of the engineer on the scale of *homo sapiens* is thus challenged. This challenge can be met through efforts to formalize practice.

## 11.5 Formalizing Practice

The formalization of engineering practice will strengthen the engineer's image as *homo sapiens*, while reinforcing his position as an agent of transformation. There are two levels at which formalization needs to evolve – at the conceptual level that deals with the engineering approach, and at the technical level that deals with practice-based knowledge.

Systems approaches can be seen as providing a formalization at the conceptual level (Dias 2008). Formalization at this level is not easy, as best expressed by David Elms (2010):

> The systems approach is not easily systematised, so to speak, partly because of the breadth of the issues involved, but more generally because there is no narrow set of applications allowing development of an easily focused theory. Structural analysis, for example, has techniques fine-tuned to dealing with structures, but the systems approach can be applied to

> anything. It has no natural boundaries. What is needed is not so much a set of immediate techniques as general principles and overarching concepts for giving the approach its power and its constraints ... The trap ... to avoid [is] being so general as to be ineffective, hence specific guides and ideas are needed.

There have however been many such frameworks proposed in the literature. The reflective practice loop is one of them, consisting of the components reflection – action – world – perception – reflection (Blockley 1992). A development of this is the Design – Build – Operate loops in the three spheres of Purpose, Process and People (Blockley 2010a). Senge (1992) has demonstrated that a number of management scenarios can be modeled with three basic elements – namely a Reinforcing Loop, Balancing Loop and Process Delay. Blockley (2010b) has proposed a framework that he calls "new process", where a process is seen as a relationship between sub-processes of questions (why) driving a set of change sub-processes (who, what, where, when) through a set of change transformation processes (how). All of these sub-processes also have their own sub-processes – they are a hierarchy of parts and wholes or holons, after Koestler (1967). Checkland and Scholes (1990) have proposed the CATWOE template for studying change management processes, the acronym covering the aspects of Customers, Actors, Transformations, Weltanschauung (Worldview), Owners and Environment. They also argue that while the world is treated in hard systems as *systemic*, and models of it as *systematic*, in soft systems the world is acknowledged as *chaotic*, and models of it as *systemic* (1990). The objective of soft systems models is not so much to *simulate* the world through systematic procedures, because such approaches will always be incomplete and lacking in real world richness; it is rather to *reflect on* the world in an integrated, systemic way, from the identification of problems to the implementation of change. In particular, such reflection could help to mitigate or even eliminate the unintended consequences associated with engineering projects.

Artificial Intelligence (AI) or more accurately Knowledge Processing can be used for or seen as providing a formalization for practice at a technical level. AI could then serve the systems approaches similar to the way in which mathematics has served the scientific method. Both AI and mathematics are formalizations at a technical (rather than conceptual) level. Dias (2007) gives examples of how some AI techniques such as neural networks, case based reasoning and interval probability theory can be used to capture, structure and process practitioner knowledge and experience. He also provides a philosophical grounding for practice based knowledge, drawing on two very diverse philosophers, namely Michael Polanyi and Martin Heidegger.

## 11.6 Conclusions

- This study of the tensions associated with technology vs. philosophy, engineering vs. science and practice vs. theory has helped to clarify some of the issues and answer some of the questions that engineers have regarding their identity in the areas of ethics, ontology and epistemology.

- In the face of the question as to whether they do more harm than good, engineers should remain proud of their contributions to society, but work at developing an acute awareness of technology's ill effects. They should also see themselves as agents of humanization as well as transformation.
- Regarding the question of whether they are managers or scientists, engineers should see themselves as holistic managers grounded in science. They should see a whole to part relationship between engineering and science, in that engineering is richer and broader than science. The engineering role also demands a sophisticated and nuanced approach in order to deal with real world complexity.
- With respect to the final question regarding the nature of engineering knowledge, we have seen that engineers largely use practical knowledge, though schooled in theory upon which they can always fall back. They should however learn to see practice as being a type of theory formation too. They also need to work at developing some formal structures, both for the engineering approach itself and for practice based knowledge.
- The adoption of "big picture" systems thinking frameworks could be useful for formalizing the engineering approach at the conceptual level. The use of knowledge processing tools such as AI may help to formalize practice based knowledge at the technical level.
- We have seen that an engineer is primarily a *homo faber*, a label of which to be proud, quite in contrast to Plutarch's reporting of Archimedes' views. However, strong arguments were made as to why an engineer should be considered as being high on the scale of *homo sapiens* too.

## References


Blockley, D. I. (1980). *The nature of structural design and safety*. Chichester: Ellis Horwood.

Blockley, D. I. (1981). Phil's eight maxims. *The Structural Engineer, 59A*(9), 292–294.

Blockley, D. I. (1992). Engineering from reflective practice. *Research in Engineering Design, 4*, 13–22.

Blockley, D. I. (2010a). *Bridges: The science and art of the world's most inspiring structures*. Oxford: Oxford University Press.

Blockley, D. I. (2010b). The importance of being process. *Civil Engineering and Environmental Systems, 27*(3), 189–199.

Blockley, D. I., & Godfrey, P. (2000). *Doing it differently: Systems for rethinking construction*. London: Thomas Telford.

Checkland, P., & Scholes, J. (1990). *Soft systems methodology in action*. Chichester: Wiley.

Dias, W. P. S. (1994). Structural failures and design philosophy. *The Structural Engineer, 72*(2), 25–29.

Dias, W. P. S. (2002). Reflective practice, artificial intelligence and engineering design: Common trends and inter-relationships. *AIEDAM, 16*, 261–271.

Dias, W. P. S. (2003). Heidegger's relevance for engineering: Questioning technology. *Science and Engineering Ethics, 9*(3), 389–396.

Dias, W. P. S. (2006). Heidegger's resonance with engineering: The primacy of practice. *Science and Engineering Ethics, 12*(3), 523–532.

Dias, W. P. S. (2007). Philosophical grounding and computational formalization for practice based engineering knowledge. *Knowledge Based Systems, 20*(4), 382–387.

Dias, W. P. S. (2008). Philosophical underpinning for systems thinking. *Interdisciplinary Science Reviews, 33*(3), 202–213.
Dias, P. (2010). *Pompeii* by Robert Harris: an engineering reading. *ICE Proceedings on Engineering History and Heritage, 163*(EH4), 255–260.
Dias, W. P. S., & Blockley, D. I. (1995). Reflective practice in engineering design. *ICE Proceedings on Civil Engineering, 108*(4), 160–168.
Ellul, J. (1948). *The technological society*. New York: Alfred A. Knopf.
Elms, D. (2010). David Blockley: An appreciation. *Civil Engineering and Environmental Systems, 27*(3), 175–176.
Feenberg, A. (1999). *Questioning technology*. London: Routledge.
Florman, S. C. (1994). *The existential pleasures of engineering* (2nd ed.). New York: St. Martin's Press.
Glaser, B., & Strauss, A. L. (1967). *The discovery of grounded theory: Strategies for qualitative research*. London: Weidenfeld and Nicolson.
Goranzon, B. (Ed.). (1995). *Skill, technology and enlightenment: On practical philosophy*. London: Springer.
Heidegger, M. (1962). *Being and time* (J. Macquarrie & E. Robinson, Trans.). London: SCM Press.
Heidegger, M. (1977). *The question concerning technology and other essays* (W. Lovitt, Trans.). New York: Harper and Row.
Koestler, A. (1967). *The ghost in the machine*. London: Picador.
Monarch, I. A., Konda, S. L., Levy, S. N., Reich, Y., Subrahmanian, E., & Ulrich, C. (1997). Mapping sociotechnical networks in the making. In G. C. Bowker, S. L. Star, W. Turner, & L. Gasser (Eds.), *Social science, technical systems and cooperative work: Beyond the great divide* (pp. 331–354). Mahwah: Lawrence Erlbaum.
Nuttgens, P. (1980). *What should we teach and how should we teach it? Aims and purpose of higher learning*. London: Gower.
Popper, K. R. (1999). *All life is problem solving*. London: Routledge.
Schon, D. A. (1983). *The reflective practitioner: How professionals think in action*. London: Temple Smith.
Senge, P. M. (1992). *The fifth discipline: The art and practice of the learning organization*. New York: Century Business.
Shallis, M. (1984). *The silicon idol: The micro revolution and its social implications*. New York: Schocken Books.

# Chapter 12
# Varieties of Parthood: Ontology Learns from Engineering

**Peter Simons**

For Brian, engineer, at 85.

*Image the whole, then execute the parts-*
*Fancy the fabric*
*Quite, ere you build, ere steel strike fire from quartz,*
*Ere mortar dab brick!*

Robert Browning

**Abstract** We survey mereology, the ontological treatment of part and whole, distinguishing its uncontroversial from its controversial principles and lamenting the excesses to which too great an attraction to formal simplicity leads ontologists. As a partial remedy we recommend greater occupation with the range and variety of uses of the concept 'part' in engineering, where artefact parts and their configurations are of vital concern. We highlight some of the major linguistic and conceptual difficulties surrounding the concept of part, and distinguish several more specific concepts of part, noting how distinctive enumeration of artefact parts at different phases in their life-cycle leads to the problem of multiple bills of materials. A related and important concept in engineering and elsewhere is that of a material feature. We discuss this and its partial affinity with the part concept.

**Keywords** Mereology • Ontology • Part-whole relations • Material features • Functional parts

---

P. Simons (✉)
Department of Philosophy, Trinity College, Dublin, Ireland
e-mail: psimons@tcd.ie

D.P. Michelfelder et al. (eds.), *Philosophy and Engineering: Reflections on Practice, Principles and Process*, Philosophy of Engineering and Technology 15,
DOI 10.1007/978-94-007-7762-0_12, 

## 12.1 Introduction

Ever since the first deliberately chipped hand-axe, humans have produced artefacts with a view to the different functions of their different parts; and ever since the first axe-head was fitted into a wooden handle, they have assembled artefacts out of functionally and structurally diverse components. In the thousands of years since then, artificers, builders and engineers have had daily currency with artefact parts, the wholes they compose, and the ways in which the parts are put together to make the whole. Philosophers by contrast have only very recently thought it worth *analysing* the concept of the part–whole relation. Of course the concept did not escape them: it is too ubiquitous for that. Plato worried about whether some abstract forms had others as parts; Aristotle pointed out that the term '*meros*' (part) has several meanings in ordinary Greek. But the concept of part did not move to centre stage in philosophical discussion until the late twentieth century. At the beginning of that century, starting with some observations of Edmund Husserl (1970), logicians, most notably Stanisław Leśniewski (1916) and Alfred North Whitehead (1919), developed formal theories of part and whole, for which theories Leśniewski coined the term 'mereology'.

In this paper I review some of the problems and controversies surrounding philosophers' formal treatments of parthood, and will conclude that their views, more prevalent than ever in the philosophical community, are simply too monocultural to account for the wide variety of part-concepts met with and required in applications, most especially in engineering. The moral drawn is that formal theories of parthood must be supplemented and if necessary corrected by empirical information about actual thought and practice outside philosophy, and that one of the best sources for such information is engineering.

## 12.2 Philosophical Mereology

Mereology was developed initially for mathematical purposes: as a nominalistically acceptable substitute for set theory (Leśniewski) or as a logical framework for geometry (Whitehead). By the late twentieth century it had become apparent that the standard formal resources of philosophers (interpreted predicate logic and set theory) were insufficient to articulate the variety of problems in ontology and metaphysics (Simons 1987), and mereology became a central instrument in the ontologist's toolkit, so that nowadays a significant proportion of metaphysical disputes turn on matters of mereology. Nevertheless, ontologists have tended to take over the strong algebraic assumptions of the early mereologists (Simons 2007). Partly as a result, a large number of the mereological problems which preoccupy metaphysicians have little or no relevance to engineering practice or theory. Despite this, the concept of part–whole in engineering is not a mere simple application, to be indicated in passing while sticking to the theoretical high road. On the contrary, the mereology of artefacts is rife with problems, for which the philosophical ontologist's mereology is of little or no use. It is the central contention of this paper that until the crucial differences between the "pure" mereology of philosophers and the "applied" mereology of engineers are more carefully articulated, there will continue to be a significant gap between their respective mereologies, rendering these mutually almost irrelevant. It is precisely the job of the *philosopher* to recognize and articulate such differences and to see that philosophical theory, no matter how abstract, does not become wholly detached from real-world considerations.

## 12.3 Uncontroversial Principles of Parthood

We use the term 'part' in its normal everyday sense, according to which a part is something less than the whole. What this means precisely can be spelled out. Firstly however what it implies is that no part is identical with its whole:

IRREFL If A is part of B, then A is not identical to B

Secondly, that a part of a whole cannot have that whole as a part:

ASYMM If A is part of B, then B is not part of A

Logically, we only need the second principle, since the first follows from it: if A were identical with B they would be alike in all respects, so then B would be part of A. But it is not, so A and B cannot be identical. A crucial formal property of the part-relation is its transitivity

TRANS If A is part of B and B is part of C then A is part of C

Many relations satisfy these principles without being part-relations, for example the less-than relation among numbers. So we need to add more to distinguish the part-relation. This takes a little more work. Here is how we do it. Firstly we define a concept of (mereological) coincidence:

Def.COIN A coincides with B if and only (Df.) A is identical with B or A and B both have parts, and they have the same parts.

This allows for the possibility that A and B have the same parts and yet are not identical.We now use this to define a notion of ingredient:

Def.INGR A is an ingredient of B iff (Df.) A is part of B or A coincides with B

So if A is identical with B, A and B are ingredients of one another, by the definition. Now we define disjointness:

Def.DISJ A is disjoint from B iff (Df.) nothing is an ingredient of both A and B

We are now in a position to say what else we need for A to be part of B: it is the principle of supplementation:

SUPPL If A is part of B then B has a part which is disjoint from A

For example, the frame is part of a bicycle, but other parts such as the wheels are disjoint from the frame.

The principles ASYMM, TRANS and SUPPL are analytically true of the part relation, and diagnostic (constitutive) of it against relations satisfying other formal principles.

## 12.4 Contentious Principles

Philosophers have put forward two mereological axioms that go well beyond the analytic principles constitutive of the part relation. These are mereological *extensionality*

EXT Coincident things are identical

And the principle of universal *composition*

UC Any collection of individuals compose a further individual, called their *mereological sum*.

Taken together these imply that the mereological sum of any collection of individuals is unique.

EXT is a thesis in keeping with standard conceptions of mathematical discourse, and in particular it is analogous to the extensionality principle of set theory, according to which sets with the same elements are identical. Although the less deleterious of the two contentious principles mentioned here, it does have one consequence that is not neutral. Sometimes an object is composed of parts yet we have reasons not to identify it with the sum of these parts. For example a casting is made of a certain consignment of metal. The sum of parts of this metal could have made other things than the casting, but the casting could not have made anything else: it is what it is. So we want to distinguish the casting from the sum of metal making it up, though they appear to have the same parts. Likewise a dry stone wall is made up exclusively of a number of stones, and it and the sum of the stones have the same parts, but they are not identical since the sum can survive scattering, whereas scattering would destroy the wall.

UC leads more obviously to ridiculous and absurd consequences, for example that there is a whole composed wholly of my left hand and Napoleon's left foot, which therefore did not exist in 1900 but did in 1820 and in 1970. Another weird whole consists of the odd-numbered breaths that Napoleon took between 1810 and 1815, another consists of Napoleon's last breath and the Tower of Pisa. Yet such bizarre ontological monsters are defended by philosophers on both pragmatic and *a priori* grounds.

To illustrate how easily the philosophical debate can become divorced from common sense, we may note that two diametrically opposed positions, both anti-common-sense, are now taken seriously in the contemporary ontological literature.

One says that there is really only one thing, and it has no parts (monism). The other says that there are only atomic (simple) things, and no complex objects (radical atomism, RAT). Such extremes have been rare since the pre-Socratics (sixth century BCE).

A related debate within the literature concerns the question under what conditions a collection of parts compose a whole. This is known as the *special composition question.* (Van Inwagen 1990). UC represents one extreme answer to this question, RAT the opposite extreme. RAT entails that mereology has no meaningful role to play in ontology, since nothing has any proper parts. Its applicability depends on the assumption that there are metaphysical atoms, which may be true, but also may not: as far as mereology and science currently tell us, there is no reason to deny that everything has proper parts without end. The extremes of UC and RAT are so attractive to philosophers that I have known at least one ontologist to simply switch his allegiance from UC to RAT without adopting any of the many possible intermediate positions, wherein, somewhere, the truth lies.

Mereology with EXT and UC is known as *classical extensional mereology* or CEM (Simons 1987, Ch. 1). Rather than enter into philosophical debate, in the spirit of (perhaps misplaced) ecumenism, let's give proponents of CEM their concept and call it that of the M-part ('M' for 'mereological'). No one to date has convicted it of inconsistency, for the simple reason that it is provably consistent. A world consisting of a single individual with no proper parts (one-element model) satisfies all the principles of CEM. The question is not whether CEM and its M-part concept are consistent, but whether they are useful, whether other concepts of part are needed and/or are preferable.

## 12.5 Ambiguities of 'Part'

One reason why the mereological part concept has been able to gain a near-monopoly of acceptance among philosophers is that there is a subtle ambiguity in the use of the term 'part', not one which was picked up by Aristotle. It turns on the distinction between 'is part of' and 'is *a* part of'. One aspect of this distinction correlates with the distinction between mass terms like 'paint' or 'steel' and count terms like 'car or 'elephant'. Part of a car is the paint on its body, another part is the steel in its body. Neither the paint nor the steel would be called *a* part of the car, unlike its engine, its windscreen, or its steering wheel. Mixtures such as alloys and solutions have different parts, in the mass sense: steel is a mixture of iron, carbon and often other elements. Bronze is a mixture of copper and tin: it has these two parts, in varying proportions depending on the type of alloy. Air is a mixture of nitrogen, oxygen, carbon dioxide, water vapour and other gases, and so on. Each of these gases is part of the mixture. On the other hand the gas in a closed container such as an aircraft in flight has parts that are not the individual components of the mixture. There is (at any one time) the part in the first-class cabin, the part in the economy class cabin, the part in the air-conditioning system, and so on.

The term 'part of' is in vernacular use more flexible that the term 'a part of' and that helps to explain why the mereological or M-part concept is liked by philosophers: it is more widely applicable. When we say '*a* part' however we typically have some more limited notion of part in mind: a selection of such concepts will be given in the next section.

Both 'is part' and 'is a part' use the grammatical singular, but 'part' can also occur with the plural: 'the women on the electoral register are an important part of the electorate' (not 'part*s* of the electorate'!) This is because when we allow plural terms as well as singular ones, the logic of 'part of' holding between plurals is the same as that for 'part of' holding among singulars. The women are part of the people but not all of them, the men are another part disjoint from the women, and so on. Very often we can interchange 'part of' with 'some of', both for plurals and for mass terms. The part of the air in the first-class cabin is some of the air in the aircraft, the women in the electorate are some of the electorate, and so on. For singulars however, instead of 'some of' we say 'one of'. The unstressed pronunciation of 'some', pronounced [sm̩], can be used with mass and plural nouns in the way that 'a' may be used with singular count nouns:

A man came to the door
Sm coffee was spilt on my keyboard
Sm people came in waving sticks

These show that English, like French, has mass and plural indefinite articles. Incidentally it is highly suggestive that the grammatical term for these in French

*Du* café a été renversé dans mon clavier
*Des* gens sont entrés en brandissant des bâtons

is 'partitive'. To date and to my knowledge no one has done a thorough investigation of the grammatical varieties of 'part' from a philosophical and linguistic point of view.

Another closely related conception is that of the constitution of an object out of matter or materials, obviously a subject of serious interest to engineers. In ancient times Aristotle proposed that all material things are compounded of two disparate elements: the form and the matter. While this hylomorphic conception of things was decisively rejected for science in the scientific revolution, in everyday terms the notion of matter or materials applies quite naturally. The engine casting is made of alloy, the dry stone wall is made of stones, more complex artefacts are composed of many disparate materials. As standardly understood, the materials of a complex whole are taken in a "mass" way: so much of this, so much of that; whereas the parts of a complex whole are understood in a "count" way: this part, that part and so on. And when we consider materials like glue, paint, weld etc., many artefacts (as indeed organisms) consist both of parts and of materials. The relationships between parts and materials are complex and by no means transparent, and have been largely ignored by ontologists (but cf. Simons 1987).

## 12.6 More Specific Part-Concepts

### *12.6.1 Physical Part*

One perhaps not wholly determinate concept, but one which is certainly worth using and trying to get more determinate, is that of a *physical* part, or P-part. Consider a metal bar. It might be cut at the centre into two pieces, but suppose it is not. Each of the two halves is a physical part of the whole, even though neither is a detached physical body. By a physical part we mean a part that could if separated from the rest be a physical object in its own right. To a first approximation, a P-part is one which is causally internally connected, but not in general a maximally connected whole. Even such arbitrary parts as the left-hand half of a car are physical parts: were such a car sliced in two (as was once portrayed in a James Bond movie) the left half would become a physical object in its own right. By contrast, the object considered by taking the sections of the bar at 1–2 cm from one end, and 3–4 cm, and 5–6 cm and so on, is not a P-part of the bar, because removing the rest does not give a physical object but several physical objects. Of course we could fuse these together somehow to give one object, but then they compose something *new*, and that's the point. Of course we may want to distinguish between *connected* and disconnected P-parts: there may be some genuine (not merely topological) basis for that further distinction. For the moment however let's stick with this first additional concept. All P-parts are M-parts, but not vice versa. M-parts need have no internal causal cohesion whatever: that's one of the things people don't like about them.

### *12.6.2 Salient Part*

There is also a somewhat vaguely delimited notion of part of something which is in some way *salient*. Call these S-parts. A part may be salient (to a given set of potential observers via one or more sensory modalities) by virtue of its geometric prominence, or its material or qualitative discontinuity from adjacent parts. An example of a salient part (which is always a physical part but not necessarily vice versa) is the lower part of an aircraft fuselage which is painted a different colour from the upper part. For example the upper part may be white and the lower part may be blue. The shape of the line separating the two parts may be deliberately chosen for example to emphasize speed, or to look elegant. Salience in this case indicates that the part is intended to be discerned by observers. But sometimes a part may be salient unintentionally or incidentally, as for example the carburetor bulges on older sports cars sometimes are (of course in time such bulges came to be associated with power and speed, so designers took pains to put them in just to advertise those connotations).

### 12.6.3 Engineering Parts: D-A-R-T

Now let's bring engineering into the picture. For any artefact that might be interesting to an engineer, some parts are more important than others. Not all P-parts are important. So call E-parts all parts that are of interest to an engineer. This is not a wholly objective demarcation so again let's try for a bit more precision, in the knowledge that improvement is incremental. Parts play different roles in engineering depending on what stage of the life-cycle of an artefact we are considering. A part which is envisaged as a unitary part during the *design* of an artefact we call a D-part. One which is manipulated as a separate individual during *assembly* we call an A-part. One which is manipulated as a separate individual during *repair* we call an R-part. And finally one which is manipulated as a separate individual during *retirement* we call a T-part ('T' as in 'reTire'). That gives DART as an acronym. It is possible for a given physical individual to play all four roles, D- A- R- and T, in the economy of a complex artefact. A door of an automobile might be an example. On the other hand, the exigencies of design, manufacture, maintenance and retirement mean that there are frequent discrepancies: what is designed as a D-part may come together only incidentally in manufacture, e.g. the braking system of a truck is never manipulated as a unitary separate object. Modular replacement and repair mean that many A-parts are never R-parts: a sealed headlamp unit in an automobile is an R-part of the automobile which has many A-parts (the unit was assembled) but no R-parts (it is replaced as a whole). Discrepancies among the different kinds of part lead to the so-called *Multiple Bill of Materials (BoM) Problem*, which is a practical hurdle facing electronic documentation of the mereology of complex artefacts across their life-cycles (Simons and Dement 1996).

We are *not* here saying there are four completely new concepts of part: what we are saying is that there are four different *roles* that parts (mostly P-parts) can fill in the life-cycle. And even parts which are not E-parts as here defined may be of at least passing interest to an engineer. Suppose a screw fails to hold a certain slightly friable material because its head is not wide enough, and the material works loose around the head. The engineer will take an interest in the screw head which is, we may suppose, a P-part but not an E-part of the screw (it was turned out of a single piece of material), in that s/he will expect the screw (*not* the head) to replaced by another with a wider head.

### 12.6.4 Functional Part

That brings me to a crucially important role for parts, the most important in regard to engineering, which it is vitally important to recognize and yet surprisingly difficult to make fully precise. That is the idea of a part which performs a unified *function* in the working of the whole artefact. Call this an F-part. For example, the screw head in the example just given is an F-part, since its function is to brace the screw against the material it is intended to hold down. We shall assume then that all F-parts are E-parts, since an engineer has to be interested in function. But as the example shows, an F-part need not be a DART-part (i.e. not any one of those). Some P-parts like the screw head are F-parts, but others are not. The left-hand half of the car is not an F-part. It will not do to invent *ad hoc* "functions" for such parts such as "holding up the right half" just to make anything an F-part. The function has to be describable independently of invoking the part in question. In this case it is not, since the right-hand half is obviously just the mereological complement of the left-hand half. By contrast a function such as "providing forward visibility while shielding occupants from the wind of forward motion" is a description of the function fulfilled by a transparent windscreen (windshield) on a vehicle, and could in principle be fulfilled by some other part or method, (e.g. without considering practical feasibility) a repulsive force-field or forceful cross-draught.

## 12.7 Material Features

There is another general concept associated with material objects (not just artefacts, but natural things as well) which is not a concept of a material part, but which is sufficiently similar and sufficiently important to require treatment here. This is the concept of a *material feature* (Simons 2002). One example is the cross-shaped recess in a screwhead, enabling it to be turned by a suitably shaped driver. Another is the helical thread on the screw with its V- or U-shaped section. Yet another is the hole in a washer or nut, which enables a bolt to pass through it. The teeth of a gear wheel are P- and F-parts of the wheel, but the recesses between the teeth, which allow it to engage with other gear wheels, are material features, not material parts. In general such features as holes (Casati and Varzi 1994), slots, grooves, recesses, cavities, edges, ledges, ridges, corners, waists, tunnels, surfaces and other interfaces are material features; and as the examples indicate, they are to be found among natural objects just as much as among artificial ones, for example in physical geography or human anatomy.

We cannot here attempt a rigorous formal ontological definition of a material feature, not least because it promises to be complicated and may require several overlapping definitions to cover different cases. But we can offer enough by the way of characterization to make the concept's distinctness and importance clear. We mention four ways in which material features are *like* material parts, and two ways in which they are *unlike* them. Firstly, a material feature is, like a material part, a *located individual*. It is not a general property, or a relation, or a mass of material. As a located individual, it can reasonably be attributed causal powers, at least of a passive nature. A hole, slot, tunnel etc. *permits* the insertion or passage of light, matter, objects, constrained by its surrounding matter. In engineering, that is often-precisely what it is there for. Secondly, like material parts, material features generally have a geometrical *shape*, whether stably or fluctuating over time. Thus engineering drawings, blueprints, and their electronic successors, CAD files, can deal with features like holes in the same way in which they deal with parts, by indicating the boundaries of material parts. Thirdly, in a quite general but intelligible sense, a material feature is *something about* a larger object in much the way that a material part is. For that reason it is tempting in various contexts to describe and think of material features as weird kinds of parts, *immaterial parts*. Of course such a conception is inherently confused, but it does signify our recognition of an affinity between parts and features, as well as our need to talk about features and give them their due. Fourthly, material features in engineering can have functions just as much as parts do. As indicated, a hole in a nut is there to allow a bolt to be inserted through it, while the thread on the inside of this hole is there to engage with the thread in the bolt and ensure a secure physical bond between them, as well as (by the threads' matched helical forms) allowing rotation to be converted into pressure exerted along the bolt's central axis in order to hold something firmly between the nut and the bolthead.

Conversely, a material feature is distinguished from a material part in two crucial ontological respects. Firstly, in general a material feature is not *made of* matter in the way in which a material part is. This applies in particular to those features which are obviously in some way concave. Obviously a hole or slot is not made of material like its surrounding matter is, otherwise it would not be there. We form a hole, slot etc. typically by *removing* matter. The hole etc. may be *filled* by something such as air or oil, fuel or hot gas, but that is different. A cavity persists as a feature despite being filled, and indeed its being suitably filled is often the point. The function of a rocket nozzle (the nozzle as material part) is to surround and define a complicatedly shaped cavity (the nozzle as material feature) through which hot expanding gas is designed to flow in a certain way.

Another kind of material feature are *boundaries*, such as a surfaces, edges, ends, tips, points: the outer surface of a sphere, the inner surface of a tube, the cutting edge of a chisel, the end of a rod, the point of a needle. There are two competing conceptions of boundary, one mathematical, one physical. The mathematical conception uses topology and geometry. In this conception a boundary has one more fewer dimensions than that which is bounds: a surface of a body has two dimensions, and edge one, a point zero. These are very apt for mathematical modeling and for simulation in computer software such as CAD systems. However such boundaries without bulk cannot account for the physical discontinuities and properties of real boundaries among material things, such as refraction, change in the velocity of sound, or surface effects such as optical films. So for engineering purposes, which is what engineering requires, a physical conception of boundary as a thin layer of material with often exceptional properties is more appropriate (Simons 1991).

Secondly, the material feature nevertheless *requires* its adjacent matter in order to be what it is: a tunnel is not nothing (ask a tunnel engineer), but it is nothing without material surrounding it. In the jargon of formal ontology, material features are *ontologically dependent* on their adjacent material. How this dependence works varies slightly from case to case.

It should be obvious even from the few simple examples given here that material features are very important in engineering, almost as important as parts. This, and the utility of CAD modeling, explain the importance of feature-based design in manufacturing engineering, despite the different senses sometimes attached to the term 'feature'. The preponderance of similarities over dissimilarities between material features and material parts also explains why we are often tempted to consider features as a sort of part. Indeed as the rocket nozzle example indicates, we sometimes use the same word for both a material feature and the material that bounds it and on which it depends, although these are ontologically speaking wholly different entities. We might even want to call material features *quasi-parts* of the objects they depend on. It is worth considering to what extent the various distinctions drawn above among different subspecies of part can be applied to quasi-parts.

## 12.8 Processes and Their Parts

Processes, apart from their ubiquity and importance in nature, are of vital interest to engineers. The operation of any artefact that entails motion or other change involves processes. These include motion of many types, chemical processes such as reactions, physical processes such as changes in temperature, pressure, shape and other parameters, as well as all the processes involved in the manufacture, running, maintenance and retirement of artefacts. Not for nothing is there a whole branch of engineering bearing the name 'process engineering'. According to some philosophers, among whom I count myself, processes are metaphysically more fundamental than enduring things or *continuants* (Simons 2000). That debate is a lively one within philosophy, albeit that for most of the history of western philosophy the idea that continuants are more basic has had the upper hand. The difference between enduring things on the one hand and processes, events and states on the other is that the latter have *temporal* parts or *phases*, whereas the former do not. So a football match or an explosion has earlier and later parts, whereas a chair or a human being does not. Their lives or careers have temporal parts. Both enduring things and processes may have spatial parts: the difference is that the parts of an enduring thing may move around and change wholesale.

There is then clearly room for a mereology of processes, and their countable associates, events. Some events are indeed minutely scrutinized and anatomized, distinguishing their various parts and their relationships: these include crucial historical events like battles and assassinations, catastrophic events like natural disasters and spectacular accidents like the sinking of the *Titanic* or the explosion of the space shuttle *Challenger*, and on a more mundane level, the critical parts of sporting events, and the myriad of crimes investigated by organs of justice. Biographers make it their trade to describe and relate the events in a person's life, itself a whole composed of countless smaller events and processes. We are thus adept, at an intuitive level, at discerning the parts of events and processes. Relatively little thought has however gone into the question whether we can simply adapt the mereology coming from mathematics and logic for processes. In many respects it appears to be easier to do so than to pursue mereology for continuants, which may change their parts over time, either naturally or by repair. There is doubtless much more to be learnt about the ontology and mereology of processes.

## 12.9 Parts at a Time

Unlike processes, which just have parts, continuants have parts at a time and over a time. For example a young man has a full head of hair, but in old age he is bald. A house may acquire an extension, new built-in cupboards. A car requires a new clutch: the old part is taken out and replaced by a new one. And so on. Some parts are permanent: the continuant has them as long as it lives, others are temporary, even intermittently parts. Some artefacts spend more time dismantled into their component parts than assembled into a whole: guns, musical instruments, some complex tools would be examples. Some parts may come into existence by the configuration of other parts, either permanently or temporarily. In general, the ontology of parts at a time and their manifold changes over time, in themselves, in their relation to one another, and in their relation to the whole, is another one of those subjects that appears to be below the interest threshold of most ontologists, while yet being of crucial concern to engineers. Yet it is not without its high theoretical interest. Plutarch, in his *Life of Theseus*, relates how

> The ship on which Theseus sailed with the youths and returned in safety, the thirty-oared galley, was preserved by the Athenians down to the time of Demetrius Phalereus. They took away the old timbers from time to time, and put new and sound ones in their places, so that the vessel became a standing illustration for the philosophers in the mooted question of growth, some declaring that it remained the same, others that it was not the same vessel.

In a twist to the story Thomas Hobbes imagines someone collecting the replaced parts until he has a complete set, whereupon he rebuilds the "original" ship; and poses the question whether the ship made of the original materials or the ship with the replacement materials is the "real" ship. Pragmatically of course a decision can simply be made on grounds of expediency, tradition or fiat, but the problem is theoretical as much as practical, and illustrates how clarity about the role of parts in a whole at and over a time can help to settle otherwise puzzling disputes.

## 12.10 Conclusion

Mereology, the formal-logical theory of part–whole and cognate relations, is around a century old. It has been thoroughly incorporated into the toolkit of modern analytic ontology, generally to the benefit of the latter. However the special purposes for which it was originally introduced, which were connected with the foundations of mathematics and physics, endowed it with a number of features which have impeded its neutral application both within and outside philosophy, and have led some ontologists to extreme and implausible positions about what there is. Since part of the point of ontology is to be able to connect with the special sciences, it is to no-one's advantage if the formal theory of part and whole interferes with the standard working assumptions of those sciences, either by unreasonably denying the existence of things everyone normally accepts (such as artefacts composed of many parts), or by propounding the existence of things no-one would normally dream of (such as bizarre mereological hybrids and monsters).

In these circumstances the obvious solution is to pare back the formal account of part and whole to cover only those logical properties that are analytically constitutive of the concept, and to leave everything else open. That allows there to be a variety of more specific part-concepts in which different features are optionally added according to the use in different contexts. That is the route we have recommended. It allows mereology to be more responsive to the needs and practices of different sciences, which benefits both sides: ontologists can bring their drive for conceptual clarity to bear on problems encountered in theory and practice outside their subject, while the rigours of encountering real problems can lend their theories greater robustness.

## References


Casati, R., & Varzi, A. (1994). *Holes and other superficialities*. Cambridge, MA: MIT Press.

Husserl, E. (1970). *Logical investigations. Investigation 3: On the theory of wholes and parts* (J. N. Findlay, Trans.). London: Routledge & Kegan Paul.

Leśniewski, S. (1916). *Podstawy ogólnej teoryi mnogości*. Moscow: Poplawski. English edition: Leśniewski, S. (1992). Foundations of the general theory of sets. In *Collected works* (trans: Leśniewski, S.) (Vol. 2, pp. 129–173). Dordrecht: Kluwer.

Simons, P. (1987). *Parts*. Oxford: Oxford University Press.

Simons, P. (1991). Faces, boundaries, and thin layers. In A. P. Martinich & M. J. White (Eds.), *Certainty and surface in epistemology and philosophical method: Essays in honor of Avrum Stroll* (pp. 87–99). Lewiston: Edwin Mellen Press.

Simons, P. (2000). Continuants and occurrents. *The Aristotelian Society Supplementary, LXXIV*, 78–101.

Simons, P. (2002). Characters and features: Individual attributes and their Kin in biology and engineering. *The Modern Schoolman, 59*, 235–252.

Simons, P. (2007). Real wholes, real parts: Mereology without algebra. *Journal of Philosophy, 103*, 597–613.

Simons, P., & Dement, C. W. (1996). Aspects of the mereology of artifacts. In R. Poli & P. Simons (Eds.), *Formal ontology* (pp. 255–276). Dordrecht: Kluwer.

Van Inwagen, P. (1990). *Material beings*. Ithaca: Cornell University Press.

Whitehead, A. N. (1919). *Enquiry concerning the principles of natural knowledge*. Cambridge: Cambridge University Press.

# Chapter 13
# Engineered Artifacts

**Byron Newberry**

**Abstract** Technical artifacts are thought to be distinct from natural and social artifacts in that they are human-made to perform functions, and their functions stem directly from their physical structure. Technical artifacts therefore have a dual nature comprising both a physical description and a functional description, which are coupled together through human intention. This paper explores the further differentiation of technical artifacts into engineered and non-engineered artifacts. Specifically, the question is addressed of what criteria might be used to distinguish an artifact that is a product of engineering, as opposed to some other type of creative or design activity. This question obviously has bearing upon related questions such as: "What is engineering?" The main argument of the paper is that the answer to the question of whether an artifact has been engineered hinges primarily on the nature of the functional requirements and specifications set out prior to the design process.

**Keywords** Engineering design • Technical artifact • Function • Specifications

## 13.1 On Social and Technical Artifacts

The goal of this essay is to explore the notion of *engineered artifacts* as distinguished from other types of made artifacts, which in turn bears on the question of how to distinguish engineering from other types of artifact design. In particular, I will make a case for the primacy of functional specifications in the determination of whether an artifact has been engineered. As a starting point, consider Kroes' (2010) discussion of the relationship between *structural* and *functional* descriptions of technical artifacts. He posits that artifacts have a dual nature, one aspect of which

B. Newberry (✉)
Department of Mechanical Engineering, Baylor University, Waco, TX, USA
e-mail: Byron_Newberry@baylor.edu

D.P. Michelfelder et al. (eds.), *Philosophy and Engineering: Reflections on Practice, Principles and Process*, Philosophy of Engineering and Technology 15,
DOI 10.1007/978-94-007-7762-0_13, 

depends on the intentions of the humans who make and use artifacts, while the other aspect depends upon artifacts' physical structure and properties.

The two types of descriptions are in one sense independent. That is, a purely structural description of some artifact need make no reference to its function; in fact, many functions might be realizable from a single artifact with a given structural description. Likewise, a functional description (the purpose for which an artifact is to be made) need not make reference to the structure of the artifact; in fact, many different physical manifestations, perhaps quite different from one another, might all perform the same function. Despite this independence, physical artifacts are in fact made to accomplish certain functions. Kroes, in a related article (2006), discusses this issue of coherence between the descriptions. That is, he discusses how designers, such as engineers, can start with a functional description (what an artifact needs to do) and reason toward a structural description (the detailed design of an artifact).

In elaborating upon this dual nature of artifacts, Kroes (2010) goes on to distinguish between *technical artifacts* and *social objects*. Kroes uses the example of money to illustrate a social object. A €10 note, for example, works—that is, allows me to purchase something—not by any physical powers that the material bill possesses, but rather by virtue of a collective agreement among people that it is money. "Whether a €10 note can perform its function as money depends on the intentions of people with regard to it," writes Kroes, adding, "In contrast to technical artifacts, there is no close connection between function and physical structure in the case of social objects." In other words, the material form of the money is somewhat arbitrary.

In the case of a technical object, such as a knife, my use of the knife is successful by virtue of its sharp edge—a physical property—independent of my social relationship to others. While this distinction is certainly clear, that a €10 note is a social object in a way the knife is not, I suggest that there is also a strong sense in which the banknote is still a technical artifact. Further, I suggest that the €10 note is a highly engineered artifact—i.e., one whose physical properties have been carefully *engineered* to be instrumental to its functionality. But what does it mean to be *engineered*?

There are several potential indicators of the engineered nature of the banknote. For example, the U.S. Bureau of Engraving and Printing, an organization that produces banknotes, employs, among others, engineers, chemists, technicians, and various types of craftspersons, whose responsibilities are typical of the types of job functions found in an engineering enterprise. Further, in the words of a European Central Bank report (ECB 2004) concerning Euro banknotes, "Banknotes are essentially products with a technical performance to deliver." The report goes on to discuss banknote technical and performance specifications in terms of such requirements as durability, security and authentication, materials and production costs, ease of handling by people, and compatibility with use in ATMs, counting machines, and vending machines. Clearly a lot of design work goes into the production of banknotes, and that work is conditioned upon having definitive performance requirements—functional specifications that, while certainly presupposing a collective agreement as to what constitutes money, are themselves primarily concerned with the physical structure of the object.

Thus, while the form money can take may be highly variable, the function of money—at least in the case of a modern banknote—cannot be wholly divorced from its physical structure. Lawson (2008) hints at this when he writes, "It is not, however, that the physical realization of social artifacts is arbitrary (as Searle seems to suggest)…Money could not be made up from water, or any other non-scarce resource, etc." Despite this non-arbitrariness of the physical form, Lawson, like Kroes, concludes that "this physical realization [of social artifacts] is inessential to their causal powers" (Lawson 2009). Here I take Lawson to mean that while the physical form may not be completely arbitrary or unimportant, it is irrelevant in the absence of the social agreement. But while such an agreement is a necessary condition, it is likely not sufficient for a modern banknote.

If I present a €10 note to buy a sandwich, there will be at least three requirements that have to be satisfied in order for the causal powers of the money to be effective—i.e., for me to receive my sandwich in return. First, the €10 note has to be something I am capable of physically exchanging with the vendor. Second, we both have to agree that the €10 note is money. Third, the vendor has to agree that the thing I hand her is actually a €10 note. We might say that the €10 note must be exchangeable, agreed-upon, and authenticable. Of those three things, two depend on the physical characteristics. The latter of the three, authentication, is something into which a great deal of design effort is put precisely for the reason that any social agreement as to what constitutes money loses its force if real money cannot be physically distinguished from fake money with reasonable accuracy.

In attempting to demonstrate the technical aspect of a €10 note, my point is not to diminish the importance of Kroes' classification of it as a social, rather than technical, object. Such distinctions are germane to answering certain philosophical questions that he and others ask about the nature of artifacts and their functions. Kroes, in fact, suggests that there is not a bright dividing line between technical and social artifacts. In an earlier work he posits an axis along which artifacts may be placed, ranging from the more technical on one end, through mixed socio-technical, to the more social at the other (Kroes 2002).

## 13.2 Engineered Artifacts

Above, in discussing the technical aspect of a €10 note, I suggested that it was *engineered*. But the main question that this article seeks to address is what, precisely, that might mean. For example, are engineered artifacts coincident with all technical artifacts? Though I will argue they are not, it is not immediately clear how to circumscribe the concept of an *engineered* artifact. The question of how to recognize an engineered artifact as distinct from other human-made artifacts, if such a distinction exists, is related to several other questions that are often asked, such as how to determine who counts as an engineer, or how to demarcate the activity of engineering, or how to distinguish engineering design from other types of design. As with these other questions, a definitive answer to the question of what counts as an engineered artifact might prove to be elusive. That elusiveness stems, at least in part, from the fact (or so I would assert) that engineering is an artificial construct. It is a category created for the sake of convenience to facilitate discourse about a particular facet of human activity, much as the category tall helps us aggregate certain people for the sake of discussion.

We may usefully, and with a relative lack of ambiguity, apply the descriptor tall to certain people. We may also generally agree that others do not warrant that label. But the dividing line between the two is arbitrary and those in the middle may be classified one way or the other depending on our specific purpose. Likewise, distinguishing what is engineering and what is not, or what is an engineered artifact and what is not, may vary depending on the objectives of our study. This is a point that has also been made by Davis (2010). To illustrate this point, consider the work of Koen (2003). He takes a rather expansive view of engineering. In fact, his view is so expansive as to be virtually coterminous with all of human activity. But this type of definition serves his goal, which is to make the claim that the engineering method is essentially the universal method of human action. In contrast, someone interested in developing legal regulations for the licensing of engineers and the practice of engineering is likely to take a much more restrictive view of what engineering is.

With due recognition of this inherent mutability of the idea, we can proceed in exploring the question of what constitutes an engineered artifact in the modest hope of gaining some useful insights. With respect to the banknote example, perhaps few would argue with the claim that a modern banknote is an engineered artifact, but I will not assume that is the case. So upon what grounds might we argue that a modern banknote has been engineered? Earlier we noted that the U.S. Bureau of Engraving and Printing employs engineers, along with others often found in engineering enterprises. In fact, many of the directors of the Bureau in the past 150 years have also been engineers (BEP 2004). Similarly, the Dutch Bank first turned over management of the technical aspects of its banknote production to an engineer in the early twentieth century (de Heij 2000). But this line of approach seems to beg the question. Defining what is engineered as that which is worked on by engineers is hardly satisfactory.

Next, we might note that scientific and technical knowledge has been used extensively in the development and production of modern banknotes. The application of such knowledge is generally thought to be a hallmark of engineering. But this alone may not be enough, since use is made of such knowledge in other activities, such as in scientific enterprises or in the process of invention, where invention may be considered distinct from engineering design (Hales and Gooch 2004). We might also draw upon the fact that the production of modern banknotes relies on the heavy use of sophisticated (and presumably engineered) technical equipment and machinery. But, a master wood worker may also use sophisticated tools and machinery to produce a piece of art, which would not itself be considered engineered, which seems to provide a counterexample.

Perhaps the most promising idea for how to justify categorizing banknotes as engineered objects is by looking at the processes used to arrive at the design and production of banknotes. In the literature on engineering and design, the methodology of the engineering design process plays a central role in defining those activities. Much effort has gone into studying engineering design methodology and trying to identify its key components and attributes. If the banknote design and production process follows the methodology of engineering, then perhaps we can rightly claim banknotes to have been engineered. Pursuing this line of thought would require further research into banknote production. Though such research would most certainly prove valuable, I have not pursued it here for the following reason. The belief that banknotes are engineered was initially an instinctive reaction on my part. Given that I had virtually no knowledge about the actual methodology used in banknote design and manufacture, it seems that my instinct must have been based on some factor other than the design methodology.

Upon further reflection, my assumptions about the functional requirements of banknotes seemed to have been instrumental to my beliefs about banknotes. Durability, portability, security, authentication, mass producibility, consistency, quality control, machine handling, and machine readability are all requirements that came immediately to mind. The simultaneous satisfaction of such diverse, and probably very specific and detailed, requirements was, to me, an indication of engineering activity, and hence that the banknote was an engineered artifact. Put another way, it seems I had instinctively used the nature of the artifact's functional requirements, and the corresponding technical specifications that might be drawn from them, as conclusive of the artifact's *engineeredness*.

This might seem like an underwhelming conclusion. After all, in discussions of engineering and engineering design methodology, the translation of perceived functional requirements into detailed technical specifications that the design of an artifact must satisfy is widely acknowledged to be a critical step in the process. "A considerable amount of modern engineering design involves working out criteria and specifications that help define how a technological system will achieve its desired function in more detail" (Nightingale 2009). But I'd like to go beyond simply stating the obvious—that the presence of significant functional requirements, and hence detailed technical specifications, correlates with engineering activity—to tentatively suggesting that the particular nature of an artifact's functional requirements/specifications is in some sense essential to whether that artifact should or should not be considered to have been engineered.

## 13.3 On Functional Requirements and Specifications

If some found object, such as a gold nugget or a gemstone, is to be used as money, we can probably agree that it is not an engineering artifact. This does not mean that there are no functional requirements for it to serve as money. At the very least it must be scarce enough to prevent everyone from being instantly rich by picking up rocks off the ground. And there are also issues of authentication, which perhaps might be settled by biting the gold nugget. But these functional requirements were not instrumental in the making of the object since it was not made at all. In the case of banknotes, which

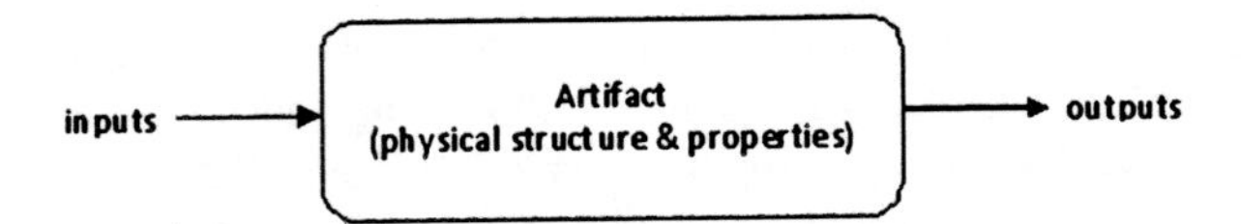


**Fig. 13.1** Black box model

are made objects, functional requirements do factor into their making. Early banknotes had little in the way of formalized functional requirements other than to have the proper information written legibly on a suitably-sized piece of paper, perhaps with signatures and seals affixed for authentication. We arguably should not consider such banknotes engineered any more than we would a garage sale sign made by writing on a piece of poster board and stapling it to a wooden stake.

But as printing technology progressed and the use of banknotes grew, technical problems of security and mass production increased. In the eighteenth and nineteenth centuries an arms race of sorts commenced between banknote printers and counterfeiters, which continues unabated today (Schell 2007). Somewhere along the way a transition occurred by which banknotes went from being merely a particular kind of paper document, with little in the way of physical functional requirements, to being artifacts whose every aspect of physical construction is carefully designed to achieve diverse, detailed, and clearly prescribed performance requirements. So even though banknotes have long been made, something apparently changed in how they were made.

Carl Mitcham (1994) has identified several types of designing with respect to artifacts. These include crafting, inventing, engineering design, and artistry. One thing all of these have in common is the dual nature of the artifacts they produce, as described by Kroes (2010) and discussed earlier in this essay. This dual nature (function/structure) may be illustrated using the classic black box with inputs and outputs, as shown in Fig. 13.1.

Here, the inside of the box represents the physical structure and properties of the artifact—the technical details of its design. The outside of the box, inputs and outputs, represent the artifact's functional requirements—what it should do and how to get it to do it. In the case of a technical artifact such as a power saw, cutting wood by pushing an electrical switch may be the functional requirements. In the case of an artistic artifact, the goal may be the evocation of a particular mental state upon visual engagement. In every case, the design task is to map the inside of the box to the outside; that is, to define the physical structure of the artifact such that the function is realized. So there is an intentional goal to be achieved. There is an end toward which the design activity is directed. But this raises important questions. What is the nature of the end, and how much and what kind of information is required to define it?

For an artist commissioned to create a memorial sculpture, the definition may be quite spare. The sculpture should elicit thoughts of that which is memorialized. The sculpture will have limits on its spatial extent. The sculpture will have limits on its costs. And so forth. But little else will be explicitly specified by way of driving or constraining the design. For another example, one of my personal hobbies is landscaping my yard. Whenever I start a project, I begin with only vague notions of the functional requirements I want to achieve: something that will add color, something that will draw people into a certain part of the yard, or something that will visually demarcate different areas. As often happens in design of all types, the target might move in the course of designing my landscape. As I develop ideas, they feed back to change my objectives. And in the case of my landscaping, my objectives—in the form of functional requirements and specifications—are predominantly qualitative, highly provisional, and rarely fixed in any medium beyond my thoughts.

In the case of invention, many authors have sought to differentiate it from engineering design. Mitcham (1994) writes, "*As opposed to designing*, inventing appears as an action that proceeds by nonrational, unconscious, intuitive, or even accidental means. Designing means rationality and planning." Although some invention may proceed in such a fashion, I am not convinced that such non-rationality and lack of planning is constitutive of invention. In his history of the development of powered flight, James Tobin (2003) chronicles the Wright brothers' methodical and meticulous efforts in achieving their inventive success. Their approach would appear to be quite rational, with very little reliance on luck or fortune. Perhaps a more appropriate juxtaposition of invention and engineering design is given by Hales and Gooch (2004, p 122), who write,

> An inventor comes up with ideas that may or may not be worth pursuing, and every now and then the chances are that a viable idea will surface. In some cases it becomes a winner. A design engineer defines a technical problem based on a set of requirements and sets off to find the most appropriate solution to the problem within defined constraints of time, money, and other resources.

This seems to capture something critical about inventing that Mitcham's account does not (while foregoing any mention of non-rationality), and that is the relative level of uncertainty in the outcome of invention when compared to engineering design. In engineering design, the functional requirements are specified in such a way that the existence of a solution—i.e., of a physical description that will realize the functional description—is not considered to be much in question (which is not to say that projects have never failed because of unrealistic requirements). For example, when the English Channel Tunnel project was initiated, there may have been some uncertainties about the costs, duration, and particular technical challenges to be faced, but there was little doubt about an eventual successful outcome, nor little doubt about what it would look like. Securing the enormous financial commitments necessary to undertake the project likely would never have been possible without a great deal of certainty about the final product. On the other hand, the success of the Wright brothers was much less assured, and was even highly doubted by much of the public. The Wrights may have been personally convinced of the possibility of achieving successful heavier-than-air, powered flight, but no one knew for sure until it happened, nor did anyone know exactly what might be required for its accomplishment.

And therein lies a critical point. For the English Channel Tunnel Project, people could specify in great detail the functional requirements the tunnel had to satisfy with respect to modes of transportation it would provide, the ventilation, the lighting, safety issues, and so forth. Rational estimates could be made for project costs,

**Fig. 13.2** Wright Brothers 1903 Flyer (NASA: www.nasaimages.org)

duration, manpower, equipment, and all manner of other resources. The requirements may have changed or drifted in the course of the project, and the resource estimates may have been more or less correct. Nonetheless, the project concluded in something close to the manner expected at the outset. On the other hand, when the Wright brothers first set out to accomplish powered flight, there was little in the way of specific functional requirements other than a few broad, qualitative objectives: to become airborne, stay airborne for some period of time, maneuver while aloft, and then land safely at or near the starting point. Any more specific functional requirements that the Wright brothers may have been working toward were likely tacit and constantly in flux. Further, at the outset of their active pursuit of flight, circa 1899, there was little basis upon which to accurately predict when success might come, what it would eventually cost, or what specific form it might take. The Wrights' now-famous 1903 Flyer (Fig. 13.2) is thus the product of an inventive process that was short on detailed functional requirements and long on uncertainty. This is certainly not to say, however, that the Flyer was not carefully crafted, with great attention to design details, and drew upon the best technical knowledge of the day.

Even though the flight of the 1903 Flyer is widely hailed as the inception of the era of powered, heavier-than-air flight, the device was not capable of useful service. The Wrights worked for the next several years, largely in secret, to incubate and improve their design to make it reliable and serviceable. In so doing, they were probably establishing many new functional requirements for the device, even if still tacitly. That is, with each trial they likely set, informally at least, more refined, and more concrete, goals for their device. But it was not until they gained sufficient confidence in their design, in the 1905–1906 timeframe, that they began in earnest

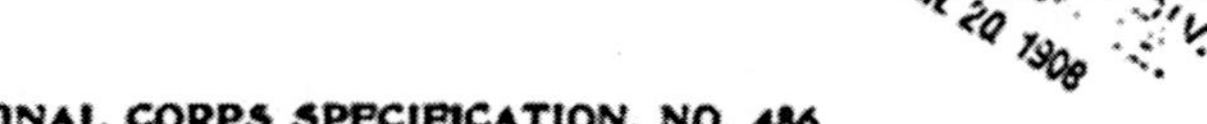

SIGNAL CORPS SPECIFICATION, NO. 486.

ADVERTISEMENT AND SPECIFICATION FOR A HEAVIER-THAN-AIR FLYING MACHINE.

TO THE PUBLIC:

Sealed proposals, in duplicate, will be received at this office until 12 o'clock noon on February 1, 1908, on behalf of the Board of Ordnance and Fortification for furnishing the Signal Corps with a heavier-than-air flying machine. All proposals received will be turned over to the Board of Ordnance and Fortification at its first meeting after February 1 for its official action.

Persons wishing to submit proposals under this specification can obtain the necessary forms and envelopes by application to the Chief Signal Officer, United States Army, War Department, Washington, D. C. The United States reserves the right to reject any and all proposals.

Unless the bidders are also the manufacturers of the flying machine they must state the name and place of the maker.

*Preliminary.*—This specification covers the construction of a flying machine supported entirely by the dynamic reaction of the atmosphere and having no gas bag.

*Acceptance.*—The flying machine will be accepted only after a successful trial flight, during which it will comply with all requirements of this specification. No payments on account will be made until after the trial flight and acceptance.

*Inspection.*—The Government reserves the right to inspect any and all processes of manufacture.

GENERAL REQUIREMENTS.

The general dimensions of the flying machine will be determined by the manufacturer, subject to the following conditions:

1. Bidders must submit with their proposals the following:
   (*a*) Drawings to scale showing the general dimensions and shape of the flying machine which they propose to build under this specification.
   (*b*) Statement of the speed for which it is designed.
   (*c*) Statement of the total surface area of the supporting planes.
   (*d*) Statement of the total weight.
   (*e*) Description of the engine which will be used for motive power.
   (*f*) The material of which the frame, planes, and propellers will be constructed. Plans received will not be shown to other bidders.

**Fig. 13.3** Signal Corps Specification No. 486 (USAF: www.ascho.wpafb.af.mil)

to seek a market for their invention. And it was not until early 1908 that they finally secured a contract with the United States Army. That contract came as a result of the Wrights' response to Signal Corp Specification No. 486, a portion of which can be seen in Fig. 13.3, which was a solicitation for bids on a heavier-than-air flying machine.

This document contained detailed, quantitative performance specifications for any proposed flying machine. These included: carrying enough fuel to cover 125 miles, spending 1 h aloft, averaging a speed of 40 miles per hour over a 5 mile course, being able to be packed for transport on an army wagon and then assembled in 1 h, and carrying two persons, sitting upright, with a combined weight of up to 350 lb. Largely in the dark about the Wrights' capabilities, the editors of the *American Magazine of Aeronautics* wrote: "There is not a known flying-machine in the world which could fulfill these specifications at the present moment…We doubt very much if the government receives any bids at all possible to be accepted" (Kelly 1943/1989). Despite this skepticism, by 1909 the Wright brothers had succeeded in meeting all the specifications and delivered a working product to the United States Army. It was designated Signal Corps Airplane No. 1 (Fig. 13.4).

It is my contention that Signal Corps Airplane No. 1 is an engineered artifact in a way that the 1903 Wright Flyer (as well as the Wrights' other, more functional, intermediaries) is not. Granted, they are extremely similar with respect to materials, components, geometry, and function. Further, much the same technical knowledge

**Fig. 13.4** Signal Corps Airplane No. 1 (USAF: www.af.mil)

and construction techniques went into their designs and development. Yet there was a fundamental difference in how the two were designed. The 1903 Flyer was the contingent result of the pursuit of rather speculative and loosely defined functional objectives, objectives for which the very possibility of attainment was in considerable doubt. Airplane No. 1, on the other hand, was the anticipated result of an effort to meet explicit, detailed, and quantitative performance requirements and specifications using known (at least to the Wrights) technological capabilities.

## 13.4 Concluding Remarks

What I would like to tentatively suggest is that whether we ought to call an artifact *engineered* or not might best be determined by looking at the nature of the functional requirements and specifications that were defined for the artifact *prior* to the commencement of its design and development (as opposed to, say, looking at the nature of the processes used along the way). All creative designing, whether artistry, crafting, inventing, or engineering must begin with some type of objective that guides the process of determining the physical structure and properties of the final artifact. Objectives are operationalized to greater or lesser extents, and more expressly or more tacitly, through functional requirements and specifications. My argument is that the likelihood that an artifact might appropriately be called *engineered* depends on the extent to which the initial functional requirements and specifications are:

- Explicit
- Quantitative
- Detailed
- Technical

- Copious
- Diverse in kind
- Coupled
- Conflicting
- Recondite
- Definitive

To what extent does each of these characteristics need to be satisfied for a resulting artifact to be called *engineered* as per my suggestion? Space does not permit much exploration of that question here, but it is an important one. And it is similar in nature to asking about the location of the boundary between short and tall. The various types of creative design activity shade into each other, and I suspect there are many artifacts whose provenances lie in gray areas.

I do not necessarily wish to discount the *process* by which functional requirements and specifications are translated into a physical description as being critical to whether some artifact is considered *engineered*. The process is of critical interest in understanding how one goes from thought to thing, and it is the process that often winds up being central in discussions of *engineering* as an activity. In fact, the development of requirements and specifications is often taken to be one of the beginning steps of the process, and may not be easily decoupled from other parts of the process. That is, often certain specifications, or even the need for certain specifications, cannot be known precisely until the design process is underway, and may continue to be in flux throughout. Rarely, if ever, are a set of *a priori* specifications complete and unchanging throughout the design process. An illustration of the focus upon *process* as being constitutive of engineering is in the field of software engineering. The advocates of that field have sought to establish it as a "true" discipline of engineering, and have tried to develop software engineering practices that emulate the perceived processes of engineering (Denning and Riehle 2009; Simons et al. 2003).

But I would argue that the process is in some sense determined by, or at least highly driven by, the nature of the requirements and specifications. That is, if the requirements and specifications satisfy the above-listed criteria to a high degree, then certain types of processes will be more effective than others in producing an artifact to satisfy those requirements. In the case of software engineering, the detailed determination of requirements and specifications is taken to be a key factor in the goal of emulating engineering design. Under my proposal, if the functional requirements and specifications of proposed software satisfy the above-listed criteria, then the resulting software has been engineered.

One upshot of taking this view is that engineered artifacts cannot be mapped one-to-one to the activity of people with the title *engineer*. That is, if the nature of the functional requirements and specifications is determinative of an engineered artifact, the question of who does the translating between requirements and physical form is left open. There may be many people who are not credentialed engineers who nonetheless are primarily responsible for creating designs to satisfy the type of functional requirements listed above. Conversely, there are many engineers whose job functions do not contribute (directly, at least) to translating between the functional requirements and physical form of any artifact.

## References

BEP (2004). *BEP history*. Washington, DC: Bureau of Engraving and Printing. http://www.moneyfactory.gov/images/BEP_History_Sec508_web.pdf. Accessed 15 Jan 2011.
Davis, M. (2010). Distinguishing architects from engineers: A pilot study in difference between engineers and other technologists. In I. van de Poel & D. Goldberg (Eds.), *Philosophy and engineering: An emerging agenda*. Dordrecht: Springer.
de Heij, H. A. M. (2000, January 27–28). The design methodology of Dutch banknotes. In *Proceedings of SPIE Vol. 3973*, San Jose, CA, USA.
Denning, P. J., & Riehle, R. D. (2009). Is software engineering engineering? *Communications of the ACM, 52*(3), 24–26.
ECB (2004). *Monthly bulletin—August*. Frankfurt: European Central Bank. http://www.ecb.int/pub/pdf/mobu/mb200408en.pdf. Accessed 15 Jan 2011.
Hales, C., & Gooch, S. (2004). *Managing engineering design* (2nd ed.). London: Springer.
Kelly, F. C. (1989). *The Wright brothers: A biography authorized by Orville Wright*. Mineola: Dover. (Original work published 1943)
Koen, B. (2003). *Discussion of the method: Conducting an engineer's approach to problem solving*. New York: Oxford University Press.
Kroes, P. (2002). Design methodology and the nature of technical artifacts. *Design Studies, 23*, 287–302.
Kroes, P. (2006). Coherence of structural and functional descriptions of technical artifacts. *Studies in History and Philosophy of Science, 37*, 137–151.
Kroes, P. (2010). Engineering and the dual nature of technical artifacts. *Cambridge Journal of Economics, 34*(1), 51–62.
Lawson, C. (2008). An ontology of technology: Artifacts, relations, and functions. *Techné, 12*(1), 48–64.
Lawson, C. (2009). Ayres, technology, and technical objects. *Journal of Economic Issues, 43*(3), 641–659.
Mitcham, C. (1994). *Thinking through technology: The path between engineering and philosophy*. Chicago: The University of Chicago Press.
Nightingale, P. (2009). Tacit knowledge and engineering design. In M. Anthonie (Ed.), *Philosophy of technology and engineering sciences* (Vol. 9, pp. 351–374). Amsterdam: Elsevier.
Schell, K. (2007). History of document security. In K. de Leeuw & J. Bergstra (Eds.), *The history of information security: A comprehensive handbook*, pp. 197–241.
Simons, C., Parmee, I., & Coward, D. (2003). 35 years on: To what extent has software engineering design achieved its goals? *IEE Proceedings-Software, 150*(6), 337–350.
Tobin, J. (2003). *To conquer the air: The Wright Brothers and the great race for flight*. New York: Free Press.

# Chapter 14
# Engineering Ethics: From Preventive Ethics to Aspirational Ethics

**Charles E. Harris Jr.**

**Abstract** An important distinction in engineering ethics is between preventive ethics, which consists of guidelines for preventing harm to the public, and aspirational ethics, which consists of guidelines and motivating considerations for using one's professional expertise to promote human well-being. Preventive ethics is stated in rules and is considered mandatory for all members of a profession. Aspirational ethics allows the professional more discretion in determining what it involves and when and how it is implemented. While preventive ethics must continue to be an important part of professional ethics in engineering, aspirational ethics should be given a more prominent place. Four types of action falling in the category of aspirational ethics can be distinguished, based on their increasingly direct focus on promoting human well-being. Four virtues can be identified as having special importance in motivating and guiding aspirational ethics.

**Keywords** Professional ethics • Aspirational ethics • Preventative ethics • Virtue ethics

## 14.1 Preventive Ethics

Engineering ethics can be divided into two areas. "Preventive ethics," which might also be called "regulatory ethics," consists of guidelines for preventing harm to the public. Preventive ethics in turn can itself be divided into two components. The first component is ethical guidelines designed to prevent specific types of professional misconduct, such as violating confidentiality when it is not justified, having an

C.E. Harris Jr. (✉)
Sue and Harry Bovay Professor of the History and Ethics of Professional Engineering, Texas A&M University, College Station, TX, USA
e-mail: e-harris@philosophy.tamu.edu

D.P. Michelfelder et al. (eds.), *Philosophy and Engineering: Reflections on Practice, Principles and Process*, Philosophy of Engineering and Technology 15,
DOI 10.1007/978-94-007-7762-0_14, 

undisclosed conflict of interest which corrupts one's professional judgment, and practicing outside one's area of professional competence. Such guidelines supply most of the content of the engineering codes of ethics. By my count, 80 % of the content of the code of the National Society of Professional Engineers (NSPE) is devoted to this type of regulation.

The second and more general component has to do with directions to exercise the proper degree of professional responsibility in one's work. Although these larger aspects of engineering responsibility are rarely mentioned in the codes, they follow from the directive present in most engineering codes to "hold paramount" the safety and health of the public, and they are often discussed in textbooks and other documents in engineering ethics. Engineers, for example, must exercise "due care" or "reasonable care" in the performance of professional duties. This requires more than merely exercising that minimal degree of responsibility necessary to avoid legal problems. Engineers must act in an anticipatory and proactive way, attempting to eliminate possible problems before they arise and even identifying and correcting problems caused by other engineers, when practically possible.

Preventive ethics has been the center of attention in the emerging discipline of engineering ethics. Much of the impetus for preventive ethics has come from the so-called "disaster cases" that have aroused public concern and demonstrated the need for protecting the public. A mining disaster in Wyoming resulted in the creation of the first state board of registration for engineers in the US, and a natural gas explosion in a school in Texas resulted in legislation setting up professional registration in the state. The Hyatt Regency walkway collapse also caused widespread concern about structural safety. The *Challenger* and *Columbia* crashes are probably the preeminent examples of disaster cases that caused public concern about safety in engineering.

Engineers can exhibit adherence to preventive ethics in various ways, some of which have been suggested already, such as avoiding conflicts of interest or anticipating and preventing events that can adversely affect the health or safety of the public. But the ultimate manifestation of preventive ethics is "whistleblowing," which often involves risking one's job or even one's career to protect the public. The best-known justification of whistleblowing, by Richard De George, holds that whistleblowing is only morally obligatory when one has evidence that would convince a responsible, impartial observer that organizational policy is wrong and strong evidence that making the information public will prevent serious harm to the public. De George's argument thus aligns itself with the preventive-ethics orientation (De George 1981).

In summary, preventive ethics, insofar as it applies to engineering, has three characteristics. First, its precepts are designed to protect the public from harm, either from technology itself or from the misconduct or lack of responsibility on the part of engineers themselves. Second, the provisions of preventive ethics are mandatory. They are ethically mandatory, because they appear in the codes or are implied by the obligation to hold paramount the health and safety of the public; they may be legally mandatory if engineers are registered by a governmental entity. Third, since the major obligations of preventive ethics are set out by the engineering profession itself, in the codes and other documents in professional ethics, they are independent of the ideals or values of individual professionals. Individual engineers do not avoid conflicts of interest simply because such conflicts would violate their personal morality, but because conflicts of interest violate the standards set out by their profession. They learn that conflicts of interest are prohibited by the profession and that they must be avoided for that reason. Hopefully conflicts of interest violate their personal morality as well, but this is not the primary reason for avoiding them.

## 14.2 Aspirational Ethics

Despite the importance to the public of preventive ethics, it is difficult to conceive of it as comprising the whole of professional ethics. One does not enter a profession merely to avoid engaging in professional misconduct or harming the public. The best way to comply with these essentially negative aims would be to avoid becoming a professional altogether. Professional ethics in its highest sense must involve something more than preventing harm to the public. Let us call this more positive aspect of professional ethics "aspirational ethics."

One way to get at the more positive dimension of professional ethics is to ask, "What is the social good that a profession promotes?" For medicine, this social good is promoting health. The last of the nine Principles of Medical Ethics of the American Medical Association (AMA) says that the AMA "supports access to medical care for all people." While not specifying how this goal is to be achieved, the code endorses health as a social good for which medicine has a special responsibility. For law, the social good is generally thought to be the promotion of justice. To be sure, many attorneys may be more interested in promoting the interests of their clients than in seeking justice, but an argument can be made that the adversary system itself promotes justice, and that the work of lawyers in advocating the interests of their clients is an essential part of the adversary system.

What should be said about engineering? What is the social good for which engineering has a special responsibility? Engineering codes suggest an answer. The complete version of the "paramountcy" statement in the NSPE code referred to earlier says: "Engineers, in the performance of their professional duties, shall hold paramount the safety, health and *welfare* of the public."[1] While the references to safety and health are essentially negative and suggest a protective or preventive function, the term "welfare" suggests a distinctly positive ideal. I propose therefore that the social good of engineering is the promotion of the welfare of the public. But what does "welfare" mean? If "safety" and "health" refer to preventing harm to the public, to what does the term "welfare" refer?

Some hints for interpreting the term "welfare" can be found in the codes themselves. When discussing the obligation of engineers to "serve the public interest," the NSPE code, in section III.2.a uses the expression "safety, health and

[1] I have added the emphasis on "welfare."

well-being" instead of "safety, health and welfare." This suggests that "welfare" and "well-being" may be synonymous, thus confirming the more positive orientation of the term "welfare."

Other codes and sources give further grounds for holding that the term "welfare" should be given a more positive interpretation. The first sentence of the code of the Institute of Electrical and Electronics Engineers (IEEE) says that members of the IEEE recognize "the importance of our technologies in affecting the quality of life throughout the world.... "It goes without saying that "affecting the quality of life" means improving the quality of life. The first of the "Fundamental Principles" of the code of the American Society of Mechanical Engineers (ASME International) commits engineers to "using their knowledge and skill for the enhancement of human welfare...." Here the more positive interpretation—*enhancing* welfare—is explicit. The first of the "General Moral Imperatives" of the "ACM Code of Ethics and Professional Conduct" of the Association for Computing Machinery (ACM) directs computing professionals to use the products of their efforts to, among other things, "meet human needs."

Finally, a statement by William A. Wulf, then President of the National Academy of Engineering (NAE), gives clear and emphatic support for a more positive aim for engineering. Commenting on the NAE's selection of the 20 greatest engineering achievements of the twentieth century, Dr. Wulf said the criterion for selection was

> *not* technical "gee whiz," but how much an achievement improved people's quality of life. The result is a testament to the power and promise of engineering to improve the quality of human life worldwide. (Wulf 2000)

Enhancing human welfare, meeting human needs, improving the quality of human life—these are clear and unmistakable references to a positive ideal appropriate to the engineering profession. But how should we understand these terms? What sense of these terms is appropriate for the engineering profession?

## 14.3 Material Well-Being

The work of economist Amartya Sen and philosopher Martha Nussbaum is helpful in answering this question. In constructing criteria for measuring progress in developing countries, these writers have proposed the following approach. Let us define "functionings" as those activities that people value and "capabilities" as the abilities to engage in these activities and thereby "to lead the kind of life they have reason to value" (Sen and Anand 2000). Nussbaum has constructed a list of ten functionings, or activities that people value, which she believes apply to most humans around the world. We can consider these to be various aspects of welfare or well-being. In abbreviated form, these functionings are the following:

1. Living a normal length of life.
2. Having clean water, food, and shelter.
3. Moving about freely and safely.

4. Using one's senses and imagination and having free expression.
5. Having love and attachments to things and other people.
6. Being able to form a conception of the good life and to plan one's life.
7. Being treated with respect and dignity.
8. Living with concern for and in relation to nature.
9. Engaging in recreational activity.
10. Being able to participate in the political process, preserve material goods, and hold property (Nussbaum 2000).

It is noteworthy that engineering contributes to most of these functionings in some way. Medical technology helps to lengthen life. The contribution of civil engineering to the production of clean water and shelter is widely recognized, and the contribution of chemical and agricultural engineering to food production is equally evident. Free movement requires roads and the means of transportation, for which engineering is crucial. Free expression and attachment to others is facilitated by communication, including the use of computers. Being able to plan one's life and carry out those plans and being treated with respect and dignity are also facilitated by a minimal level of material well-being, which is facilitated by all branches of engineering. Being able to live in relation to nature and enjoy recreational activity are facilitated by transportation and other benefits of engineering. Finally, material goods cannot be preserved until they are first possessed, and engineering contributes to the production of material goods.

This enumeration points to an important fact, namely that engineering is especially associated with the material or physical factors that are important in enabling people to achieve a high quality of life or well-being. Therefore we can say that *the social good of engineering is the promotion of the material basis of human well-being or quality of life*. I propose that this is the good in view in aspirational ethics in engineering. In the next section, I suggest four ways in which engineers can promote the material basis of human well-being or quality of life, listed in terms of the increasing centrality of the goal of promoting human well-being. Let us refer to them as "aspirational acts."

## 14.4 Four Types of Aspirational Acts

Let us call the first category *Acts Exhibiting Exemplary Professional Excellence*, that is, actions that manifest the highest level of professional expertise and achievement. While preventive ethics may require minimal levels of professional competence, aspirational ethics advocates professional expertise and achievement that goes as far beyond this minimum level as the professional's capabilities allow. Although the direct and immediate focus is on attaining the highest level of professional excellence rather than promoting human well-being, the indirect result can be the production of engineering works of outstanding merit that increase human well-being.

The second category I call *Supererogatory Preventive Acts*. These are actions that are concerned with preventing harm to the health and safety of the public, but that go beyond what is required by preventive ethics. They are actions, like all supererogatory actions, that are praiseworthy, but not required. Richard De George's justification of whistleblowing, cited earlier, illustrates the distinction between a required action of preventive ethics and a supererogatory preventive action. For De George, if the evidence for the harm is overwhelming and if making the information public will almost certainly prevent the harm, the action is required and therefore falls into the category of (mandatory) preventive ethics. If, on the other hand, one can only say that the harm is serious, the concern has been reported to superiors, and the organizational channels have been exhausted, taking action to prevent the harm is supererogatory and falls in the category of aspirational ethics. Protesting the emission of a chemical from one's plant whose harmfulness is in dispute is also an example of a supererogatory preventive action.

Another example of a supererogatory preventive action is given in an opinion of the NSPE's Board of Ethical Review (BER). In case 82–85, the Board defended the right of an engineer to protest what he believed were excessive costs and time delays on a defense contract on the part of his employer. The BER's judgment was that, although the engineer was not ethically required to protest his employer's actions, he had "a right to do so as a matter of personal conscience." The reason cited by the Board to justify this right was that, in being concerned about the responsible expenditure of public funds, the engineer was looking after the welfare of the public. Here the welfare of the public is interpreted in terms of protecting the financial interests of taxpayers. Unlike actions in the category of preventive ethics, however, this action is described as non-mandatory. Furthermore, the action is described as deriving from the "personal conscience" of the engineer rather than strict professional obligations, as in the case of preventive ethics. Protecting the financial well-being of taxpayers when a threat to health and safety is not involved falls into the category of aspirational ethics.

The third category is what Michael Pritchard has called *Good Works* (Pritchard 1992). Professional activities in this category might be considered no different from any other type of engineering work, except that the public good is more clearly in mind, they are often highly innovative, they are frequently performed with a high degree of enthusiasm, and they sometimes involve an element of self-sacrifice. James is excited about being put on a project to develop an experimental automobile that has many recyclable parts, is lightweight, is unusually safe, and gets at least 60 miles per gallon of fuel. He works with unusual intensity and energy and is willing to put in overtime hours without pay to achieve the goals of the project. Students in a senior design class build an auditory visual tracker for use in evaluating the training of visual skills in children with disabilities. The students meet the children for whom the equipment is being designed, and this encounter so motivates them that they work overtime and even when the course is over to complete the project (Harris et al. 2009). A chemical engineer devotes his career, with some risk, to developing a highly efficient engine, a biomass conversion system, and other projects in "green engineering" (Harris et al. 2009, 191–192). In the 1930s a group of General Electric engineers, acting against considerable skepticism, worked overtime with no pay to develop a sealed beam headlight, which greatly reduced the number of accidents caused by night driving (Meese 1982).

I designate the fourth category as *Altruistic Engineering Acts.* Actions in this category are characterized by a still more direct focus on promoting public well-being, perhaps a deviation from a normal career path, and a special concern to utilize one's professional expertise to help those who are disadvantaged or in distress. At age 27, Frederick C. Cuny, who attended engineering school but was not a degreed engineer, founded the Interact Relief and Reconstruction Corporation. He organized relief efforts, involving engineering work, in Bosnia after the war and in Operation Desert Storm (Pritchard 1998). The work of engineers in Engineers Without Borders also falls into this category.

## 14.5 Characteristics of Aspirational Ethics

The above discussion suggests three characteristics of aspirational ethics. First, the provisions of aspirational ethics have a distinctly positive and idealistic element. Their orientation is not toward simply protecting the public from harm, but achieving the highest rungs on the ladder of professional excellence. In fact, the ideal of professional excellence is central in aspirational ethics.

Second, the provisions of aspirational ethics are non-mandatory, in that how and to what extent one implements them is a matter for personal discretion. While holding the welfare of the public paramount may be mandatory, it is left to individual engineers to determine how they will implement this provision. By contrast, the provisions of preventive ethics are more specific and ethically mandatory—even legally mandatory if the engineer has professional registration. Engineers may be condemned ethically and perhaps legally sanctioned for engaging in such practices as having undisclosed conflicts of interest or inappropriately revealing confidential information. These requirements are firmly grounded in the codes and other literature of engineering ethics. Aspirational ethics is different. Even failing to embrace aspirational acts altogether would not be cause for professional or legal reprimand, although it would involve an ethical failure of a lesser sort.

Third, the motivation for aspirational ethics, as well as the determination of how it is implemented, is in personal ideals, although these ideals may be importantly related to one's professional work. The BER ruling cited earlier hints at the personal grounding of aspirational ethics when it says that the engineer's decision to protest his employer's misuse of taxpayer funds was "a matter of personal conscience."

Mike W. Martin has even more clearly recognized the personal grounding of the aspirational aspects of professional ethics. Discussing the intersection of professional ethics with personal ideals, Martin says:

> Personal commitments motivate, guide, and give meaning to the work of professionals...I seek to widen professional ethics to include personal commitments, especially commitments to ideals not mandatory for all members of a profession. (Martin 2000)

One of Martin's favorite examples is Dr. David Hilfiker who "left a comfortable medical practice in rural Minnesota to work in a ghetto in Washington, D.C." According to Dr. Hilfiker's own testimony, his reason for doing this was to achieve a closer relationship with God (Martin 2000, 3). The examples in the category of altruistic engineering bear an obvious similarity to the example of Dr. Hilfiker.

As Martin stresses, aspirational acts are non-mandatory in nature. They are not grounded in rules promulgated in codes of ethics that are ethically (and perhaps legally) required. Rather, they are grounded in what the BER calls "personal conscience" and what Martin calls "personal commitments." They are grounded, that is, in traits of character. This means that they are grounded in what have traditionally been called virtues. I turn now to the nature of virtues and how they can serve as a grounding for aspirational ethics.

## 14.6 The Virtues

To begin, we can still profitably call upon Aristotle's definition of a virtue. For Aristotle, "…virtue or excellence is a characteristic [H. Rackham translates: "settled disposition of the mind"] involving choice, …that…consists in observing the mean relative to us, a mean which is defined by a rational principle, such as a man of practical wisdom would use to determine it." (Aristotle 1962) A virtue is a character trait which determines action, but not in a mechanical way. The determination of action must always be made "relative to us," i.e. relative to the circumstances of a particular situation. Moral judgment is necessary, for example, to discern what honesty requires in a particular situation. Further, as the definition also indicates, a virtue is something stable and abiding. Being courageous on one occasion is not enough to make one courageous, just as being cowardly on one occasion is not enough to make one cowardly.

Another important characteristic of a virtue is that it pervades the entire personality. Rosalind Hursthouse depicts the complexity of a virtue:

> A virtue such as honesty is a disposition which is well entrenched in its possessor, something that, as we say, "goes all the way down," unlike a habit such as being a tea-drinker—but the disposition in question, far from being a single track disposition to do honest actions for certain reasons, is multi-track. It is concerned with many other actions as well, with emotions and emotional reactions, choices, values, desires, perceptions, attitudes, interests, expectations and sensibilities. To possess a virtue is to be a certain sort of person with a certain complex mindset. (Hence the extreme recklessness of attributing a virtue on the basis of a single action.) (Hursthouse 2012)

The complexity and depth of the virtues is often overlooked, and it is an important consideration in determining how the virtues are to be taught.

A final point about the virtues that has been emphasized by contemporary research in social psychology may be contrary to Aristotle's understanding of the virtues. Aristotle appears to assume what Martha Merritt calls the "motivational self-sufficiency" of the virtues: that character is sufficient to motivate action (Merritt 2000). A vast body of social psychological research, however, casts doubt on "the Aristotelian certainty that a good upbringing, together with an accumulation of practical experience, is sufficient to secure virtuous dispositions as firm and unchangeable under normal circumstances" (Merritt 2000, 376). Instead, Merritt finds in social psychological research strong validation for "*the sustaining social contribution* to character" (Merritt 2000, 374). Effective transfer of the virtues as character traits to actual behavior appears to require social support. Without this support, individuals, influenced by the contingencies of the situation, may fail to consistently manifest the virtues in behavior. In the professions, this social support should come from professional societies and the professional community itself. Now I want to suggest four virtues that are of special importance in motivating and guiding the aspirational acts described earlier.

## 14.7 Four Virtues for Aspirational Ethics

The first virtue is *aspiration to professional excellence*, the disposition to achieve at the highest possible level in one's area of professional competence. Professional excellence can be linked to the more general Greek concept of excellence (*arête*), which is the quality that enables its possessor to perform his own particular function well. For the Greeks, it is the quality that enables a shoemaker to make good shoes or a warrior to be a good fighter. Excellence results in pride and satisfaction in a job well done, a job performed to the highest standards of the activity in question. Accordingly, an excellent engineer is one who performs to the highest standards of his or her profession. Minimal standards of competence are enforced by law and required by codes of ethics, but the aspiration to achieve the highest of which one is capable is not, and cannot be, mandated.

Since ancient times, many advocates of virtue ethics have maintained that the virtues can be taught. Teaching the virtues that motivate and guide aspirational conduct should be, therefore, an important aspect of moral education in engineering. Teaching the virtues has often been facilitated by the use of exemplars. While exemplars, such as Roger Boisjoly, have often been cited in engineering ethics for praiseworthy conduct in protecting (or attempting to protect) the public, it is also important to identify engineers for excellence in engineering work itself. Many exemplars could be cited in this category. Charles Steinmetz was important in the development of alternating current that made possible the expansion of the electric power industry in the U.S. Paul MacCready, inventor of the Gossamer Penguin, the first successful completely solar-powered aircraft, was cited by the Academy of Achievement as Engineer of the Century.

The second virtue is what Paul Taylor has called *respect for nature*, a disposition to appreciate and care for the natural world (Taylor 1986). It is a virtue that is important in motivating many good works, such as engineering projects devoted to protecting the environment. Engineering has more direct effect on the natural world than any other profession, so responsibility for environmental impact is a special obligation of engineers.

Rosalind Hursthouse has suggested that respect for nature is a "new" virtue. As she is the first to admit, however, inculcating a virtue is no simple matter, because a virtue involves a range of emotions, sensibilities, perceptions and in fact "a way of being human" (Hursthouse 2007). It might even involve "a complete transformation of character" (Hursthouse 2007, 163). One cannot simply decide to have a virtue, because its acquisition ordinarily (though not always) begins in childhood, before conscious decisions of this type are made. Training in the virtue of respect for nature is no exception. Its inculcation should begin in childhood and continue through adulthood.

An engineer might manifest the virtue of respect for nature in various ways, but the most obvious way would be a commitment to environmentally friendly engineering projects. Examples of engineers who have committed themselves to environmentally friendly project are also important (Harris et al. 2009).

How can the virtue of respect for nature be nurtured? For engineering students, exposing them to readings in environmental philosophy and literature, encouraging them to take courses in biology, and encouraging engineering professors to consider issues of environmentally friendly engineering and sustainable engineering come to mind. With young children, parents can encourage them to respect the lives of wild animals and not to kill them unnecessarily and to appreciate the beauty and intricacy of nature. "See that spider web? Isn't it beautiful? Don't tear it up when you do not have to."

The third virtue is also perhaps a "new" virtue, which I shall call *techno-social sensitivity*, a disposition to be aware of the effects of technology on society and to insure that these effects are as humane as possible. Hursthouse has reminded us that a virtue includes "sensibilities," which for our purposes can be taken as synonymous with "sensitivity" or even "awareness." This is an aspect of a virtue that is especially important here.

Even more than respect for nature, techno-social sensitivity is a virtue that students probably did not learn early in life. Furthermore, acquiring this virtue appears to be especially difficult for engineering students, as a recent study has indicated (Kuhn 1998). The primary vehicle for inculcating this virtue is probably exposure to the history of technology and, especially, exposure to the disciplines of Science and Technology Studies (STS) and the philosophy of technology. From these disciplines students learn about the effects of technology on human life and our perception of the world and other people. Some of these effects are salutary, but some are not. The increasing ability to dominate nature may have diminished our ability to experience the transcendent, and the effect of computer networking on the development of social skills may not always be to the better.

The fourth virtue is *benevolence*, the disposition to do good to others. Unlike respect for nature and techno-social sensitivity, benevolence is a long-recognized virtue. In the engineering context, benevolence is especially associated with supererogatory protection of the public from harm and promotion of the material well-being of the public, including the least advantaged.

Probably the best way to encourage benevolence is to encourage empathy (actually feeling the distress of others) or sympathy (having a compassionate or caring attitude towards the suffering of others). In a series of experiments, Batson showed that empathy/sympathy does indeed lead to genuinely altruistic motivation, and that it is best induced by imagining how one would feel in the situation of another (not how the other feels). Batson has probably also shown that empathy/sympathy is a causal factor in bringing about actual helping behavior (Batson 1991).

Encouraging benevolence in engineering students is probably best accomplished by means of service learning, such as the design project at Texas A&M mentioned earlier, and participating in projects sponsored by Engineers Without Borders.

## 14.8 Conclusion

Preventive ethics has been and will continue to be an essential aspect of engineering ethics, because the public must be protected from threats to health and safety and from the misuse of professional expertise by engineers. But preventive ethics does not, at least for the most part, connect with the highest professional ideals, or the personal motivations that give one's work as a professional their deepest meaning. Aspirational ethics should be given a larger place in the thinking of engineers and in the teaching of engineering ethics.

## References

Aristotle. (1962). *Nicomachean ethics* (M. Ostwald, Trans., 1107a). Indianapolis: Library of Liberal Arts.

Batson, C. D. (1991). *The altruism question: Toward a social psychological answer*. Hillsdale: Lawrence Erlbaum.

De George, R. T. (1981). Ethical responsibilities of engineers in large organizations. *Business and Professional Ethics Journal, 1*, 1–14.

Harris, C. E., Jr., Pritchard, M. S., & Rabins, M. J. (2009). *Engineering ethics: Concepts and cases* (pp. 191–192). Belmont: Wadsworth.

Hursthouse, R. (2007). Environmental virtue ethics. In R. L. Walker & P. J. Ivanhoe (Eds.), *Working virtue* (Vol. 160). Oxford: Oxford University Press.

Hursthouse, R. (2012). *The Stanford Encyclopedia of Philosophy*. Virtue ethics. The stanford encyclopedia of philosophy. (2012 edition). E. N. Zalta (Ed.). URL http://plato.stanford.edu/archives/spring2012/entries/ethics-virtue/

Kuhn, S. (1998). When worlds collide: Engineering students encounter social aspects of production. *Science and Engineering Ethics, 4*, 457–472.

Martin, M. W. (2000). *Meaningful work* (p. vii). New York: Oxford University Press.

Meese, G. P. E. (1982). The sealed beam case. *Business and Professional Ethics, 1*, 1–20.

Merritt, M. (2000). Virtue ethics and situationist personality psychology. *Ethical Theory and Moral Practice, 3*, 365–383.

Nussbaum, M. (2000). *Women in human development: The capabilities approach*. Cambridge: Cambridge University Press.

Pritchard, M. S. (1992). Good works. *Professional Ethics, 1*, 155–177.

Pritchard, M. S. (1998). Professional responsibility: Focusing on the exemplary. *Science and Engineering Ethics, 4*, 215–233.

Sen, A., & Anand, S. (2000). The income component of the human development index. *Journal of Human Development, 1*, 83–106.

Taylor, P. (1986). *Respect for nature*. Princeton: Princeton University Press.

Wulf, W. A. (2000). *Great achievements and grand challenges*. Revised version of a lecture given on October 22 at the 2000 annual meeting of the National Academy of Engineering

# Chapter 15
# Making the Case for the Inclusion of Lay Persons on Engineering Accreditation Panels: A Role for an Engineering Hippocratic Oath?

**William Grimson and Mike Murphy**

**Abstract** Different professions have been tarnished through ethical lapses on the part of their members. This can partly be explained by the profession seeking to protect itself when things go wrong. In the engineering profession, the education of engineers is subject to scrutiny through the process of accreditation. While there are well documented learning outcomes associated with engineering programmes, the members of accreditation panels are invariably engineers. Later in the engineer's career, in the evaluation stage for professional recognition, the candidate engineer must demonstrate a number of competences. The interview board members evaluating these competences are engineers. The involvement of non-engineers in these activities would be beneficial. There is also considerable benefit, during the education of the engineer, in reflection on an engineering oath, similar to the Hippocratic Oath.

**Keywords** Hippocratic oath • Trust • Professional ethics • Universal code of ethics • Lay persons on engineering program accreditation panels

## 15.1 Introduction

Baroness Onora O'Neill in her 2002 BBC Reith Lecture noted that Confucius told his disciple Tsze-Kung that three things are needed for government: weapons, food and trust (O'Neill 2002). If a ruler can't hold on to all three, then give up the weapons first and the food next, but trust should be guarded to the end: *without trust we cannot stand.* O'Neill went on to say "it isn't only rulers and governments who prize and need trust. Each of us and every profession and every institution need trust.

W. Grimson (✉) • M. Murphy
College of Engineering and Built Environment, Dublin Institute of Technology, Dublin, Ireland
e-mail: william.grimson@dit.ie; mike.murphy@dit.ie

D.P. Michelfelder et al. (eds.), *Philosophy and Engineering: Reflections on Practice, Principles and Process*, Philosophy of Engineering and Technology 15,
DOI 10.1007/978-94-007-7762-0_15, 

We need it because we have to be able to rely on others acting as they say that they will, and because we need others to accept that we will act as we say we will". How professions, particularly engineering, address the question of trust is of interest not only to the professions themselves but more importantly to society in general. Accepting a broad definition of what constitutes a profession, there can be little doubt that trust has been questioned or eroded in recent years across a spectrum of professions. Examples of loss of trust abound for religious, banking, political, and healthcare institutions and it would be foolish to imagine that science and engineering are immune to potential loss of trust. Of course not every institution or everyone is untrustworthy but trust lost can have far-reaching consequences and trust lost is not easily regained.

The issue of trust is not just one concerned with the professional providing the service they say they will: there is another dimension to be considered. In his play *the Doctor's Dilemma,* Shaw claimed that the professions are conspiracies against the laity (Shaw 1906/1946). He felt that the professions could be accused of the charge of 'hiding of shortcomings'. Peer review by fellow professionals is perhaps the most widely adopted approach by professions to protect their standards of conduct and service. However, peer review does not guarantee that professional shortcomings will be either discovered or acknowledged. There are a number of compelling examples that illustrate this point, ranging from clerical and institutional abuse to medical negligence and disasters in chemical plants. Just one case is presented here to demonstrate what Shaw alluded to in his writings.

## 15.2 Case Study: Medical Profession, Ireland

In 2006 'The Lourdes Hospital Inquiry' report was published by Judge Maureen Harding Clark S.C on the subject of peripartum hysterectomy at Our Lady of Lourdes Hospital, in Drogheda, Ireland. The full report should be read carefully to realise the complexity of the story and the lessons to be learnt but, in summary, what unfolded is as follows. A midwife brought to the attention of hospital management that unnecessary surgical procedures were being carried out. However, the general ethos of the hospital and the seniority of the consultant resulted in no action being taken. But the underlying issue did not go away and subsequently a three-member medical panel reviewed the matter. They found that there was no case to answer. However, the statistics for medical procedures conducted pointed clearly to something being wrong and the regional health authority decided a further investigation was required. It was eventually determined that many surgical procedures had been wrongly carried out. In hindsight, this should have been clear from the start, given that one consulting doctor had carried out twice as many procedures as three other consultants combined over the period of the report.

What went wrong? The ethos of the hospital meant it was difficult to challenge a senior consultant. Midwives and junior doctors felt they could not influence what happened. And senior hospital management, initially at least, took no responsibility.

There was an issue of inadequate training and a lack of team meetings both of which could have helped avoid the unnecessary outcomes. There was also the fact that relevant patient files had been lost or possibly removed. Interestingly, Judge Clark noted in her report that Lourdes hospital was "a relatively small but very busy hospital which operated by a separate and unique set of rules, and was accountable to a religious community rather than to objective medical standards". But it is hard to avoid the conclusion that the profession protected itself from admitting there were shortcomings and that it required legal involvement and a judicial review to reveal the whole story.

For engineering as a profession and engineering educators, what are the appropriate measures to help build and safeguard as far as possible the trust amongst engineers themselves, between the engineering profession and other professions and more significantly between engineers and society? It should not be overlooked that as citizens, engineers are the product of society, their own circumstances, their education and a plethora of social interactions that make up daily life. Hence the role of the engineering profession and engineering educators in respect of ethical behavior is to build on what has been developed in the individual before that individual embarks on becoming an engineer.

## 15.3 The Challenge

An article in the medical journal Lancet had as its sub-title *the sources of professional ethics: why professions fail*. Robert Veatch wrote that 'we can quickly see the problem if we ask what the source of professional ethics ought to be (Veatch 2009). The classic answer was that it comes from the professional group'. He also went on to note that 'Hippocratic physicians pledge not to reveal their knowledge of medicine to lay people (not here pledging to keep patients' information confidential, but rather promising not to disclose the secret knowledge of remedies or healing theories)'. This is hardly a position that engenders trust! It might be concluded that an ethical framework should be established externally from the profession based upon a sound philosophy. But for engineering there is the view, expounded perhaps most clearly by the philosopher Carl Mitcham, that engineering as a profession is philosophically weak, which extends to the ethical positioning of engineering (Mitcham 2008). The situation is not helped by a strictly utilitarian view of engineering. There is a view that teaching the professions has shifted the function of the university from that of providing students with an opportunity for education to that of acquiring employability skills. The philosopher Robert Paul Wolff has argued that such a shift is detrimental to the fundamental role of the university. He questioned whether the university should serve as a training camp for professionals, and consequently that the education of the professions should not even reside within the modern university (Wolff 1992). With the ideal type of a university in mind, Wolff directed his criticism against the professions and towards their lack of intellectual inquiry and critique. He viewed the relationship between professional bodies and academic professionals as being inherently in conflict with the independent pursuit of knowledge within the ideal university.

If one accepts Wolff's point, there is an interesting conflict here regarding what is good for the education of the individual as opposed to what is good for the university. One imagines that a general university education provides a breadth of learning that should succeed in developing the young engineer, which is all the better to yield a desired pattern of professional behavior. Instead, there is the exhortation that the professions ought to be sent to what are essentially boot-camps for their professional education. By which is meant that the professions are not to be trusted and worse there is the risk they might contaminate the rest of the intellectual community in the university.

So the challenge is to determine what response is required of those responsible for the formation of the engineer, when one bears in mind that the education of professional engineers is likely to remain within the realm of the university, while taking the views of Veatch, Mitcham and Wolff into consideration?

## 15.4 Formation of the Engineer

In Ireland, the United Kingdom and the United States the formation of the engineer consists of two phases. The first phase involves an undergraduate academic engineering programme. The second phase consists of relevant professional work experience gained over a period of normally 4 years or more. The 'standard' established for the first phase is expressed in terms of programme learning outcomes and the second phase is described as a set of competences.

To illustrate, consider the Irish engineering professional society, Engineers Ireland, which is also the accrediting body for engineering education programmes. Engineers Ireland specifies the following programme outcomes which apply to all honours Bachelor degree engineering programmes aimed at satisfying the education standard for the title of Chartered Engineer (Engineers Ireland 2007). Programmes must enable graduates to demonstrate:

(a) The ability to derive and apply solutions from a knowledge of sciences, engineering sciences, technology and mathematics;
(b) The ability to identify, formulate, analyse and solve engineering problems;
(c) The ability to design a system, component or process to meet specified needs, to design and conduct experiments and to analyse and interpret data;
(d) An understanding of the need for high ethical standards in the practice of engineering, including the responsibilities of the engineering profession towards people and the environment;
(e) The ability to work effectively as an individual, in teams and in multi-disciplinary settings together with the capacity to undertake lifelong learning;
(f) The ability to communicate effectively with the engineering community and with society at large

The educational standard for engineering programmes is a result of a complex process that includes historical best practices as interpreted by engineering academics, input from professional bodies, and the demands of industry and to some extent society. The outcome of these processes has resulted in engineering programme accreditation criteria being established with ABET taking a leading role in the USA and the European Network for Accreditation of Engineering Education (ENAEE) acting as an umbrella organisation in Europe. In Europe EUR-ACE developed the Framework Standards for the accreditation of engineering programmes (EUR-ACE 2008). Further, there has been a degree of harmonisation of accreditation criteria globally through the mechanism of the Washington Accord. It can be argued that the underlying objectives embedded in the learning outcomes of the accreditation criteria, at least to some extent, address the issues raised above. On the other hand, in the spirit of Wolff's argument, what is in place are self-referencing sets of criteria generated from a self-selecting body of like-minded professionals thus bolstering the arguments about an inherent conflict of interest. The proof, in principle, that this is not the case is evident in modern engineering curricula where there is a discernable re-focus and re-alignment with the agreed accreditation criteria.

But there are concerns: specifically have the non-technical societal and ethical objectives of the accreditation criteria been absorbed in a meaningful way into engineering programmes? Consider the 'need for high ethical standards' and the 'ability to communicate effectively with the engineering community and with society at large'. Two issues immediately arise. First, who sets the ethical standards and who adjudicates compliance with respect to these standards? Second, in what sense is communication with society at large a true dialogue rather than a unilateral monologue, often coded in technical language, in the direction of the profession towards society? To the first point, it can hardly be questioned that self-regulation with respect to compliance has not been without its difficulties across a wide range of professions and engineering is no exception. To the second point, the views of Samuel Florman are as valid today as when they were published some quarter of a century ago in respect of the poor communication between engineers and society (Florman 1976). The spectre of loss of trust arises because of a perceived gap between intent and practice.

Let us now turn our attention to the second phase in the formation of the engineer: professional work experience. This is characterised as development competences to be achieved during the early stages of professional employment. A typical set of engineering competences are framed as follows:

1. Use a combination of general and specialist engineering knowledge and understanding to optimize the application of existing and emerging technology
2. Apply appropriate theoretical and practical methods to the analysis and solution of engineering problems.
3. Provide technical, commercial and managerial leadership.
4. Use effective communication and interpersonal skills.
5. Make a personal commitment to abide by the appropriate code of professional conduct, recognising obligations to society, the profession and the environment.

Professional bodies complying with the above framework normally have their own subsidiary requirements. So, for example, to become a Chartered Engineer in Ireland competence 5 above is elaborated to state that an individual must (Engineers Ireland 2011):

1. Place responsibility for the welfare, health and safety of the community at all times before responsibility to the profession, to sectional interests, or to other engineers;
2. Comply with the Code of Ethics of Engineers Ireland;
3. Apply professional skill in the interests of employer or client, for whom they act in professional matters, as a faithful agent or trustee;
4. Give evidence, express opinions or make statements in an objective and truthful manner and on the basis of adequate knowledge.

Repeating a point made earlier the intention is good and perhaps even honourable but, in practice, what is the outcome? Here again, there is a degree of harmony amongst those who licence professional engineers or award chartered engineer status. Could the system be criticised as inward-looking and self-referencing?

## 15.5 Accreditation as a Peer Process

In most jurisdictions an accreditation panel is composed only of engineers with both academic (university) and practicing engineers as its members. Individual panel members are largely chosen because of their expertise in the subject matter comprising the engineering programme under review. Generally no particular consideration is given to whether the panel members are in fact competent to make a judgement as to whether the societal and ethical criteria described earlier have been met. In turn this offers licence to curriculum designers to give less than full attention to requirements concerning ethical and societal matters. To put it succinctly: on the one hand the team developing or modifying a programme do not insist on the same degree of rigour when it comes to ethical and societal material as they would to the mathematical or technical content. On the other hand, the accreditation panel members may not have real expertise to highlight shortcomings in the ethical and societal aspects of the programme under review.

Let us summarise the role that the relevant accreditation body has in this process. First, it has adopted and promoted the accreditation criteria. Second, it has selected the members of the accreditation panel. Third, it will have an Accreditation Governance Board that will consider panel reports from all accreditation visits and seek fair and hence uniform decisions. The salient point to note is the total domination of the process by members of the profession (all engineers) and the lack therefore of any independent voice. This in itself does not necessarily mean that a programme so accredited is inadequate, but it does leave the profession open to the charge that it is self-serving and inward looking. The counter argument that has been used is that engineers are first and foremost members of society and therefore, almost by definition, they can take the societal aspects into account. Suggested remedies are considered later.

## 15.6 Professional Review of Engineers Seeking Professional Standing

At the end of their formation period candidates are eligible to apply to become licensed engineers (US) or chartered engineers (UK, Ireland) and thus be designated as professional engineers. In Ireland the process involves the candidate writing a report outlining their career followed by an in-depth interview. The interview board consists of a peer group with at least one member having expertise in the relevant area as demonstrated in the candidate's report. In the interview heavy emphasis is placed on the technical and management competences, with the result that the ethical and societal dimensions are given far less scrutiny. In many respects this is understandable. Nevertheless it runs counter to what the professional institutions profess. Further, who briefs the interviewers as to how they should conduct interviews? Well, other engineers of course and thus the circle is closed.

The end result of the two processes (education plus work experience) followed by the professional review by and large is good and often excellent. Is there a case then to modify the process? Consider what happens when things go wrong. Most institutions have a disciplinary or fit-to-practice process. In the case of Engineers Ireland its Council established an Ethics and Disciplinary Board and stipulated that the Board shall have a maximum membership of 16 persons including the Chair and shall include up to 4 persons who are not members of the engineering profession. The overall implications of this are clear: it is only when things have gone wrong or might have wrong or been seen to go wrong is it time to involve non-engineers. The key point being made in this chapter, notwithstanding the earlier comment that processes are essentially sound, is that the involvement of non-engineers is too little and too late.

## 15.7 Comparison with the Medical Profession: A Hippocratic Oath for Engineers?

The medical and engineering professions have much in common. This is not surprising as they both aim to improve the conditions of mankind and address needs that are largely physical. They are amongst the oldest of professions with long histories. Their knowledge bases are equally eclectic and draw on science, technology, craft and heuristic practices. Not everything found to work had a rational explanation when cures or devices were first tried. Science often played a catch-up role in the sense of explaining after-the-fact how a cure or device worked. Medicine and engineering are now much more scientific than they were previously but one only has to consider treatments and engineering approaches of less than 100 years ago to realise that the ongoing processes for both professions continues to rely more heavily on science. There is one aspect of the medical profession that has attracted much attention over many centuries, namely the Hippocratic Oath. Could or should there be an equivalent oath for engineers and if so what form should it take?

A modern version of the Hippocratic Oath is as follows (Lasagna 1964):

- I swear to fulfil, to the best of my ability and judgment, this covenant: I will respect the hard-won scientific gains of those physicians in whose steps I walk, and gladly share such knowledge as is mine with those who are to follow.
- I will apply, for the benefit of the sick, all measures [that] are required, avoiding those twin traps of overtreatment and therapeutic nihilism.
- I will remember that there is art to medicine as well as science, and that warmth, sympathy, and understanding may outweigh the surgeon's knife or the chemist's drug.
- I will not be ashamed to say "I know not," nor will I fail to call in my colleagues when the skills of another are needed for a patient's recovery.
- I will respect the privacy of my patients, for their problems are not disclosed to me that the world may know. Most especially must I tread with care in matters of life and death. If it is given to me to save a life, all thanks. But it may also be within my power to take a life; this awesome responsibility must be faced with great humbleness and awareness of my own frailty. Above all, I must not play at God.
- I will remember that I do not treat a fever chart, a cancerous growth, but a sick human being, whose illness may affect the person's family and economic stability. My responsibility includes these related problems, if I am to care adequately for the sick.
- I will prevent disease whenever I can, for prevention is preferable to cure.
- I will remember that I remain a member of society, with special obligations to all my fellow human beings, those sound of mind and body as well as the infirm.
- If I do not violate this oath, may I enjoy life and art, respected while I live and remembered with affection thereafter. May I always act so as to preserve the finest traditions of my calling and may I long experience the joy of healing those who seek my help.

Much of the language could be adopted by engineers and we have underlined some key phrases that seem particularly relevant. We suggest that the very least that might be attempted in our engineering schools is for senior engineering classes to critically discuss the Hippocratic Oath and to learn and take from it what is appropriate. But more could be envisaged.

A number of different oaths have been proposed by various prominent members of the scientific community. Sir Joseph Rotblat in his Nobel Prize acceptance speech said "the time has come to formulate guidelines for the ethical conduct of scientists, perhaps in the form of a voluntary Hippocratic Oath" (Rotblat 1995). Sir David King, the UK government's chief scientific advisor, laid out a *universal code of ethics* for researchers across the globe (King 2007):

- Act with skill and care in all scientific work. Maintain up to date skills and assist their development in others.
- Take steps to prevent corrupt practices and professional misconduct. Declare conflicts of interest.
- Be alert to the ways in which research derives from and affects the work of other people, and respect the rights and reputations of others.

- Ensure that your work is lawful and justified.
- Minimise and justify any adverse effect your work may have on people, animals and the natural environment.
- Seek to discuss the issues that science raises for society. Listen to the aspirations and concerns of others.
- Do not knowingly mislead, or allow others to be misled, about scientific matters. Present and review scientific evidence, theory or interpretation honestly and accurately.

Even a quick reading of the above universal code of ethics shows how the societal aspect is strongly addressed. Again, would student engineers not benefit from a thorough discussion of this code and its relevance to their chosen engineering profession? A somewhat different oath in character was introduced as a graduation ceremony oath in the University of Toronto for its Medical Scientists (Institute of Medical Science 2007):

> I have entered the serious pursuit of new knowledge as a member of the community of graduate students at the University of Toronto. I declare the following:
>
> - **Pride**: I solemnly declare my pride in belonging to the international community of research scholars.
> - **Integrity**: I promise never to allow financial gain, competitiveness, or ambition cloud my judgment in the conduct of ethical research and scholarship.
> - **Pursuit**: I will pursue knowledge and create knowledge for the greater good, but never to the detriment of colleagues, supervisors, research subjects or the international community of scholars of which I am now a member.
>
> By pronouncing this Graduate Student Oath, I affirm my commitment to professional conduct and to abide by the principles of ethical conduct and research policies as set out by the University of Toronto.

The tone is somewhat self-centred with correspondingly less concern for society at large but the intentions are clear and worthwhile.

As a further example, the Institute for Social Invention Oath proposed the following (Codling):

> I vow to practice my profession with conscience and dignity; I will strive to apply my skills only with the utmost respect for the well-being of humanity, the earth, and all its species; I will not permit considerations of nationality, politics, prejudice, or material advancement to intervene between my work and this duty to present and future generations. I take this Oath solemnly, freely, and upon my honor.

This short and elegant statement perhaps says all that needs to be said and its implications could be teased out within well directed workshops for engineering students.

Finally, in this brief review of the Hippocratic Oath and similar oaths, the following has been proposed as an Engineering Oath (Susskind 1973):

> I solemnly pledge myself to consecrate my life to the service of humanity. I will give to my teachers the respect and gratitude which is their due; I will be loyal to the profession of engineering and just and generous to its members; I will lead my life and practice my profession in uprightness and honor; whatever project I shall undertake, it shall be for the good of mankind to the utmost of my power; I will keep far away from wrong, from corruption, and from tempting others to vicious practice; I will exercise my profession solely for the

benefit of humanity and perform no act for a criminal purpose, even if solicited, far less suggest it; I will speak out against evil and unjust practice wheresoever I encounter it; I will not permit considerations of religion, nationality, race, party politics, or social standing to intervene between my duty and my work; even under threat, I will not use my professional knowledge contrary to the laws of humanity; I will endeavour to avoid waste and the consumption of non-renewable resources. I make these promises solemnly, freely, and upon my honor.

Nobody expects that merely taking an oath will avoid all wrong-doing or bad behaviour. But the act of discussing and understanding an oath in whatever form it takes can only lead to a more responsible attitude by the professors of that oath in their subsequent careers. What is clear is that the programme learning outcomes expected in engineering programmes and the competences required of a professional engineer, especially as they relate to ethics and societal matters, have a relationship with the sentiments contained within the various oaths above. A medical doctor does not become a good doctor merely by taking the Hippocratic Oath; but in working amongst a peer group who also have taken the oath there is the reasonable expectation that behaviour patterns will be of a high ethical standing. Of course there have been and will always be exceptions but a common framework for ethical and society-aware conduct is to everyone's benefit.

## 15.8 The Role of Laypersons

There are in essence three junctures wherein a layperson could influence the formation of an engineer. First, it can occur in the design stage of an engineering curriculum. A modern trend appears to be, under the pressure of shoehorning much into a crowded curriculum, that ethical and societal material is embedded in technical subjects. Whilst this provides the advantage of providing contextual learning, it might dilute the overall general principles that need to be established. Further the teacher of the engineering technical subject may not be sufficiently versed in the general area of ethics and societal issues. The use of expert non-engineers (what we term a layperson) could only help in devising sound subject matter covering ethical and societal material. Ethics and the impact of engineering on society are sufficiently important to warrant dedicated space in the curriculum alongside technical subjects. Thus it would be desirable that such persons teach the subject as a standalone class, perhaps with an enriching mix of students aspiring to graduate in a range of professional programmes.

The second point at which laypersons could make an important contribution is at the programme accreditation stage. Here they could act as guardians of society to help ensure that the 'rules of engagement' are being complied with by the engineering school proposing their programme. These rules are essentially the learning outcomes provided by ABET, ENAEE or equivalent body. The third clear opportunity for the involvement of a layperson is at the interview stage where a candidate is applying to become a professional engineer (known as a Chartered Engineer in

Ireland and the UK). The candidate would be expected to respond to the layperson on a range of probing questions evaluating their competences, particularly Competence 5 previously discussed.

One issue that immediately arises is the category or type of layperson required at each of the three opportunities just discussed. In the first instance an academic in the university delivering ethics lectures and holding tutorials and workshops is well suited to developing that part of the curriculum. Occasionally a member of staff within an engineering school might be suited to this role. Further a sharing of material across regions or clusters of universities would provide a useful impetus. The major task is to facilitate the inclusion of relevant material within the curriculum. Creating time in the curriculum for any important subject material requires strong leadership at dean level to counter opposition from staff.

This leads to the second opportunity for making use of a layperson, namely, in the accreditation stage where compliance with programme learning outcomes can be judged. The assertion is that an independent person is more likely to objectively question the extent to which the ethical and societal learning outcomes have been adequately addressed. Judgement would, in most accreditation systems, depend not solely on a paper study or interviews with staff but would also involve meeting graduates of the programme. Who might such an independent layperson be? The common model used in the peer review process is that academics from engineering schools within the general region act on accreditation panels together with representatives from the relevant industry. In line with this approach, the simplest solution is to use academics from other universities who are responsible for teaching the ethical and societal material. Occasionally as part of ensuring calibration with other jurisdictions, visitors from other institutions would sit and observe accreditation proceedings. The rationale for this lies with the similarity of specified programme learning outcomes, which affords a valued comparison of solutions with respect to delivering on those outcomes.

Finally, at the professional review stage, there is a case to be made that a non-academic and independent layperson should be involved. This is perhaps the hardest role to pin down as the person needs to be an all-rounder, have experience and insight into how professionals interact with society, and have strong contextual expertise that is grounded in empirical evidence or historical perspective. There are a wide range of potential candidates. These include: former senior managers of companies that have/had wide ranging interaction with society, senior managers of NGOs, civil servants who have had dealings with large scale projects involving complex planning permissions, allied professions in areas such as environmental protection, and representatives of citizens groups. With such a scheme in place there would be an onus on companies providing a rounded experience for their newly graduated employers as it would not reflect well on that company if the candidate failed their professional interview. There might be a degree of exaggeration in this last point but an overall push at all stages, from initial education to early experience, to ensure what is considered ethical behaviour, is bound to impact on all concerned.

## 15.9 Conclusions

A fundamental principle of any profession must be that its members are fit to serve the needs of the public and society in general. In turn it follows that engineers must not only aspire to be trusted by society but to have, through their educational and experiential backgrounds, the necessary means to generate that trust. The prevailing emphasis on the purely technical side of the profession is of course understandable but professional institutions should be obliged to ensure that the non-technical aspects of engineering are equally addressed. After all, the institutions have signed up to a generally agreed set of learning outcomes and so they must follow through and ensure compliance. The idea of involving laypersons is not novel as illustrated by their deployment on professional Disciplinary Boards. At all stages of the formation of engineers, the introduction of new 'thinking' and new 'observing' that laypersons can offer, must be worth considering. Further a Hippocratic-like Oath for engineers should be considered not as an end point but rather as a means of discussing the inherent issues throughout the education of engineers and beyond, In short the oath is an outward expression of a set of deeply held values arrived at after careful deliberation. Finally, it might be said whether correctly or otherwise that engineering does not have a trust problem. A good conservative engineering principle would be to take steps to help ensure that trust is not lost, for once lost it is hard to regain.

## References

Clark, M. H. (2006). *The Lourdes Hospital inquiry: An inquiry into peripartum hysterectomy at Our Lady of Lourdes Hospital, Drogheda*. Dublin: Stationery Office, Government Publications. Available at http://www.stateclaims.ie/ClinicalIndemnityScheme/publications/lourdes.pdf

Codling, P. (attributed). *The Hippocratic Oath for scientists, engineers, and executives*. London: Institute for Social Inventions. http://www.onlineethics.org/cms/7331.aspx

Engineers Ireland. (2007). *Accreditation criteria for engineering education programmes*. Available at http://www.engineersireland.ie/media/engineersireland/services/Download%20the%20accreditation%20criteria%20(PDF,%20240kb).pdf

Engineers Ireland. (2011). *Chartered Engineer: Regulations for the title of Chartered Engineer*. http://www.engineersireland.ie/EngineersIreland/media/SiteMedia/membership/professional-titles/Chartered-Engineer-Regulations-2012.pdf

EUR-ACE. (2008). *The European quality label for engineering degree programmes at Bachelor and Master level*. Available at http://www.enaee.eu/wp-content/uploads/2012/01/EUR-ACE_Framework-Standards_2008-11-0511.pdf. Accessed 14 Mar 2012.

Florman, S. C. (1976). *The existential pleasures of engineering*. New York: St. Martin's Press.

Institute of Medical Science. (2007). University of Toronto. Available at http://www.ims.utoronto.ca/current/oath.htm. Accessed 22 Mar 2012

King, D. (2007, September 12). The great beyond: 'Hippocratic Oath for scientists'. *Nature News Blog*. Accessed 22 Mar 2012. http://blogs.nature.com/news/2007/09/hippocratic_oath_for_scientist.html

Lasagna, L. C. (1964). *Hippocratic Oath—Modern version*. Available at http://www.pbs.org/wgbh/nova/doctors/oath_modern.html. Accessed 4 Mar 2012.

Mitcham, C. (2008). *Philosophical weakness of engineering as a profession. WPE-2008 workshop on philosophy and engineering*. London: Royal Academy of Engineering.

O'Neill, O. (2002). *A question of trust.* Reith Lectures 2002. Available at http://www.bbc.co.uk/radio4/reith2002/. Accessed 14 Mar 2012.

Rotblat, J. (1995). *Remember your humanity.* Available at http://www.nobelprize.org/nobel_prizes/peace/laureates/1995/rotblat-lecture.html

Shaw, G. B. (1946). *The doctor's dilemma.* London: Penguin Books (Original work published 1906).

Susskind, C. (1973). *Understanding technology* (p. 118). Baltimore/London: The John Hopkins University Press.

Veatch, R. M. (2009). The sources of professional ethics: Why professions fail. *The Lancet, 373*(9668), 1000–1001.

Wolff, R. P. (1992). *The ideal of the university.* New Brunswick: Transaction.

# Chapter 16
# Ethical Awareness in Chinese Professional Engineering Societies: Textual Research on Constitutions of Chinese Engineering Organizations

**CAO Nanyan, SU Junbin, and HU Mingyan**

**Abstract** Common ethical awareness is a key for the establishment of a profession. To examine the professional ethical awareness embodied in the constitutions of Chinese engineering societies, this chapter analyzes the texts of the articles of constitution of 48 national engineering societies in Mainland China up-to-date. In comparison to the model code of ethics of the World Federation of Engineering Organizations (WFEO) and the constitution of the Institute of Electrical and Electronics Engineers (IEEE), we find that the universal principles of justice, fairness and concern for human beings are not given full attention and that "do the right thing" is under-emphasized in comparison to "do the thing right" in the ethical awareness of Chinese engineering societies.

**Keywords** Ethical awareness • Professional ethics • Codes of ethics • Moral ideal • Engineering society of mainland China

## 16.1 Introduction

Michael Davis has asked an interesting question: Is there a profession of engineering in China? He has expressed the view that a profession is a number of individuals in the same occupation voluntarily organized to earn a living by openly serving a

N. CAO (✉)
Institute of STS, Tsinghua University, Beijing 100084, China
e-mail: caonanyan@gmail.com

J. SU
School of Journalism and Communication, Xiamen University, Fujian 361005, China

M. HU
Department of Philosophy, Party School of the Central Committee of C. P. C,
Beijing 100091, China

D.P. Michelfelder et al. (eds.), *Philosophy and Engineering: Reflections on Practice, Principles and Process*, Philosophy of Engineering and Technology 15,
DOI 10.1007/978-94-007-7762-0_16, 

common moral ideal in a morally-permissible way beyond what law, market, morality, and public opinion would otherwise require (Davis et al. 2007). Do the engineering societies of China[1] have such a moral ideal? Engineering ethical awareness can be reflected in the constitutions or ethical codes/creeds of engineering societies. A constitution of an engineering society is an important aspect of the institutionalization of engineering professional ethics (Cong 2006). The purposes and beliefs expressed in the tenets of the constitutions of engineering societies often embody their ethical consciousness; while the professional code of ethics is a statement of principles by which the practitioner may calibrate his personal attitude and conduct to the model approved by his peers (Schaub et al. 1983). Codes of ethics state the moral responsibilities of engineers as seen by the profession and as represented by a professional society (Martin and Schinzinger 2005).

The ethical awareness of engineering societies of China and its historical changes will be studied in this paper, using the means of textual research on those societies' constitutions and ethical codes.[2] The authors investigate the constitutions of 48 out of 68 societies under the jurisdiction of the China Association for Science and Technology. Firstly, we analyze the text of the constitutions of these 48 societies. Secondly, we compare the constitutions and ethical codes of engineering societies in mainland China with IEEE's and the model code of ethics of the WFEO (World Federation of Engineering Organizations) which belongs to UNESCO. Then, we select a special engineering society and investigate a number of amendments to its constitution throughout its history which reflect the changes in its ethical consciousness. Finally, we sum up the characteristics of the constitutions of Chinese engineering societies, and explain the factors shaping these characteristics.

## 16.2 Ethical Awareness in the Constitutions of Contemporary Engineering Societies in Mainland China

To enhance the image of the profession, to clarify rules of conduct within the profession, and to promote the public good (Vesilind 1995), international engineering societies generally have their own ethical codes. The written codes of ethics, either statutorily independent or included in the society's constitutions, demonstrate self-conscious professional ethical awareness.

All nationwide engineering societies in Mainland China are uniformly managed by the China Association for Science and Technology (CAST).[3] Up to the end of 2009,

[1] In this paper, an engineering public organization – a society, federation, or association – is called an engineering society.

[2] This paper does not comprehensively study the constitutions of engineering societies and the system of registered engineers in Hong Kong, Macau or Taiwan.

[3] China Association for Science and Technology (CAST) is a non-governmental organization of Chinese scientific and technological workers, funded by public finance. "Through its member

CAST had 68 affiliated societies of engineering and technology (Only some of these societies define themselves as engineering societies, such as the China Machinery Engineering Society. The rest are both engineering societies and trade associations). We investigated the constitutions of 48 engineering societies (see Appendix), which reflect the basic situation of engineering societies of mainland China.

Generally speaking, the constitutions were framed when these societies were established, and they are revised on a regular basis. Unfortunately, for most societies, we could not find a complete history of their constitutional revisions. What we have is only the latest version of these constitutions. Of the 48 constitutions we collected out of the 68 nationwide societies related to engineering in mainland China, 2 were published in 2001, and 46 were published afterwards. In 1998, the Ministry of Civil Affairs of the People's Republic of China (MCA), which is responsible for managing public organizations including CAST, issued a "model text of a public organization's constitution." All the existing constitutions were stipulated or revised according to this model text.

After investigating the ethical awareness reflected in the articles of the 48 constitutions of engineering societies of Mainland China, we find that:

1. The "tenet" of those public organizations, in which ethical awareness is reflected in general, includes four primary aspects: to comply with the State Constitution and laws; to serve the economic construction of the country and the development of science and technology; to enhance national and international academic communion; and to insist on democracy in managing the societies.
2. With very few exceptions, hardly any engineering societies have formulated separate written ethical codes of engineering.[4]
3. The ethical consciousness, indirectly reflected in most of the constitutions of the societies we studied, is still vague and pays little attention to important ethical principles, such as public safety, health and well-being, and environment protection.[5]

---

societies – nearly 200 in number and local branches all over the country, the organization maintains close ties with millions of Chinese scientists, engineers and other people working in the fields of science and technology" (http://english.cast.org.cn). In fact, CAST, managed by the Chinese Ministry of Civil Affairs, is a quasi-governmental organization with some administrative responsibilities for scientific and technological societies in Mainland China.

[4] In this study of 48 engineering societies, the China Computer Federation (CCF) is the only society dedicated to formulating an ethical code for engineers. The code of ethics includes five more specific criteria, namely, respecting intellectual property rights, respecting facts, evaluating work objectively, keeping impartiality in peer review, forbidding duplicate publication, and so on. As its title "China Computer Federation academic ethical code" shows, the code of ethics of the CCF is also a scientific moral code, rather than the ethical code of engineering profession.

[5] For example, the Chinese Hydraulic Engineering Society clearly requires adherence to the "harmony between man and nature" (2004); the China Mechanical Engineering Society put "people-oriented, seeking social welfare" as the purpose (2006). However, even in these societies' constitutions, the ethical responsibility for public well-being and the environment is still in a subordinate position.

4. The universal principles of justice, fairness and concern for human are underemphasized. The ethical awareness of the current engineering societies in mainland China remains at the stage of considering the positive impacts of engineering and requiring engineers to "do the thing right," rather than "do the right thing."
5. The constitutions of engineering societies in various fields mainly just replicate the "model text of a public organization's constitution" issued by the MCA, and barely emphasize their own characteristics of engineering societies. The ethical awareness of the current engineering societies in mainland China merely expresses the government's requirements for engineering societies, and not the engineers' consciousness within professional groups.
6. Ignoring the specific characteristics of engineering fields, the constitutions of engineering societies in mainland China neither reflect the ethical demands of contemporary society nor attach importance to the maintenance of public interest. Even in the trades of coal excavating, food processing, paper making etc., where accidents have frequently occurred in the past few decades, ethical principles, such as public safety, health and environmental protection, are rarely mentioned in the constitution of the relevant societies.[6] By contrast, the constitution of the Institute of Electrical and Electronics Engineers (IEEE 2006) stresses, "In recognition of the importance of our technologies in affecting the quality of life throughout the world, and in accepting a personal obligation to our profession, its members and the communities we serve, do hereby commit ourselves to the highest ethical and professional conduct and agree: to accept responsibility in making decisions consistent with the safety, health and welfare of the public, and to disclose promptly factors that might endanger the public or the environment; etc."

## 16.3 Comparison to the WFEO Model Code of Ethics

The World Federation of Engineering Organizations (WFEO), which belongs to UNESCO, is composed of engineering societies from more than 90 countries in the world. Since 1990, the WFEO has worked to prepare a Code of Ethics. It was expected that this model code would be used to define and support the creation of codes for its member institutions, and that it would be adopted in the near future. The WFEO model code of ethics expresses the ethical considerations of the

[6] Even engineering disasters arousing public attention have not urged the leaders of those engineering societies to upgrade ethical standards for engineering professional behavior. We cannot find any changes in the purpose of the China National Coal Association about safety and health of mineworkers after so many mine disasters; while foodstuff safety and public health are not mentioned in the purpose of China Cereals and Oils Association; environment protection is not mentioned in the purpose of China Paper Making Association; energy conservation is not mentioned in The Architectural Society of China, etc.

international engineering society. We believe that it manifests the moral ideal that the WFEO members should openly serve.

The final version of the WFEO model code of ethics, adopted in 2001, consists of four sections. Based on clarifying the concept of ethics, the first section indicates the orientation of a professional ethical code of engineering: "A code of professional ethics is more than a minimum standard of conduct; rather, it is a set of principles which should guide professionals in their daily work." (World Federation of Engineering Organizations 2001) The second section lists nine practical criteria of ethics. It prescribes that professional engineers should "hold paramount the safety, health and welfare of the public and the protection of both the natural and the built environment in accordance with the principles of sustainable development." (World Federation of Engineering Organizations 2001) The third section lists seven separate environmental ethical guidelines. The fourth part summarizes further the ideals of engineering professional ethics.

The WFEO model code of ethics concludes that engineers should "always remember that war, greed, misery and ignorance, plus natural disasters and human induced pollution and destruction of resources, are the main causes of the progressive impairment of the environment and that engineers, as active members of society, deeply involved in the promotion of development, must use our talent, knowledge and imagination to assist society in removing those evils and improving the quality of life for all people" (World Federation of Engineering Organizations 2001). It emphasizes that the responsibility of "do the right thing" has precedence over "do the thing right." This provides the possibility for engineers to shoulder social responsibilities and moral obligations beyond their professional limits.

In short, the WFEO model code of ethics discusses the role and orientation of engineering ethics, proposes that engineers should take nine professional ethical responsibilities and seven environmental responsibilities, and finally sublimates to the comprehension of public morality. Thereby, it carries out the process from moral ideals to the professional code of ethics, and returns to moral ideals.

CAST joined the WFEO in 1981, and became a national member. Since then, delegations composed of CAST leaders and scientists have attended all general assemblies. China's famous scientists held several important leadership positions in WFEO, such as Vice-Chairman and member of the Executive Committee, and a number of scientists are serving as members of the professional committees. China plays an important role in the WFEO and expands its influence in the engineering world. Through participating in WFEO activities, CAST promotes exchanges and cooperation with engineering organizations around the world. In 2004, CAST, Chinese Academy of Engineering (CAE) and WFEO successfully cooperated and hosted the World Engineers Conference in Shanghai, China.

Nevertheless, the engineering societies affiliated with CAST have made no response to the revised model code of ethics of WFEO (2001). Engineering societies in mainland China devote little concern to preparing written codes of ethics, and even if they have such a code (as do the CAE and the China Computer Federation), it focuses only on academic research ethics instead of engineering ethics.

CAE, as the highest level honorary and consultative academic organization in the Chinese engineering technological world, was set up in 1994. CAE stipulated academicians' scientific moral codes in 1998, and some self-discipline regulations of academicians' scientific moral codes in 2001. The scientific moral codes and self-discipline regulations came on in response to allegations of scientific misconduct. In both CAE constitutions and codes, we can find only obligations to scientific activities, e.g. to enhance the spirit of scientific inquiry, to spread abroad scientific thoughts, to promote advanced scientific culture, to maintain the dignity of scientific morality, and to popularize scientific and technological knowledge. There is no mention of obligations to engineering practice. The content of codes and regulations contains merely scientific moral items, such as authorship, intellectual property, peer review, academic critics, commercial propaganda, fighting against pseudoscience and superstition, etc. without any items of engineering ethics.

Moreover, even for all of the constitutions of the engineering societies revised after 2001, their tenets absorb little of the core content of the WFEO model code of ethics, such as the care for public health, safety and well-being and protection of the ecological environment. Among the 48 constitutions that we have investigated, there are 38 constitutions on which the WFEO model code of ethics has almost no impact. Thus we can infer that engineering societies in mainland China lack awareness of the professional ethics of engineering.

## 16.4 The Evolution of the "Chinese Engineers Creed" from 1933 to 1996

Is it true, then, that China's engineering societies have never clearly expressed their ethical consciousness? In order to clarify this point, we now examine the Chinese Institute of Engineers (Zhong Guo Gong Cheng Shi Xue Hui), the earliest engineering society in China.

In 1912, three engineering societies, the Chinese Engineering Society (Zhong Hua Gong Cheng Xue Hui), the Chinese Institute of Engineering (Zhong Hua Gong Xue Hui) and the Road Workers Colleague Masonic Organization (Lu Gong Tong Ren Gong Ji Hui), were set up successively in Guangzhou, Shanghai etc. Before long, the three societies were merged and renamed as the Chinese Engineers Society (Zhong Hua Gong Cheng Shi Hui), with Tianyou Zhan (the English name is Jeme Tien Yow), who graduated from Yale University in 1881 with a bachelor's degree in Civic Engineering, elected as its president. The society was then renamed China Institute of Engineers (Zhong Hua Gong Cheng Shi Xue Hui) in 1914. In 1917, Chinese students studying in the United States initiated the Chinese Engineering Society (Zhong Guo Gong Cheng Xue Hui). In 1931, the China Institute of Engineers and the Chinese Engineering Society combined and officially changed their name to the Chinese Institute of Engineers (Liu et al. 2002; Mao et al. 1987).

In 1933, the Chinese Institute of Engineers formulated the "Chinese Engineer Credendum," which is the earliest written Chinese engineering professional code of ethics (Liu et al. 2002). Consisting of the following six criteria, it stipulates that an engineer:

1. Shall not give up or be disloyal to duty;
2. Shall not give or accept undue reward;
3. Shall not clash with one another or exclude colleagues;
4. Shall not directly or indirectly harm the reputation or professional work of colleagues;
5. Shall not compete for business or position by despicable means;
6. Shall not disseminate false information, or behave in a manner undignified for the profession.

Out of the above six criteria, four of them are related to engineers' responsibilities to their colleagues. They are made for the emerging situation of engineering societies, with apparent tendency of self-regulation within the occupation; and they are also the products of reference to the precedents of other countries (Mao et al. 1987). In terms of its content, this engineer credendum is not differentiated from the ethical credenda of other industry associations, which shows that the emerging occupational groups of Chinese engineers were not yet clearly aware of their differences from other occupational groups.

In 1941, during its 10th Annual Conference, the Chinese Institute of Engineers adopted an amendment to rename the "Chinese Engineer Credendum" as the "Chinese Engineers Creed", broaden engineers' responsibility to their country and nation, and modify the content of the original credendum (Mao et al. 1987). The revised eight criteria stipulate that an engineer is to:

1. Comply with economic development policies of the country and the national defense policies, to realize the industrial plan of the national Father Sun Yat-sen.
2. Recognize that the national interest is above all else, and be willing to sacrifice freedom and make contributions.
3. Promote the industrialization of the country, strive for self-sufficiency of essential goods and supplies.
4. Implement industry standardization, meet national defense needs and people's livelihood.
5. Seek no fame, resist the lure of material, maintain professional dignity and comply with service ethics.
6. Seek truth from facts, keep improving for the sake of excellence, strive for independent creation, and focus on collective achievements.
7. Have the courage to take responsibility, be devoted to duty; moreover, one should cooperate sincerely with the spirit of mutual aid and fraternal love.
8. Be strict with oneself, generous to others, and develop clean, simple, prompt and faithful life habits.

Compared to the "Chinese Engineer Credendum" of 1933, this creed, on the one hand, raises the height of spiritual philosophy by emphasizing national interests with stronger political overtones, while on the other hand, decreasing the directivity to the professional groups and thus lacking real binding.

After the founding of the People's Republic of China in 1949, the headquarters of the Chinese Institute of Engineers was relocated to Taipei, China. In 1976, the Chinese Engineers Creed was modified. For example, "freedom" in the second criterion was changed to "small ego." It was amended again in 1996; this creed is still in use in Taiwan. The creed of 1996 contains four rules and eight sub-rules, which summarize engineers' responsibilities to society, the profession, clients, employers and colleagues (Chinese Institute of Engineers 1984). The details are as follows:

1. **Responsibility to society**

*Law-abiding and dedication*: adhere to statutes and regulations, protect public safety, and promote public well-being.

*Respect for nature*: maintain ecological balance, conserve natural resources, and preserve cultural assets.

2. **Responsibility to the profession**

*Devotion to work and duty*: apply professional knowledge, do one's duty strictly, follow engineering practices properly.

*Innovation and enhancement*: study new science and technology, progress toward excellence, improve product quality.

3. **Responsibility to clients and employers**

*Sincere service*: work with full capacity and wisdom, provide the best service, and achieve the work goals.

*Mutual trust and mutual benefit*: establish mutual trust and build a win-win consensus, create project success.

4. **Responsibility to colleagues**

*Division of labor and cooperation*: implement the division of expertise, focus on coordination and cooperation, and increase operational efficiency.

*Connecting the past and future*: self-motivate and encourage one another, inherit and carry forward technical experience, and train junior talent.

Compared to the 1941 version of the Chinese Engineers Creed, the new creed highlights the responsibility of engineers, puts social responsibility in first place, and broadens engineers' responsibility to the profession, clients, and employers. Moreover, based on the Chinese Engineers Creed, the Chinese Institute of Engineers formulated the "Implement Rules of the Chinese Engineers Creed," embodying the engineers' responsibility to make it more operational. The revised version of the Chinese Engineers Creed (1996) is a set of fairly complete, systematic and operational codes of engineering professional ethics.

From the formulation of the Chinese Engineers Creed and its history of revisions, we can see that as early as 1933, the written code of ethics for Chinese engineers had already been set. After modifications in 1941 and 1976, it has gradually matured till 1996. However, due to various historical reasons, to some extent, the engineer's ethical awareness in mainland China has stayed on the level of the Chinese Engineers Creed from 1941 to 1976.

## 16.5 Conclusion

Through the above analysis, we can achieve the following understanding about the engineering ethical awareness embodied in the constitutions or ethical norms of Chinese engineering societies.

Although the ethical obligations of engineers have become an important part of the qualification standard for Chinese registered engineers, engineering societies of mainland China still lack clear and comprehensive recognition about the ethical responsibilities of engineers, and lack a moral ideal, beyond civic morals, exclusively belonging to engineers.

At present, with respect to the concept of ethical consciousness, engineering societies of mainland China are still focused on "do the thing right" rather than "do the right thing." The latter is subjected to the interests of the whole society. Moreover, it lacks important connotations, such as the responsibilities of "informing" and "reporting." In addition, there is a certain kind of ambiguity in the orientation of the ethics of engineering societies in mainland China. The ethical codes tend to be too abstract when it comes to ethical principles and too concrete when it comes to conduct requirements.

We can see that the ethical awareness of engineering societies in mainland China still remains grounded in the idea from World War II. Historically speaking, when the state and the nation's survival was challenged by foreign aggressors, engineers, as a group of people with mastery of the power of technology, treated the state and national interests as a priority and dedicated themselves to the development of the national economy and science and technology, which was especially essential at a time of domestic and international war. However, in the era of globalization, when sustainable development is promoted, and in a context where implementing a scientific outlook on development and building a harmonious society is advocated in China, the requirements for engineers are much more than just "do the thing right." In view of this, the ethical awareness of the Chinese engineering society lags far behind.

In fact, both CAE and public engineering organizations regard themselves as academic groups instead of groups of engineering practitioners. While we could say that, to some extent, the Chinese government, the scientific community and the public have attached importance to scientific research integrity/ethics, we must say that there is much more work to be done about the ethics of engineering. The primary step is to revise the texts of constitutions and formulate codes of ethics. In addition, it is also necessary to reform the institution of engineering activity and improve engineering education.

Indeed, the engineering societies' constitutions and the written texts of their codes of ethics do not completely represent the ethical awareness of these groups and their members. Ethical awareness might not be in concert with the ethical demands of the actual situation. Nonetheless, the constitutions and codes of ethics do guide and constrain individuals and groups, and affect their actual behavior. As N. Fairclough has mentioned, in the social and cultural environments, the relationships of language and values, religious beliefs and power are interactional. Language is a social practice. It is a kind of timeless intervention force of social order, which reflects reality from various perspectives, and manipulates and influences social process through the reproduction of ideology. Written items would affect people's values and also their behavior. Therefore, the use of language can promote changes in discourse and transformations in society (Fairclough 1989, 1992, 1995).

In recent decades, Chinese engineering has played a great role in the development of the Chinese society and economy, which is obvious to all over the world. But it is undeniable that disasters endangering public safety, health and well-being occur frequently in Chinese engineering activities. The underlying cause of these disasters is the lack of ethical awareness of engineering societies in mainland China. It merits attention from our domestic engineering theorists and engineering practitioners.

**Acknowledgments** The authors appreciate Prof. Philip J. Chmielewski and Prof. Bocong Li for their helpful suggestions on this research. We are also grateful to Miss Liying Wu of National Central University in Taiwan and Miss Xuefen Jiang of Chinese Institute of Engineers in Taiwan. Miss Wu and Miss Jiang have kindly helped in finding historical documents about CIE in Taiwan.

## Appendix: Directory of 48 Engineering Societies Under the China Association for Science and Technology

| Name of societies | The time of societies establishment | The time of the constitutions stipulated or revised[a] |
|---|---|---|
| China Civil Engineering Society | 1912 | 2002 |
| Chemical Industry and Engineering Society of China | 1922 | 2007 |
| China Textile Engineering Society | 1930 | 2005 |
| Chinese Hydraulic Engineering Society | 1931 | 2004, 2009 |
| Chinese Society for Electrical Engineering | 1934 | 1989,2009 |
| Chinese Mechanical Engineering Society | 1936 | 2006 |
| The Chinese Society of Naval Architects and Marine Engineers | 1943 | 2006 |
| The Architectural Society of China | 1953 | 2005 |
| The Chinese Society for Metals | 1956 | 2006 |

(continued)

(continued)

| Name of societies | The time of societies establishment | The time of the constitutions stipulated or revised[a] |
|---|---|---|
| Chinese Society for Geodesy, Photogrammetry and Cartography | 1959 | 2009 |
| The Chinese Ceramic Society | 1959 | 2004 |
| International Measurement Confederation | 1961 | 2009 |
| Chinese Association of Automation | 1961 | 2008 |
| The Chinese Institute of Electronics | 1962 | 2006 |
| China Computer Federation | 1962 | 2004, 2008 |
| Chinese Society for Agricultural Machinery | 1963 | 2006 |
| Society of Automotive Engineers of China | 1963 | 2006 |
| China Ordnance Society | 1964 | 1964, 1978, 1988, 1993, 1997, 2003 |
| China Technical Association of the Paper Industry | 1964 | 2008 |
| Chinese Association of Refrigeration | 1977 | 2004 |
| China Association for Standardization | 1978 | 1979, 2004 |
| China Highway and Transportation Society | 1978 | 2004 |
| China Railway Society | 1978 | 2002, 2008 |
| China Engineering Graphics Society | 1979 | 2009 |
| Chinese Society of Agricultural Engineering | 1979 | 2004 |
| China Institute of Communications | 1979 | 2000, 2007 |
| China Instrument and Control Society | 1979 | 1991, 2003, 2008 |
| Chinese Society of Astronautics | 1979 | 2004 |
| Chinese Vacuum Society | 1979 | 1999, 2009 |
| Chinese Society for Corrosion and Protection | 1979 | 2006 |
| Chinese Nuclear Society | 1980 | 2008 |
| Chinese Institute of Food Science and Technology | 1980 | 2001 |
| China Electro-technical Society | 1981 | 2004, 2010 |
| China Energy Research Society | 1981 | 2002, 2009 |
| China Occupational Safety and Health Association | 1983 | 1983, 1988, 1993, 2002, 2008 |
| China Society of Motion Picture and Television Engineers | 1984 | 2005 |
| China Tobacco Society | 1985 | 2009 |
| Chinese Society of Particuology | 1986 | 1986, 1996, 2002, 2006 |
| China Illuminating Engineering Society | 1987 | 2007 |
| Chinese Society of Biotechnology | 1993 | 2001 |
| China Water Engineering Association | 2005 | 2005 |
| The Chinese Society of Rare Earths | 1979 | 2004 |
| China Coal Society | 1962 | 2007 |
| Chinese Cereals and Oils Association | 1988 | 2004 |

(continued)

(continued)

| Name of societies | The time of societies establishment | The time of the constitutions stipulated or revised[a] |
|---|---|---|
| The Nonferrous Metals Society of China | 1984 | 2005 |
| Chinese Society of Inertial Technology | 1987 | 2006 |
| Chinese Society For Vibration Engineering | 1985 | 2007 |
| Chinese Association for Artificial Intelligence | 1981 | 2005 |

[a]Most societies have been amended several times, but the authors did not find the amendment of those constitutions on the whole history, the table lists the modified time is not complete

## References


Chinese Institute of Engineers. (1984). *A memorandum of the operation of Chinese Institute of Engineers in the past three decades after moved to Taiwan*. Taipei: Chinese Institute of Engineers.

Cong, H. (2006). The status in quo and prospect of engineering ethics. *Journal of Huazhong University of Science and Technology (Social Sciences), 4*, 76–81.

Davis, M. (2007). How is a profession of engineering in China possible? In C. Du & B. Li (Eds.), *Engineering studies* (Vol. 3, pp. 132–141). Beijing: Beijing Institute of Technology Press.

Fairclough, N. (1989). *Language and power*. London: Longman.

Fairclough, N. (1992). *Discourse and social change*. Cambridge: Polity Press.

Fairclough, N. (1995). *Critical discourse analysis: The critical study of language*. London: Longman.

IEEE. (2006). *Code of ethics*. Retrieved June 11, 2010, from http://ethics.iit.edu/ecodes/node/3248

Liu, H. (2002). *The establishment and the development and their historical status of the Chinese Engineering Societies*. Master degree dissertation, Tsinghua University, Beijing.

Mao, Y. (1987). Brief history of Chinese Institute of Engineers. In Committee of the Research on History and Culture of the CPPCC National Committees (Ed.), *Historical data selections* (100th series, p. 117). Beijing: Chinese Literature Press.

Martin, M. W., & Schinzinger, R. (2005). *Ethics in engineering* (p. 44). New York: McGraw-Hill.

Schaub, J. H., & Pavlovic, K. (1983). *Engineering professionalism and ethics* (p. 28). New York: Wiley.

Vesilind, P. A. (1995). Evolution of the American Society of Civil Engineers Code of Ethics. *Journal of Professional Issues in Engineering Education and Practice, 121*(1), 4–11.

World Federation of Engineering Organizations.(2001) *The WFEO model code of ethics*. Retrieved 25 August, 2001, from http://www.wfeo.net/wp-content/uploads/WFEO_MODEL_CODE_OF_ETHICS_Final.pdf

# Chapter 17
# Engineering for Peace: An Obligation of Professional Capabilities

**W. Richard Bowen**

**Abstract** This chapter aims to contribute to a reprioritisation of engineering in the pursuit of peace. Some of the devastating effects of modern weapons and the temptations of military technology are outlined. A philosophical grounding for the proposed reprioritisation is presented based on a description of engineering as a *practice*, in Alasdair MacIntyre's sense, combined with an *obligation of professional capabilities* developed from the work of Amartya Sen. The concept of *sustainable security*, the cooperative and peaceful resolution of the root causes of conflict, is outlined and its surprisingly rapid incorporation into UK government security strategy is described. Building on this philosophical analysis and government policy, some ways that individual engineers, university engineering departments, commercial engineering enterprises and professional engineering associations can contribute to the promotion of genuine peace are proposed.

**Keywords** Engineering as a practice • Peace engineering • Professional capabilities • Professional obligations • Sustainable security

## 17.1 Introduction

In November 2010, two sisters, Paeng, 15, and Piou, 10, were returning from school in central Laos when the younger girl picked up a small object to show her sister. She then threw it to the ground where it exploded. Both girls were taken to hospital in the capital, 3 h away. The younger girl bled to death 30 min after arrival in hospital. Her sister had severe fragmentation wounds in her neck, hand and hip (Buncombe 2010).

W.R. Bowen (✉)
i-NewtonWales, 54 Llwyn y mor, Caswell, Swansea SA3 4RD, UK
e-mail: wrichardbowen@i-newtonwales.org.uk

D.P. Michelfelder et al. (eds.), *Philosophy and Engineering: Reflections on Practice, Principles and Process*, Philosophy of Engineering and Technology 15,
DOI 10.1007/978-94-007-7762-0_17, 

The younger sister had picked up a cluster munition bomblet, probably dropped by US forces more than 30 years before.[1] The design of these submunitions is such that many cause immediate and indiscriminate injury and death, but others remain unexploded until subsequently disturbed. These quiescent submunitions, which are small and often brightly coloured, are especially attractive to children. They have caused the injury and death of tens of thousands of civilians. The death of Piou and the severe injuries sustained by Paeng were typical for these weapons, but their case was especially poignant as it occurred whilst the first meeting of states party to the Convention on Cluster Munitions, which prohibits all use, stockpiling, production and transfer of such weapons, was taking place in Vientiane, also in Laos.

The design, manufacture and use of cluster munitions require the application of sophisticated engineering across the range of the discipline. Hence, in a real sense, Paeng and Piou were victims of engineering. They were two of the many such victims, for the greatest tragedy of the engineering profession is that during the twentieth and early twenty-first centuries generations of the most able engineers have worked on the development, manufacture and use of many types of weapons of indiscriminate effect and huge devastation potential. War has become the normal business of engineering: almost a third of engineers in the US are employed in military-related activities (Gansler 2003) and the largest single employer of engineers in the UK is an arms-producing company. The resources used are enormous, with world military expenditure in 2010 exceeding US$1,630 billion (SIPRI 2011).

Such military engineering can be accompanied by an astonishing degree of ethical detachment on the part of individuals, commercial engineering enterprises and professional engineering associations. At the individual level, a leading exponent of "nuclear deterrence" in the UK has described that issue as "intellectually congenial perhaps because of its combination of complexity and abstractness" whilst advising that to reach the Soviet Union's "threshold of horror" would require up to ten million dead (Mottram 2009; Edwards 2010). Again, a senior engineer has described his early work as a weapons engineer dealing with sonar, radar, guns and missiles as the "fun hands-on part" of his career (Duckett 2008). Commercial engineering enterprises usually take great care to fully assess and make known the effect of their activities on persons, communities, the environment and the economy. However, the annual reports of arms companies do not record the number of civilians injured or killed by their products, and requests for such information are declined on the grounds of not commenting, on principle, on individual customers or individual contracts. Nevertheless, such information has been recorded: for example, analysis of Wikileaks documents shows that 201 civilians have been killed and 498 civilians injured in Iraq by weapons with components from Norwegian arms companies (Skille et al. 2010). Finally, professional engineering associations can

[1] A cluster munition is a means of delivering and scattering a large number of explosive submunitions (bomblets). A single cluster munition may scatter submunitions over an area of 1 $km^2$. The number of submuntions delivered may be enormous: during the 1991 US operation "Desert Storm" in Iraq it is estimated that 11,000,000 were fired from rockets, of which 220,248 were fired in the first 5 min (McGrath 2004).

play an important role in leading informed debate about the role of engineering. However, their contribution to discussion of the suffering caused by military engineering is notably absent.

The present chapter aims to contribute to a reprioritisation of the use of engineering by challenging engineers to consider how they can best use their skills in the pursuit of peace. It begins with some examples of the beguiling temptations of military technology. Secondly, to provide a philosophical basis for the proposed challenge, a description of engineering in terms of Alasdair MacIntyre's concept of a *practice* is summarised. Thirdly, this description is supplemented with the concept of the *obligation of professional capabilities*, building on a proposal of Amartya Sen. Fourthly, recent analyses of the root causes of conflict, approaches to security, and their incorporation into government policy are outlined. Finally, building on the philosophical analysis and government policy, ways in which individual engineers, commercial engineering enterprises, university engineering departments and engineering associations can contribute to the promotion of genuine peace are suggested.

## 17.2 The Temptations of Military Technology

Certain types of weapons have been considered so horrendous that their use has been proscribed by international law. Important examples of such restrictions include the Biological and Toxin Weapons Convention (1972), the Chemical Weapons Convention (1993) and the recent Convention on Cluster Munitions (2010).[2] These conventions have made valuable contributions to the protection of life. However, they also have limitations. For example, major producers and users of cluster munitions, including Israel, Russia and the United States, have not signed the CCM. Further, there are always temptations to find ways around such legislation or to develop entirely new types of weapons. Two of the many such possibilities will be noted here: the use of drugs as weapons and the use of drones (unmanned aerial vehicles).

Considerable work is currently being undertaken by governments, industries and universities on military applications of new biological knowledge. Such approaches are often euphemistically described by their proponents as "drugs as weapons" or "non-lethal weapons" and typically target neurological activity, with possibilities for ethnic selectivity. There is ambiguity as to whether they are covered by the BTWC and CWC. Engineering skills are needed for the scale-up of the production, purification and encapsulation of the active ingredients of such weapons, and to provide the theoretical and practical basis for their deployment, probably as dispersed aerosols.

---

[2] Biological and Toxin Weapons Convention (BTWC), *Convention on the prohibition of the development, production and stockpiling of bacteriological (biological) and toxin weapons and on their destruction*; Chemical Weapons Convention (CWC), *Convention on the prohibition of the development, production, stockpiling and use of chemical weapons and on their destruction*; Convention on Cluster Munitions (CCM).

As the development of such weapons also requires medical skills, the British Medical Association (BMA), which represents doctors in the UK, has published a detailed assessment of the topic (BMA 2007). The overall conclusion is that "the BMA is fundamentally opposed to the use of any pharmaceutical agent as a weapon", with key reasons including the need to uphold existing law unequivocally (BTWC and CWC) and the multiple, and probably insurmountable, difficulties that will prevent the use of pharmaceuticals as weapons without causing innocent deaths and disability. This conclusion is consonant with the BMA's overall guidance on the involvement of doctors in weapons development:

> …the BMA considers that doctors should not knowingly use their skills and knowledge for weapons' development…through their participation doctors are lending weapons a legitimacy and acceptability that they do not warrant. Doctors may consider that they are, in fact, reducing human misery through their involvement, but in reality the proliferation of weapons shows this to be untrue. (BMA 2001)

This authoritative analysis should also give cause for concern to any engineer approached with a proposal for work in this area.

One of the key promoters of ethical action is proximity. Indeed, Levinas (1961/1969) has defined an ethical act as "a response to the being who in a face speaks to the subject and tolerates only a personal response". Correspondingly, it is known that even highly trained soldiers are averse to killing at close range. However, a major current technological priority is the development and use of sophisticated weapons that allow remotely-controlled killing at great distances, particularly aerial drones but also land and sea equivalents. These are an attractive option for the military due to their relative cheapness in comparison with manned equipment and because there is essentially no risk to their operators.

Drones are widely used in Afghanistan whilst being controlled from Nevada, USA. Some are used for surveillance, but others are equipped with bombs and missiles. The latter are reported to cause many civilian casualties, though quantitative data is difficult to obtain as such drones are often used in remote areas with inadequate monitoring of effects. Great concern has been expressed about their use by well-informed and expert analysts. Thus, a report to the United Nations General Assembly Human Rights Council (UN 2010) has described such weapons, which are operated through computer screens, as giving rise to a risk of a "Playstation" mentality to killing. Again, one of the most senior UK judges has compared drones to internationally forbidden weapons such as land mines and cluster bombs, "so cruel as to be beyond the pale of human tolerance" (Bingham 2010). A further concern is that the development of drones has facilitated targeted killings ("state-sanctioned assassinations") outside of war zones. For example, there were more than 110 missile strikes by US drones in Pakistan during 2010, and drones have been used in other states outside war zones, such as Yemen. Such use is authoritatively regarded as being in most circumstances illegal under international law (UN 2010). Most drones are currently controlled by human operators, but increasing automation is in progress with the aim of computer-controlled selection and destruction of targets. This creates a further distance between the initiator of the action and the victim, and makes less clear the allocation of ethical and legal responsibility for the action.[3]

The United Nations Foundation (2008) estimates that 90 % of those killed, wounded or displaced in violent conflict are (civilian) women and children. An argument is sometimes used by arms producers along the lines that more technically-sophisticated weaponry can reduce civilian casualties. The indiscriminate and disproportionate injury and death caused by many modern weapons suggests that such an argument cannot be entirely true. Indeed, a detailed and careful study of casualties in Iraq in the period 2003–2008 shows that "sophisticated" weaponry used at a distance resulted in a far greater proportion of indiscriminate civilian deaths of women (46 %) and children (39 %) than more primitive techniques used at close range (Hicks et al. 2009). Experts advise that the patterns found in Iraq are likely to be replicated wherever similar weapons are used. "Sophisticated" weaponry may also have very long-term detrimental effects on civilians, as exemplified by the high incidence of birth defects in Fallujah, Iraq since 2003 (Alaani et al. 2011) and the many deaths caused by cluster munitions.

Drugs as weapons and drones are just two of the very many types of weapons which are currently under engineering development, with potential for hugely deleterious effects on human wellbeing. A challenge to such development may begin with a summary of a philosophical approach to the nature of engineering.

## 17.3 Engineering as a *Practice*

The overall nature of engineering may be clarified by considering it as a *practice*, "a coherent and complex form of socially established activity", of the type first proposed by MacIntyre (1981/1985). The UK Royal Academy of Engineering has provided a cogent and challenging basis for a description of what might be considered *the practice of engineering*:

> Professional engineers work to enhance the welfare, health and safety of all whilst paying due regard to the environment and the sustainability of resources. They have made personal and professional commitments to enhance the wellbeing of society through the exploitation of knowledge and the management of creative teams. (RAE 2007)

Practices have a number of key features, here defined using MacIntyre's terminology with descriptions of their engineering application (Bowen 2009):

1. *Internal goods* – For engineering these are particularly those associated with technical excellence: the accurate and rigorous application of scientific knowledge combined with imagination, reason, judgement and experience. Such goods

[3] Proponents of such technology may refer to "autonomous" systems able to make "decisions". Such terminology seems intended to imply mind-like properties, possibly to distract from the responsibility of manufacturers and operators. However, such "autonomous" drones remain machines (Lucas 1961).

are best recognised by participation in the practice, and characteristically directly benefit all who participate in the practice, and less directly all those affected by the practice.

2. *External goods* – For engineering these include considerable economic benefits to society, but more particularly technological artefacts. Such goods are typically the possession of an individual or group.
3. *Ends (or goals)*[4] – For engineering this may be described as the promotion of the flourishing of persons in communities through contribution to their material wellbeing.
4. *Virtues* – These facilitate the success of a practice, and those particularly necessary in the case of engineering are: accuracy and rigour; honesty and integrity; respect for life, law and the public good; and responsible leadership – listening and informing.
5. *Institutions* – These sustain practices and in the case of engineering include university departments, professional associations and commercial enterprises.
6. *Systematic extension* – Successful practices will seek to continuously develop their internal goods, external goods, ends, virtues and institutions.

Two characteristics of the practice of engineering are especially noteworthy in the present context. Firstly, the practice is described as being concerned with the welfare, health and safety of *all*. This is a very demanding aspiration, which includes communities beyond our usual boundaries and the individual persons in those communities. Secondly, a successful practice pays appropriate attention to *all* of its key constituent features. A cautionary note is required here. MacIntyre noted the dangers of too great a focus on external goods such as wealth, fame or power. In the case of engineering there is an additional and particular danger of focusing too greatly on the external goods of technological artefacts. Too great a prioritisation of the development of technically ingenious artefacts can lead to mistaking the external goods of the practice for the real end of the practice.

Many engineers are attracted to work for arms companies by the opportunities offered for involvement with the development of highly sophisticated technological artefacts. However, when engineering is considered as a practice, technological artefacts are only contingent products, external goods, in the pursuit of the flourishing of persons in communities. Furthermore, the prioritisation of technological artefacts of a type designed to cause great human suffering is a very perverse approach to engineering. Nevertheless, concern for the welfare, health and safety of all should naturally include consideration of actions that promote security and peace. Here a further feature of a practice is important: that its goods and ends should be systematically extended. The following sections will lead to proposals for a reprioritisation and extension of the role of engineering in the pursuit of peace.

[4] Philosophers use the term "end" to describe what an engineer might describe as a "goal".

## 17.4 The Obligation of Professional Capabilities

The technical aspects of engineering can bring great satisfaction. However, this leads to the danger of becoming so absorbed in the technical aspects that the ethical dimension, the effect of the technology on others, is neglected or lost. The avoidance of this danger can be stated in terms of a positive challenge to engineers: *can the great possibilities for technical innovation in engineering be matched by a corresponding innovation in the expression and acceptance of ethical responsibility*? That is, can engineers adopt a truly aspirational approach to their work? In particular, in the present context, can engineers promote peace and security by non-military means?

Development of such an aspirational approach may benefit from aspects of the work of the philosopher and economist Amartya Sen, as most recently expressed in his book *The Idea of Justice* (2009). Sen is concerned with the removal of injustice and the promotion of justice in the world. A central feature of his approach is that it does not seek to identify some ideal state, a task adopted by many philosophers, but rather seeks practical improvements to whatever circumstances presently exist. Hence, the approach has affinities with engineering at its best. A key concept in Sen's analysis is the *obligation of power*:

> …if some action that can be freely undertaken is open to a person (thereby making it feasible), and if the person assesses that the undertaking of that action will create a more just situation in the world (thereby making it justice enhancing), then that is argument enough for the person to consider seriously what he or she should do in view of these recognitions. (Sen 2009)

This obligation could be considered as a generalisation of the "rule of rescue": the compelling motivation to save endangered human life wherever possible. It should also be noted that this obligation is practical rather than idealistic, for it concerns the serious consideration of feasible options and thus recognises that there may be situational constraints on the action (at least initially).

The obligation certainly refers to a type of situation in which many engineers may find themselves, for they have at their disposal a range of knowledge, skills, techniques and technologies of uniquely powerful potential. However, engineers rarely have the type of (political) power referred to by Sen. It is, therefore, proposed here to retain the definition but refer instead to an *obligation of professional capabilities*.

'Capabilities' is a term which Sen uses to build an approach to social justice in terms of the various things that a person manages to do or be in leading a life. Such capabilities he describes in terms of both *wellbeing* and *agency*, the latter being the possibility to advance whatever goals and values a person has reason to advance. Wellbeing is particularly useful in assessing issues of distributive justice. Agency gives attention to the person as a doer (Sen 1987). The specific inclusion of agency is a characteristic feature of Sen's work and allows for a much richer description of the benefits of social justice than the consideration of wellbeing alone. However, in the present chapter the term *professional capabilities* is taken to refer specifically to the professional actions which an engineer can undertake to remove injustice and to promote justice. The civilian deaths and injuries caused by military engineering are clearly instances of injustice. Hence, in these terms there is a clear obligation of professional capabilities to refrain, where practically possible, from activities resulting in such injustice. However, the adaptable skills of engineers provide a possibility for the further expression of such an obligation of professional capabilities: to contribute to the removal of the underlying root causes of violent conflict and hence to promote genuine peace.

## 17.5 The Root Causes of Conflict, Approaches to Security and UK Government Strategy

Independent organisations such as the Oxford Research Group have provided perceptive analyses of current threats to peace and of the most effective responses (ORG 2006). The Group identifies four factors as the likely root causes of possible future insecurity and conflict: (1) climate change – leading to loss of infrastructure, resource scarcity and the mass displacement of peoples, giving rise to civil unrest, intercommunal violence and international instability; (2) competition over resources – including food, water and energy, especially involving unstable parts of the world; (3) marginalisation of the majority world – increasing socioeconomic divisions and the political, economic and cultural marginalisation of the vast majority of the world's population; and (4) global militarisation – the increased use of military force as a security measure and the further spread of military technologies, including chemical, biological, radiological and nuclear weapons. The Group characterises the predominant current responses as a power projection *control paradigm* – an attempt to maintain the existing state of affairs through military means. It proposes that a more effective approach is a *sustainable security paradigm* – to cooperatively resolve the root causes of these threats using the most effective civilian means available (ORG 2006, 2010).

Despite the modest size of its population and its peaceful geographical location, the UK has the third highest military budget in the world in cash terms (after the USA and China), and the world's second largest arms producing company is also UK-based (SIPRI 2011). UK security strategy therefore has global significance,[5] and it was first clarified in a single document by a recent government (CO 2008). That publication made clear that "The broad scope of this strategy also reflects our commitment to focus on the underlying drivers of security and insecurity, rather than just immediate threats and risks". It further recognised that climate change, competition for energy and water stress are "the biggest potential drivers of the breakdown of the rules-based international system and the re-emergence of major inter-state conflict, as well as increasing regional tensions and instability".

[5] The present analysis will hence focus mainly on the UK, though similar developments are taking place in other countries.

The consonance of these aspects of the strategy document with the Oxford Research Group's analysis is striking. However, two new motivating factors arose in May 2010: (1) an election resulting in a coalition government with a broader view of security, and (2) the financial necessity of substantially reducing overall government spending so as to ensure a balanced national budget. An early initiative of the new government was the creation for the first time of a National Security Council with high-level representation across the full range of government departments. Then, in October 2010 the government published two key documents: *The National Security Strategy* and *The Strategic Defence and Security Review* (HM Government 2010a, b).

The *National Security Strategy* sets out two core objectives: (1) ensuring a secure and resilient UK, and (2) contributing to shaping a stable world. It describes a commitment to a "whole government" approach based on "a concept of security that goes beyond military effects". The document reports the National Security Council's judgement of the four highest priority risks over the next 5 years: (1) international terrorism, (2) cyber attacks, (3) international military crises, and (4) major accidents and natural hazards. Eleven less likely risks are also identified, categorised in two further tiers of priority. The document gives high priority to tackling the root causes of instability, identifying such causes as competition for resources, marginalisation, environmental factors and climate change. The *Strategy* suggests a strong commitment to change: "we have inherited a defence and security structure that is woefully unsuitable for the world we live in today. We are determined to learn from those mistakes, and make the changes needed".

The *Strategic Defence and Security Review* provides more detail on the implementation of the *Strategy*. Overall, although wider security is given significant attention, the emphasis and budget allocations still prioritise military solutions. Thus, although only one of the four highest priority risks (international military crises, and this is expressed vaguely)[6] could be clearly addressed by the sophisticated weaponry that engineers have developed in recent years, the *Review* nevertheless prioritises expenditure on exactly that sort of military equipment: aircraft carriers, "hunter-killer" submarines, naval destroyers, combat jets and nuclear weapons. These represent a continued commitment to an outdated "Cold War mindset" which the Strategy elsewhere criticises: it recognises that "we face no major state threat at present and no existential threat to our security, freedom or prosperity". The only specified major change in expenditure that could benefit the *Strategy's* core objective of contributing to shaping a stable world is a proposed increase of Official Development Assistance to 0.7 % of Gross National Income over the next 3 years, with 30 % of this being used "to support fragile and conflict-affected states and

[6] The use of conventional military force to address the threat of terrorism is regarded by key experts as counter-productive. Thus, the Director General of the UK security service MI5 between 2002 and 2007 has advised that "the invasions of Iraq and Afghanistan radicalised parts of a generation of Muslims who saw the military actions as an 'attack on Islam'...Arguably, we gave Osama bin Laden his Iraqi jihad" (Manningham-Buller 2010). The Chief of the UK Defence Staff regards military victory against al-Qa'ida and the Taliban as not possible (Richards 2010).

tackle drivers of instability". In short, the *Review* does not adequately implement the analysis of the *Strategy*.

Neither *The National Security Strategy* nor *The Strategic Defence and Security Review*, which together run to 113 pages, uses the word "engineering" even once. However, science and technology are mentioned, including an important role for the National Security Council to "provide focus and overall strategic direction to the science and technology capability contributing to national security". These factors provide a challenge to engineers to make known to the Council the ways in which engineering can make unique contributions to fulfiling the core security objectives through civilian means.

## 17.6 Engineering for Peace

As discussed in the preceding section, the analysis of the Oxford Research Group has provided a convincing case for a move towards a sustainable security paradigm, which seeks to cooperatively resolve the root causes of conflict using the most effective peaceful means available. In the UK, successive governments have incorporated such a concept of sustainable security as a core feature of security strategy with surprising, but welcome, rapidity. However, the practical implementation of this strategy is inadequate: the present UK government continues to give priority to funding the development and commissioning of large-scale, complex weapons systems of a type best suited to military power projection. Such lack of consistent political commitment is puzzling but undoubtedly reflects the strong political influence of arms companies and the military hierarchy in the UK.

It will be noted that engineers can play a major role in resolving the root causes of conflict identified by the ORG and the UK government. To give some illustrative examples: development of renewable energy sources and transition to low carbon energy economies can reduce climate change; improved efficiency, better recycling, and the introduction of innovative processes and materials can reduce resource competition; generation of wealth through the introduction of appropriate engineering processes in impoverished societies can diminish marginalisation; reducing or halting weapons development and reducing trade in arms can limit militarisation. However, the UK government appears unaware of such potential contributions of engineering.

Engineers have often been attracted to work for arms companies by the opportunities offered for working on the development of advanced technological artefacts. However, philosophical analysis shows that technological artefacts, such as sophisticated weapons systems, are only part of the practice of engineering, examples of external goods. Engineers need also to consider in a balanced way the other key constituent features of their practice, including internal goods, ends, virtues and the systematic extension of the practice. Advanced engineering will, in particular, seek to balance these constituent features in a way that seeks to enhance the welfare, health and safety of all. *Hence, a crucially important point is: advanced engineering is* not *synonymous with advanced technology*. For example, advanced engineering may involve the application of an ingenious, but technically simple, means of meeting a genuine human need. It is particularly important for engineers to avoid the danger of becoming so absorbed in technical wizardry that the ethical dimension, the effect of the technology on others, is neglected or lost. On the contrary, the versatile range of knowledge, skills and techniques at the disposal of engineers may be seen as leading to an obligation of professional capabilities to use such skills for the removal of injustice and the promotion of justice.

The many deaths and injuries to civilians caused by a control paradigm of security sustained by military engineering represent clear instances of injustice. Hence, there is a clear obligation of professional capabilities to refrain from such military engineering. Furthermore, and more generally, absence of peace results in much injustice in the world. Absence of violent conflict is a necessary but not sufficient condition for sustainable peace. Peace is additionally characterised by relationships between individuals, and social groupings of all sizes, based on honesty, fairness, openness and goodwill (Bowen 2009). It may thus be considered that there is an obligation of professional responsibilities to use engineering in ways that can promote peace as understood in this way. Such an obligation may be expressed at the levels of the individual engineer, commercial engineering enterprises, university departments, and professional engineering associations.

Arguably the most important professional decision that an engineer can make is his or her choice of first job, for this may play a crucial role in determining an entire career. In making such a decision, or in subsequently changing career direction, engineers have the opportunity to consider how the intended work can contribute to the flourishing of persons in communities in a manner consonant with an obligation of professional capabilities. Here engineers are fortunate in often having considerable freedom of choice, for the practical skills and numeracy which are characteristic of an engineering education give rise to great versatility. Each individual engineer can seek in their work to achieve a balance of internal goods, external goods, ends, virtues and the systematic extension of their professional practice. A further important aim for each individual engineer should be to seek continuity and coherence of ethical values across personal and professional life. Past development of weapons technology has often taken place in a context of "ethical bracketing" of professional activities. A useful protection against such culpable ignorance are questions such as, "What would my family and friends think about this activity?"

A key reason for the surprisingly rapid emergence of sustainable security as an important feature of UK government security strategy is undoubtedly that levels of financial spending on increasingly complex and expensive weapons technology have themselves become unsustainable. Reductions in government arms expenditure are at present relatively modest, due as already noted to a lack of political coherence and the effect of influential lobbies. However, if the change in strategic thinking is genuine, as it seems to be, and if economic difficulties persist, as is likely, then there is a clear message to commercial engineering enterprises such as arms companies: it would be wise to realign with the new strategy if business is to succeed. Arms companies are in fact well placed to make such a change for they already employ many of the country's most able engineers. Additionally, the new security strategy may lead to business opportunities for other engineering companies with expertise that can genuinely lead to the amelioration of the root causes of conflict.

University engineering departments can make an important contribution to engineering for peace, broadly understood, by developing an integrated approach to teaching societal and technical aspects of engineering, indeed by teaching engineering as a practice promoting the flourishing of persons in communities. Some of the most innovative departments already have such an approach, and this is leading to the recruitment of new types of able and socially aware students who were previously discouraged by the nerdy image of the discipline. More specifically, university engineering departments can teach the obligation of professional capabilities and ensure that students are aware of the framework of international law concerning armed conflict (legal teaching is at present mostly limited to national health and safety regulations). Graduating students will then be better able to make informed and responsible career choices.

Professional engineering associations have a particular responsibility for making government aware of the possibilities of engineering and for leading informed public debate. The lack of reference to engineering in the *National Security Strategy* and *Strategic Defence and Security Review* documents shows that much work needs to be done in informing government of the potential contribution of engineering to sustainable security. Such associations also need to make a specific effort to make such potential contributions publicly known. However, sustainable security transcends national boundaries, so professional associations also need to inform and support international initiatives. The UN initiative for a Culture of Peace (UN 1999, 2006) is especially relevant and has identified eight areas of action to: foster a culture of peace through education; promote sustainable economic and social development; promote respect for human rights; ensure equality between men and women; foster democratic participation; advance understanding, tolerance and solidarity; support participatory communication and the free flow of information and knowledge; and promote international peace and security. All of these areas can benefit from engineering involvement (Bowen 2009).

Finally, it may be hoped that individual engineers, commercial engineering enterprises, university engineering departments and professional engineering associations will aspire to create a Culture of Peace *within* engineering. Given the hugely deleterious effects that military engineering has on human wellbeing, it might even eventually be considered appropriate for professional engineering associations, say the UK Royal Academy of Engineering and the US National Academy of Engineering, to provide advice following the pattern of the British Medical Association:

> …the UK Royal Academy of Engineering and the US National Academy of Engineering consider that engineers should not knowingly use their skills and knowledge for weapons' development…through their participation engineers are lending weapons a legitimacy and acceptability that they do not warrant. Engineers may consider that they are, in fact, reducing human misery through their involvement, but in reality the proliferation of weapons shows this to be untrue.

It is hoped that the present chapter provides a contribution to a challenge to the engineering profession to move in such a direction, hence promoting genuine peace.

**Acknowledgment** I thank Iselin Eie Bowen for perceptive comments during the development of this chapter.

## References

Alaani, S., Savabieasfahani, M., Tafash, M., & Manduca, P. (2011). Four polygamous families with congenital birth defects from Fallujah, Iraq. *International Journal of Environmental Research and Public Health, 8*, 89–96.
Bingham, T. (2010). Interview with the British Institute of International and Comparative Law (Reported in Verkaik, R., Top judge: Use of drones intolerable. *The Independent*, 6 July).
Bowen, W. R. (2009). *Engineering ethics: Outline of an aspirational approach*. London: Springer.
British Medical Association. (2001). *Recommendations for the medical profession and human rights: Handbook for a changing agenda*. London: BMA.
British Medical Association. (2007). *The use of drugs as weapons*. London: BMA.
Buncombe, A. (2010, November 13). Cluster bombs kill another child as nations reach accord. *The Independent*.
Cabinet Office. (2008). *The national security strategy of the United Kingdom*. London: CO.
Duckett, A. (2008, September). A view from the top. *The Chemical Engineer*, 20.
Edwards, R. (2010, November 26). Secret files from 70s reveal Trident strike needed to kill 10m Russians. *The Guardian*.
Gansler, J. S. (2003). Integrating civilian and military industry. *Issues in Science and Technology*. http://www.issues.org/19.4/updated/gansler.html
Hicks, M. H.-R., Dardagan, H., Serdan, G. G., Bagnall, P. M., Sloboda, J. A., & Spagat, M. (2009). The weapons that kill civilians – Deaths of children and non-combatants in Iraq, 2003–2008. *The New England Journal of Medicine, 360*, 1585–1588.
HM Government. (2010a). *The national security strategy*. London: TSO.
HM Government. (2010b). *The strategic defence and security review*. London: TSO.
Levinas, E. (1969). *Totality and infinity*. Pittsburgh: Duquesne University Press. (Originally published as *Totalité et infini*. The Hague: Martinus Nijhoff, 1961).
Lucas, J. R. (1961). Minds, machines and Gödel. *Philosophy, 36*, 112–127.
MacIntyre, A. (1985). *After virtue* (2nd ed.). London: Duckworth. (First published 1981).
Manningham-Buller, E. (2010, July 20). Evidence to the Chilcot enquiry (Reported in Siddique, H., Iraq inquiry: Saddam posed very limited threat to UK, ex-MI5 chief says). *The Guardian*.
McGrath, R. (2004, November 11). *Cluster munitions – Weapons of deadly convenience? Reviewing the legality and utility of cluster munitions*. Presentation at Meeting of Humanitarian Experts, Geneva.
Mottram, R. (2009, March 2). Sir Michael Quinlan: Leading strategist in the MoD during the cold war and defender of nuclear deterrence. *The Guardian*.
Oxford Research Group. (2006). *Global responses to global threats*. Oxford: ORG.
Oxford Research Group. (2010, May). Reviewing Britain's security. In *International Security Monthly Briefing*. Oxford: ORG.
Richards, D. (2010, November 15). (Reported in Sengupta, K., Head of armed forces says victory over al-Qa'ida is not possible). *The Independent*.
Royal Academy of Engineering. (2007). *Statement of ethical principles*. London: RAE.
Sen, A. (1987). *On ethics and economics*. Oxford: Blackwell.
Sen, A. (2009). *The idea of justice*. London: Allen Lane.
Skille, Ø. B., Andersen, E., Krekling, D. V., Westhrin, V., & Imrie, G. (2010, October 28). Norske våpen involvert i drap på sivile under Irak-krigen. *Norsk Rikskringkasting*. http://www.nrk.no/nyheter/verden/1.7355609
Stockholm International Peace Research Institute. (2011). *SIPRI yearbook 2011: Armaments, disarmament and international security*. Oxford: Oxford University Press.

United Nations Foundation. (2008). *Conflict prevention and peace building*. New York: UNF.

United Nations General Assembly. (1999). *Declaration and programme of action on a culture of peace* (A/RES/53/243). New York: UN.

United Nations General Assembly. (2006). *Culture of peace* (A/61/175). New York: UN.

United Nations General Assembly. (2010). *Report of the special rapporteur on extrajudicial, summary or arbitrary executions, Philip Alston* (A/HRC/14/24/Add.6). New York: UN.

# Chapter 18
# Roboethics and Telerobotic Weapons Systems

**John P. Sullins**

**Abstract** A technology is used ethically when it is intelligently controlled to further a moral good. From this we can extrapolate that the ethical use of telerobotic weapons technology occurs when that technology is intelligently controlled and advances a moral action. This paper deals with the first half of the conjunction; can telerobotic weapons systems be intelligently controlled? At the present time it is doubtful that these conditions are being fully met. I suggest some ways in which this situation could be improved.

**Keywords** Ethics of technology • Moral agency • Telerobotic weapons systems • Telepistemological distancing • Normalization of warfare • Military Ethics • Robotic Weapons • Philosophy of Engineering • Telerobots ethical concerns • Automated warfare

## 18.1 Introduction

A technology is used ethically when it is intelligently controlled to further a moral good. The philosopher Carl Mitcham explains that the intelligent control of technology requires:

> (1) Knowing what we should do with technology, the end or goal toward which technological activity ought to be directed; (2) knowing the consequences of technological actions before the actual performance of such actions; and (3) acting on the basis of or in accord with both types of knowledge—in other words, translating intelligence into active volition. (Mitcham 1994)

J.P. Sullins (✉)
Department of Philosophy, Sonoma State University, Rohnert Park, CA, USA
e-mail: john.sullins@sonoma.edu

D.P. Michelfelder et al. (eds.), *Philosophy and Engineering: Reflections on Practice, Principles and Process*, Philosophy of Engineering and Technology 15,
DOI 10.1007/978-94-007-7762-0_18, 

We can easily extrapolate that the ethical use of telerobotic weapons technology occurs when that technology is intelligently controlled and advances a moral action (Sullins 2009). This paper will not attempt to decide the question of when warfare is just or ethical, for now it will be assumed that there are at least some cases where it might be. Instead we will look at the first half of the conjunction and decide if telerobots can indeed be intelligently controlled in the manner that Mitcham requires. At the present time it is doubtful that these conditions are being fully met. I suggest some ways in which this situation could be improved.

## 18.2 Problems with the Intelligent Control of Telerobotic Weapons Systems

### *18.2.1 Telepistemological Distancing*

***Telerobotic systems influence how the controller of the telerobot sees the situations within which he or she is trying to navigate the robot***. The first insufficiently addressed effect of telerobotic weapons systems is telepistemological distancing—the removal of the operator from the location of military activity. The main function of military robotics is to extricate precious human agents from the direct harm encountered on the battlefield.

There are at least two distinctive types of robots used for this purpose: autonomous robots and telerobots. Telerobots are to be distinguished from autonomous robots in that a telerobot has one or more human agents who have some direct control over the activation of several, or all, of the systems, motors and actuators in the machine. The famous NASA Mars Rovers are good examples of telerobots in that they receive commands from their controllers on Earth and then they execute those commands remotely on Mars. Telerobots come in many forms such as robots that are in direct radio control of a human operator who determines each and every action of the machine to semi-autonomous machines that are only under intermittent human control but perform some action autonomously.

Fully autonomous robots, on the other hand, would have little or no direct input from their owner/operators and would be able to make important operational decisions on their own. Today, it is arguably not the case that there are any such things as fully autonomous robots equivalent to human or even animal natural agents. The machines that may be developed in the future which might have robust AI (artificial intelligence) and ALife (artificial life) functions are fascinating to contemplate and raise many intriguing ethical issues (Sullins 2005, 2006, 2008a, 2009). However, currently, they are not being deployed to a battlefield so I will not cover their ethical status in this paper.

The autonomy of telerobots is more subtle. Even though there are obvious human operators interacting with the machine during its operation, there are also many autonomous systems in the machine over which the operator has minimal control. In some cases the operator sole control may only be an abort-action button (Sullins 2008b). Since robots used on a battlefield and other the hostile do not always afford their operators the luxury of giving them the time necessary to make deliberate decisions about the robot's behavior, the tendency is to add more and more autonomy to telerobots—even in situations where we might prefer the full human control of these machines (Arkin 2007).

The operators of telerobots necessarily see the world a little differently when they look at it through the sensors and cameras mounted on the machine and this may impact their ability to make ethical decisions or at least influence the kinds of ethical decisions they chose while operating the machine. When one is experiencing the world through the sensors on a robot one is experiencing the world telepistemologicaly, meaning that the operators are building beliefs about the situation that the robot is in even though the operator may be many miles away from the telerobot. This adds a new wrinkle to traditional epistemological questions. In short, how does looking at the world through a robot color one's beliefs about the world?

Epistemology is no trivial subject and this paper is not meant to be a full treatise on the subject. But suffice it to say that a useful epistemology will provide some sense of assurance that the propositions one believes about the world are true or at the very least useful to the agent that possesses them. As anyone who has tried to program autonomous robots can attest, getting a machine to reliably discern useful information about their environments turns out to be fiendishly difficult. Even just getting a robot to autonomously recognize a soda can in a lab environment is tough. One solution is to have a human agent help the machine make these determinations telerobotically by having the human operator analyze the data coming in from the machine to help it determine if an object is a soda can or some other object. If we move the robot out of the lab and onto a battlefield, and task it to not just looking for innocent soda cans but for enemy agents who are actively trying to deceive the machine, and then added to all this complexity we also have to distinguish between the enemy and friendly or neutral agents who are also present at the scene, then we must realize that this is obviously a monumental problem that will tax our telepistemological systems design to the limit.

Thus, the first requirement for the intelligent control of telerobotic weapon systems must be that the view of the world that the robot provides to its operators must be one that is epistemologically reliable. In order to be successful, let alone ethical, a telerobotic weapons system must provide a telepistemological view of the situation that the machine is in, which is accurate enough that given some agents *A* (a military telerobot/human team), the telerobot provides true knowledge of an event, meaning that some proposition P (e.g. "There is an enemy in that house and there are no civilians in the house") is believed to be true if and only if that proposition is indeed true given some small margin of error. Even in a noisy environment filled with smoke and low light, the Agents *A* are provided with accurate and meaningful information allowing for the operators to use the telerobot effectively to advance some ethically positive course of action. The agents can't just believe P to be true by pure luck, gut feeling, or happenstance; there have to be good reasons to support the belief.[1]

---

[1] I am well aware that this is a quick gloss over the Reliable-Indicator theory of epistemology and that there are well know paradoxes that can occur, such as beliefs that ensure their own truth self referentially. That detail is unimportant here and for a full exploration of this point I refer the reader to Armstrong (1973).

We might begin by noticing that many mundane technologies have a similar function. For instance, binoculars or other vision aids must provide a soldier with accurate information about the world from a distance. How does a soldier know what she sees through the binoculars is a true representation of the world? She knows what she sees is true because the binoculars operate under the physical laws of optics and she knows, or can know, the laws of optics and check the results herself. This is a reliable chain of causes so the soldier is justified in believing what she sees.

The question now is whether or not a telerobot provides a reliable chain of causes for its operator(s). The answer is not as simple as it was for the simple optical binoculars. This is due to the fact that the images the operator(s) see on their screen as they operate a telerobot are digitally enhanced or altered computationally, alterations which may be epistemologically suspect (Goldman 2001). Also, we have ignored the fact that even with the simple binoculars, once the images enter into the mind of the operator or soldier, myriad social, political, and ethical prejudgments may color the image that has been perceived with epistemic noise. As this sociological prejudgment is a problem in the simple case of the binoculars, it will also occur in telerobotic systems.

We can now see that there are two loci of epistemic noise in the use of telerobots; (1) the digitally enhanced medium in which the message is contained and (2) the sociological preconditioning of the human agent receiving the message transmitted by the robot. Let's now look at each of these conditions as they apply to telerobotic weapons systems.

(1) To know "that P" about some remote location through a technological medium, there must first be an actual fact to be known, and the receiving agent has to correctly believe it to be true. There is an additional requirement that the process by which the agent acquired her belief should be accurate so that if the fact were not true, then the agent would not be fooled and correctly disbelieve it. Is it reasonable to believe that telerobotic weapons now in use actually fulfill this requirement?

Today many of the Telerobots in use are drone aircraft. Flying them is not easy due to the telepistemological difficulties the pilot encounters trying to develop an accurate situational awareness of the conditions the aircraft is operating in P. F. Singer reports in his book "Wired for War" that:

> The use of drones has increased significantly…There are so many UAV's buzzing above Baghdad, for instance, that it is the most crowded airspace in the entire world, with all sorts of near misses and even a few crashes. In one instance an unmanned Raven drone plowed into a manned helicopter. (Singer 2009)

As Singer's example illustrates; there is reason to be worried about the efficacy of the telepistemological value of current technologies already in use. Unless this changes there is little hope of controlling these weapons intelligently and thus diminished chances that they can be used ethically.

(2) Another important location of epistemic difficulty can be found in the preconceived notions that the operators of telerobots bring to the equation. The most important factor we must focus on now is what I will call *telepistemological*

*distancing*. The operators of these machines are typically many miles, sometimes many thousands of miles, away from the military missions that the telerobots are accomplishing. Arguably, this can be seen as a moral good which these machines provide, in that the operator may be very safe from harm. But this moral good also has a few unfortunate consequences, one of which is that it makes the use of military force more likely. Given that few, if any ethical theories consider the state of war a moral good, anything that propagates war instead of seeking its speedy end cannot be used ethically. Telerobots provide an opportunity for military adventures that cost fewer lives and resources, thus helping make these actions much more politically palatable. Telepistemological distancing also helps facilitate political arguments that propagate the impression that modern warfare is a surgical affair. The compelling videos posted on YouTube of these machines in action also help foster the idea that warfare can be clean and surgical.

These images provide compelling anecdotal evidence that we can eliminate our enemies from the air with ease and precision, fostering an illusion of military omnipresence and omnipotence. The videos we see are highly selective and focus on the big successes. This selective sample will obviously result in a skewed political opinion of the technology and make leaders more likely to use them (Sullins 2009).

The perceived accuracy and omnipotence of military operations provided by telepistemological distancing is obviously a powerful disincentive towards accountability and media scrutiny of military affairs because this technology provides video of its own operation which can be prescreened and then given to reporters, the ultimate imbedded reporter. The most compelling of these videos can be selected for release to the public, assuaging any arguments that are critical of violent political action. Without this scrutiny it will be more likely for wars to be waged, which is obviously not an ethical outcome.

I am emphatically not arguing for placing human soldiers in harm's way when a telerobot could accomplish the mission with minimal risk. I am simply pointing out that in doing so we must also acknowledge that it might make us more readily turn to force as a solution to our political problems.

### *18.2.2 The Normalization of Warfare*

***Telerobots contribute to the acceptance of warfare as a normal part of everyday life.*** As a consequence of the telepistemological distancing that telerobots provide, there is a growing tendency for the operators to be located great distances from the field of battle, sometimes even thousands of miles away (Singer 2009; Kelly 2005). In fact, one of the major bases of operations for the US Air force's unmanned aircraft is located just outside Las Vegas. The pilots of these aircraft commute to work and then operate telerobots on military missions for a 3 h shift, after which they return home to their normal lives (Knapp 2005). For these pilots, fighting the war is just a normal part of their lives. Is there something ethically wrong with the normalization of warfare and the creation of shift-work military telerobot operators? The problem is that operating one of these machines is not just any old job; it is a job that requires the use of deadly force and the witnessing of the effects of that force on a regular basis. Imagine the mental gymnastics required to compartmentalize one's life to be at war one moment and then a few hours later to be watching TV with one's family. To use this technology ethically we must be certain that we are not psychologically damaging the operators of these machines or their friends and family.

Regardless of the ethical status of the use of these machines, politically, there is a strong motivation to pursue extending this practice to as many military operations as possible. As this process continues, then more and more military operations would be accomplished by telerobots located in the air, at sea and even on the ground. The result would be far fewer casualties. In fact, there would not even be all that much of a lifestyle loss for the military personnel that operated these telerobots. As long as the wars that are propagated with these technologies remain targeted at countries that cannot retaliate in kind, then these wars might go almost unnoticed by the general public (Sullins 2009).

As these systems become more autonomous and less technically demanding to fly or operate, then the need for military professionals to operate them will diminish and the job might eventually migrate to military contractors to realize cost cutting measures. There is already a tendency for unmanned aircraft to be flown by younger enlisted men rather than drawing on expensive pilots from the typical pool of trained officers (Singer 2009).

These trends will be likely to distort the special ethical terrain that warfare inhabits. If the conduct of warfare becomes equivalent to a day at the office, then we might lose interest in its speedy conclusion. Again, this suggests that telerobotic weapons resist intelligent use and as they stand will propagate unethical situations.

### *18.2.3 The Perceived Antiseptic Layer of Telerobotics*

***Telerobots contribute to the myth of surgical warfare and limit our ability to view our enemies as fellow moral agents***. Telepistemological distancing is designed to place an impenetrable barrier between the aggressor and the targets of that aggression.

Telerobotic weapons systems place a tremendous antiseptic layer of technology between the combatants, and that may help each side to more effectively dehumanize each other. The operator of the machine will see her enemy as little more than thermal images on a screen and the human targets of these machines will see only the teleoperated mechanical weapons of their foe. This type of warfare could intensify the hatred that is already fostered by current modes of armed conflict (Sullins 2009).

This type of warfare is likely to produce a disregard for the moral agency of one's enemies and may even promote a deeper hatred than that already caused by current modes of armed conflict. Nearly every ethical theory demands that moral agents must be given special regard, even when they are one's enemy. If it is possible to fight a just war, then it can only be fought in a way that seeks to reach a quick end to the conflict, treats enemy soldiers as moral agents, and gets both sides of the conflict back to a peaceful political relationship where they can again fully respect one another's moral worth.

Telerobotic warfare will make this much more difficult to achieve. Already, the victims of the many telerobotic attacks that have occurred over the past few years have expressed their belief that these weapons are cowardly and that the weapons are also inflict devastating civilian casualties (Singer 2009). Whether or not these perceptions are true, they are the image that telerobotic weapons cultivate and will inhibit the return to peaceful relations.

If telerobotic weapons are enhancing intergenerational hatred between peoples, then they are not a technology that is being intelligently controlled and their use is unethical.

## 18.3 Mitigation Strategies

So far we have seen some very serious problems that block the intelligent control of telerobotic weapon systems, which we argued was a necessary condition for their ethical use. I must admit that this is not a universally held claim. Ronald Arkin argues that it is possible to develop these systems in ways that actually enhance the possibility for just conduct in warfare and has presented his arguments in a technical report funded by the U. S. military (Arkin 2007). Wendel Wallach and Collin Allen argue that for pragmatic reasons that, "…if the proponents of fighting machines win the day, now will be the time to have begun thinking about the built-in ethical constraints that will be needed for these and all (ro)botic applications," and they have offered some ideas on how to accomplish this (Wallach and Allen 2009). Michael and Susan Anderson also argue that machine morality is of paramount concern and have offered some ideas on how to program ethical decision making (Anderson et al. 2006). And there are many more books and papers being written on the subject. Still, many of these efforts are centered on autonomous robots, and little focus has been paid to the much more common telerobotic systems. It is easy to think that since a moral agent is controlling the telerobot, then the telerobot's actions must be moral, but I hope to have shown here that this is a suspect argument. The problem is due to the fact that the design of telerobots limits our ability to intelligently control them, whether or not the controlling agent is indeed attempting to act ethically.

I have argued that telepistemological distancing has been shown to be the root cause of many of the issues preventing intelligent control of the machine, yet I do have to admit that this very same distancing also has an ethically positive ability to reduce casualties for at least some of the combatants. It is hard to say whether the positive outweighs the negative at this point in time.

It may be impossible to remove all of the negative factors surrounding telepistemological distancing but there might also be ways of mitigating their most pernicious effects. I would like to conclude with my modest recommendations for more ethically designed telepistemological weapons systems.

- Constant attention must be paid to the design of the remote sensing capabilities of the weapon system. Not only should target information be displayed but also information relevant to making ethical decisions must not be filtered out. Human agents must be easily identified as human and not objectified by the mediation of the sensors and their displays to the operator. If this is impossible, then the machine should not be operated as a weapon.
- A moral agent must be in full control of the weapon at all times. This cannot be just limited to controlling an abort button. Every aspect of the shoot or don't shoot decision must pass through a moral agent. Note, I am not ruling out the possibility that that agent may not be human. An artificial moral agent (AMA) would suffice. It is also important to note that AMAs that can intelligently make these decisions are a long ways off. Until then, if it is impossible to keep a human in the decision loop, then these machines must not be used as weapons.
- Since the operator herself is a source of epistemic noise, it matters a great deal whether or not that person has been fully trained in just war theory. Since only officers are currently trained in this it follows that only officers should be controlling armed telerobots. If this is impossible, then these machines should not be used as weapons.
- These weapons must not be used in any way that normalizes or trivializes war or its consequences. Thus shift-work fighting should be avoided. Placing a telerobotic weapons control center near civilian populations must be avoided in that it is a legitimate military target and anyone near it is in danger from military or terrorist retaliation.
- These weapons must never be used in such a way that will prolong or intensify the hatred induced by the conflict. They are used ethically if and only if they contribute to a quick return to peaceful relations.

**Acknowledgements** I would like to thank my students and colleagues for their many stimulating discussions on this topic, which helped formulate my thoughts as they are expressed in this paper. I would also like to particularly thank Dr. George Ledin who has graciously supported my efforts in this field of study and who has given me many opportunities to present my ideas to his students and colleagues in the computer science department at Sonoma State University. Finally I would like to thank my research assistant Jennifer Badasci who was instrumental in the completion of this work.

## References

Anderson, M., Anderson, S., & Armen, C. (2006). An approach to computing ethics. *IEEE Intelligent Systems, 21*(4), 56–63.

Arkin, R. (2007). *Governing lethal behavior: Embedding ethics in a hybrid deliberative/reactive robot architecture* (Technical Report GIT-GVU-07-11). College of Computing, Georgia Institute of Technology, p. 57. Available: http://www.cc.gatech.edu/ai/robot-lab/online-publications/formalizationv35.pdf

Armstrong, D. M. (1973). *Belief, truth and knowledge*. London: Cambridge University Press.

Goldman, A. (2001). Telerobotic knowledge: A reliabilist approach. In K. Goldberg (Ed.), *The robot in the garden: Telerobotics and telepistemology in the age of the Internet* (pp. 126–143). Cambridge, MA: MIT Press.
Kelly, M. L. (2005, September 16). *The Nevada home of the predator aircraft.* National Public Radio, All Things Considered. Available: http://www.NPR.org
Knapp, G. (2005). *Predator UAV 'Battle Lab' just north of Las Vegas*. Available: http://www.klas-tv.com/Global/story.asp?S=3001647CBS
Mitcham, C. (1994). *Thinking through technology: The path between engineering and philosophy.* Chicago: University of Chicago Press.
Singer, P. W. (2009). *Wired for war* (p. 202). New York: Penguin Press HC.
Sullins, J. P. (2005). Ethics and artificial life: From modeling to moral agents. *Ethics and Information Technology, 7*, 139–148.
Sullins, J. P. (2006, December). When is a robot a moral agent? *International Review of Information Ethics, 6*, 23–30. Available: http://www.i-r-i-e.net/inhalt/006/006_Sullins.pdf
Sullins, J. P. (2008a). Friends by design: A design philosophy for personal robotics technology. In P. E. Vermaas, P. Kroes, A. Light, & S. A. Moore (Eds.), *Philosophy and design: From engineering to architecture* (pp. 143–158). Dordrecht: Springer.
Sullins, J. P. (2008b). Artificial moral agency in technoethics. In R. Luppicini & R. Adell (Eds.), *Handbook of research on technoethics* (pp. 205–221). Hershey, PA: Information Science Reference, IGI Global.
Sullins, J. P. (2009) Telerobotic weapons systems and the ethical conduct of war. *The American Philosophical Association Newsletter on Computers and Philosophy, 8*(2). Available: http://www.apaonline.org/publications/newsletters/index.aspx
Wallach, W., & Allen, C. (2009). *Moral machines: Teaching robots right from wrong* (p. 21). Oxford: Oxford University Press.

# Chapter 19
# Normative Crossover: The Ethos of Socio-technological Systems

**Rune Nydal**

**Abstract** This chapter investigates normative dimensions of technologies based on an understanding of technologies as socio-technological systems. As such systems display technologies as both technical and social, they call for a corresponding clarification of the relationship between epistemic and ethico-political activities. The notion of the ethos of socio-technological systems is suggested to denote the immanent worth of the system. Analysis of the formation of the ethos brings forward how normative concerns we recognise respectively as epistemic and ethico-political are intertwined. The notion of the ethos of socio-technical systems is presented and discussed with reference to a Norwegian controversy on the ultrasound screening programme for pregnant women.

**Keywords** Ethics of technology • Obstetric ultrasound • Normative crossover • Normative controversy • Socio-technical norm

## 19.1 Normative Crossover

Technology, as advocated in Thomas Hughes' classical work on the electrification of western societies, needs not only to be understood in terms of its systems of material constituents, but also in terms of systems of social shaping and control. These societies, as Hughes (1983, 1986) described, became electrified through a "seamless web" of interactions of social and technical systems, or so-called socio-technological systems. This notion emphasizes what many scholars, including myself, have come to take as their theoretical point of departure. Technologies need to be analysed in a social context where the technical and the social needs to be

R. Nydal (✉)
Department of Philosophy and Religious Studies, Norwegian University of Science and Technology (NTNU), 7491 Trondheim, Norway
e-mail: Rune.nydal@ntnu.no

D.P. Michelfelder et al. (eds.), *Philosophy and Engineering: Reflections on Practice, Principles and Process*, Philosophy of Engineering and Technology 15,
DOI 10.1007/978-94-007-7762-0_19, 

analysed as intertwined, co-evolving or co-produced (Jasanoff 2004). This chapter discusses normative aspects of such an understanding.

Hughes' story revealed a complex dynamics between technical and social activities. Such a blurring of the technical-social distinction implies a critique of well-established analytical and institutional separation of epistemic and ethico-political concerns. As socio-technological systems are understood as both social and technical, it is difficult to see how one can purify domains of legitimate authority of existing technical and political institutions. The engineering activity of constructing robust material and social orders (that, for instance, make electrification possible) is intrinsically linked to activities we think of as ethico-political: the choices of how to electrify our societies and how to live in an electrified world.

'Socio-technological systems' is a concept that, in Latour's (1999: 174–215) vocabulary, "crossover" the social and the technical. Latour borrowed the crossover notion from genetics, having chromosomal crossover in mind. Such crossover concepts, given Latour's analogy, are to capture the exchange, mixing and mutual blending of the social and the technical. Normative crossover concepts would then correspondingly capture a mutual blending of normative concerns we recognise as epistemic and ethico-political.

I draw on the work of Charles Taylor's philosophical anthropology, in suggesting it makes sense to think of the notion of the ethos of a socio-technological system. 'Ethos' is a moral term that goes back to the Greek discussion of the moral character of man. The ethos of a socio-technological system is to refer to the moral character of such a system. Paying attention to how the ethos is shaped brings to the forefront ways in which ethical and epistemic activities are intertwined.

Human agents, as analysed by Taylor (1985a, b), are self-evaluating beings. Their self-evaluations may be what Taylor describes as weak or strong: humans evaluate their actions, but display more or less integrity, are more or less trapped in traditions, self-conceit and so forth, but are never morally indifferent to their actions. Humans, given Taylor's anthropology, cannot escape evaluating their own actions; and moral terms are consequently needed in order to account for individual as well as collective human agency.

Given that moral self-evaluations are weak or strong, more or less reflected and reasoned, humans are not necessarily fully aware of why they act as they do individually or collectively. Even crucial decisions we live by may, to a varying degree, be well deliberated or well argued. To describe a practice is, from this perspective, to address and engage the identity of the practitioners individually and collectively. By discussing the goals of a practice we simultaneously discuss who we are and what we want to be.

In classical reference to the "ethos of science" (e.g. Merton 1996) 'ethos' refers to science's characteristic trait as truth seeking, or to technological traits of safety and reliability. The worth of research activities, what could for instance substantiate the practitioner's motivations and pride, is exclusively expressed in epistemic terms. As John Ziman (1998) put it, "[t]he official ethos of academic science" nurtures a "'no-ethics' principle" as guiding professional norms and commitments revolve around the task of ensuring conditions for truth seeking research activity (like expressed in Merton's CUDOS).

The ethos of a socio-technical system refers to the moral trait of the socio-technical system under discussion, i.e. what is valued. It is the system that forms the unit of analysis of a technology's ethos. The system is both technical and social, shaped by humans in a dynamic interplay with nonhumans. Such a system has an ethos *qua* human practice, although the system is not exclusively determined by humans. The reference to ethos provide a short-cut reference to evaluations of worth that is seen as embedded in the socio-technological system as it play a role in mediating, maintaining, as well as destabilising the system. The reference to ethos expresses the realities to which human self-evaluative agents respond, as they are affecting or affected by the system.

I have found a medical context, obstetric ultrasound, useful for presenting and discussing the notion of ethos that I seek to articulate.[1] Obstetric ultrasound investigations are offered to all pregnant women, an offer which is conditioned by technical and social orders. The story presented in this chapter do not aim to highlight the socio-technological system as such, but explore how the notion of the ethos of the system may draw attention to the specific way human (moral) agencies blend into these systems.

The following is a story of a normative controversy that took place at a national level in Norway in 1999–2000.

## 19.2 The Ethos of an Ultrasound Screening Programme

Every pregnant woman in Norway is offered an ultrasound test in the 18th week of pregnancy, and most women have since the mid 1980s accepted the offer (Bergsjø 1997). Such diagnosis of foetuses has however been controversial in Norway from the start (Sætnan 1995). The controversy has been difficult to resolve as goals crossover medical and social reasons. The conclusions of a national consensus conference (of medical experts) held in 1995 are illustrative. Although it was determined that the screening programme had little or no medical effect, it was also concluded that the programme should nevertheless continue because "women want it" (NFR 1995). The explicit reference to women's desires reflects an uncertainty of what women appreciated and raises a question of why this appreciation could simultaneously articulate and legitimize the goals of the medical program.

Such conferences arise when the stability and the further development of a practice are challenged, and the consensus conference was one important arena where a verdict was reached about what the ethos of the programme had been and what it should be in the future. The conclusions of the conference suggest a robust ethos at the time had been formed within a fairly broad process of deliberation. The ethos

[1] Following the original presentation of this paper, Peter-Paul Verbeek published a work (2011) in which he found obstetric ultrasound to be a rewarding case for moral reflection on technology and raises questions similar to my own. While Verbeek investigates the moral dimension of materiality by turning to Michel Foucault's work as well as the phenomenological tradition, I turn to the philosophical anthropology of Charles Taylor.

was nevertheless too fragmented to be captured by a single value judgment. It seems more reasonable to think of it in terms of a spectrum of interwoven evaluative judgments. These judgments are historical and need from time to time to be subjected to re-evaluation as the program evolves. The conference did not only to assess formal objectives (determining multiple embryos, the position of the placenta, pregnancy date and deviations from normal developmental patterns (Helsedirektoratet 1991)). They also included evaluations describing the expectations, comfort and excitement on the part of users who sustained the programme by choosing to participate in it. These evaluations, originally emerging as responses to the program, had now become part and parcel of it, since they made a difference for the programs' further maintenance and performance. Even if the program did not satisfactorily meet traditional medical goals of preventing or curing illness, women had other reasons for participating in it that outweighed such goals. Diagnostics could, for instance, make a difference for how well parents could be materially and mentally prepared for the birth of a disabled child; or a medical confirmation that everything looked alright could ease substantial stress on the part of parents who for some reason or another were worried about the well-being of the foetus.

The overall ethos of the program then, was articulated at the conference in medical objectives as well as end-users' experiences, reported in evaluations of how it affects a parent's giving birth. Moreover, the overall assessment of the worth of the screening program has many components as screening is big business, being mediated by economic, industrial, scientific and political interests. Given its immanent and historic character, the program's ethos was not necessarily fully articulated and deliberated. This is not only due to matters of representation, such as there being stakeholders misrepresented or ignored in the process and who potentially could have threatened the stability of the programme. It is also about difficulties in articulating evaluations by which we live as we respond to socio-technological systems in constant transition.

The conclusions of the 1995 consensus conference might have been well-grounded. But even if they were, as time goes by, the screening practice as well as the diagnostic powers of the technology could develop and cause tensions between the actual practice and the grounds for it being judged as a good one. Even if the outcome of the Norwegian consensus conference in 1995 was robust by normative measures, it would not necessarily remain robust. In the story to follow, I suggest destabilising elements were put in motion, resulting in a heated public controversy over a particular ultrasound research project.

## 19.3 The Controversy

This controversy, evolving around a group of researchers at the Norwegian National Competence Centre for Ultrasound Research at NTNU, arose as the centre was poised to evaluate novel diagnostic tools that had been developed and used abroad. A "nuchal translucency thickness" of the foetus, best visible in the foetus from 11 to 14 weeks, had been reported to be correlated with a set of different disorders. The best correlation value had been shown for embryos with Down's syndrome. These developments had already received media attention in August 1998.[2] The researchers in Trondheim told the reporter that these methods were about to become routine in many places in countries like England, Germany, the Netherlands and Switzerland, and further investigations of both methodological quality and ethical ramifications of early ultrasound investigations were needed before such methods should be considered for implementation in Norway.

These developments had the potential to transform the screening program in ways that could destabilise its ethos As Berge Solberg analysed the situation in a public meeting,[3] there were two different issues of concern among stakeholders, which could be grouped under the headings of eugenics and medicalisation. Concerns about eugenics followed from the increased diagnostic powers of detecting disorders. The fact that one would need to consider moving the ultrasound investigation to the 12th week of pregnancy, in order to exploit the potentials of the tools, reinforced the possible eugenic effect, since the 12th week is the legal abortion limit in Norway. These developments might further increase a woman's experience of having her pregnancy "medicalised" due to the possible increase in the power of the diagnostic tools, as well as in the increase in the number of tests performed.

It was naturally expected that the Norwegian National Competence Centre for Ultrasound Technology should be updated on the limitations and possibilities of these new applications. In order to evaluate these applications, scientists argued they needed to acquire first-hand information of the technology of the sort one acquires through systematic research work. "We don't govern, politicians govern, we give advice. In order to give advice, we need data", the researchers later argued in a feature article in a national newspaper.[4] For example, researchers needed to acquire the observation skills necessary to understand the diagnostic significance of the degree of nuchal thickness, along with the possible implications of implementing these new techniques in a Norwegian setting.

It was difficult, however, to get an exact picture of what the focus of the proposed study was, especially when it came to research on the implications of implementation. When the research project reached newspaper headlines, the researchers refused to hand over their research protocol upon request; and they were likewise reluctant to be specific about the matter in public meetings. The university paper (*Universitetsavisa*), a national and a local newspaper (*Aftenposten* and *Adresseavisa*), as well as locally arranged debates are the main sources for this story, since none of the documents in question were official.

---

[2] Lene Skogstrøm. "Engelsk ultralydstudie med oppsiktsvekkende resultater: Avdekker Downs syndrom tidligere." *Aftenposten* (3.8.1998).

[3] Berge Solberg. "Hvor går ultralydforskningen ved NTNU?" Vitenskapsteoretisk forum, MTFS at NTNU, Trondheim (16.05.2000).

[4] Sturla Eik-Nes, Kåre Molne, Harm-Gerd Blaas, Kjell Salvesen. "Ultralyd tidleg i svangerskapet – styring eller trussel?" *Aftenposten* (01.11.1999).

The research objective seemed to have two focal points. The project was designed to investigate the potentials of the new diagnostic marker, and to carry out a comparative analysis of what could be gained by introducing the new tools in the broader context of a screening programme by moving this programme from the 18th to around the 12th week of pregnancy. As they explained in their feature article, the researchers set out to include 6,000 pregnant women in the local region in the project. Half the group was to be offered a test in the 12/13th week of pregnancy in addition to the one in the 18th week. The project was approved by the regional ethics committee that reviews biomedical projects on a regular basis.

The controversy was sparked by the Norwegian Minister of Health, who asked the researchers to put the project on hold as he publicly questioned the desirability of the research project on the 15th of October 1999. At the outset, this action appeared to be a clear-cut case of politically motivated intervention into knowledge acquisition processes. Already, in the year before, the Minister had commented on the international trends towards early ultrasound diagnostics, signalising his scepticism due to the possible eugenic consequences.[5] It did not come as a surprise that the Minister was highly criticized in the first phase of the debate. The criticism calmed down however, as the debate came to focus on individual and social ramifications of obstetric ultrasound.

The Minister did not try to stop the project through official channels. It was naturally beyond his formal authority to do so. Being sensitive to the political impact of the research project, he publicly requested the research group in Trondheim to delay it, referring to the ongoing work of a public committee revising a law that was, among other things, about to draw general guidelines for the scope and limits of foetal diagnostics.[6] Local politicians, having political responsibility for the hospital in question, immediately responded by putting pressure on the researchers to put the project temporarily on ice. After all, the cabinet Minister of Health had become involved, what was at stake?[7]

A public debate was triggered during which expressions of anger against the cabinet minister erupted. A majority of delegates of the National Parliament, the director of the Research Council of Norway, the Rector of NTNU, as well as the dean of the Faculty of Medicine attacked the cabinet minister for his non-legitimate political intervention in the research process. The tone was harsh, and we were reminded that the freedom of research had to be safeguarded.[8] The message was supported by reports, broadcasted nationwide, of medical researchers' lives being threatened by anti-abortion extremists, as well as positive interviews with the "outspoken" researchers in Trondheim.[9]

---

[5] "Tidlig ultralyd er betenkelig." An NTB text printed in *VG* (3.8.1998).

[6] "Vil stanse tidlig ultralyd." An NTB text printed in *VG* (15.10.1999).

[7] "Bøyde av for helseministeren." *Adresseavisen* (16.10.1999).

[8] "Politikerne støtter Eik-Nes." *Adresseavisen* (22.10.1999); "Ville utsette ultralydprosjekt: NTNU-leder kraftig ut mot helseministeren." *Aftenposten* (1.11.1999); "NTNU støtter Eik-Nes." *Adresseavisen* (2.11.1999); Emil Spjøtvoll. "Vitenskap og Politikk"; "Vil ha friere tøyler." *Universitetsavisa* (4.11.1999); "Fingrene av fatet, statsråd!" *Universitetsavisa* (18.11.1999).

[9] "Mot Veggen." *Universitetsavisa* (21.10.1999); "Frittalende forsker." *UKE-Adressa* (13.11.99).

A counter-reaction appeared as critical perspectives and negative experiences of foetal diagnostics in general popped up in the papers.[10] The focus of attention drifted from the science-politics interface to a more substantial discussion on obstetric ultrasound. One could also sense old conflicts lurking in the background, and even mutual distrust between the political and research communities in question reflecting earlier controversies. The cabinet minister now gradually gained substantial public support through different politicians in opposition.[11]

Due to the heated nature of the debate, the members of the local ethics committee announced they would reconsider their decision. The work of the committee had mainly been associated with the principles of the Helsinki declaration, which aims to secure individual rights. On this basis, the quality of the research project could be questioned. One should not distress people if important insights are an unlikely outcome of the experiments. This time, the committee chose to test the quality of the project by submitting their research protocol for peer review by an epidemiologist and statistician (Gulbrandsen 2000).

This marked a turning point. The committee, discussing the matter on the 17th of December 1999, now found it unacceptable that patients were to be confronted by results where, according to the reviewers, 45 % of the diagnoses made would be false positives (Gulbrandsen 2000). The result leaked to the press where the quality of the project was portrayed as unsatisfactory.[12] Suddenly the research community found itself without general support in the local press, or from university leaders. Critical articles, attacking the integrity of the researchers in Trondheim, now followed in the papers as well.[13] For their part, the researchers in Trondheim questioned the rationale for the new decision made by the ethics committee. They found the expert judgment questionable (claiming it would not stand the test of an international review), and indicated that the fact that the report had leaked to the press displayed unwarranted scientific and political power struggles.[14]

The new evaluation of the ethics committee, I would say, functioned as a catalyst for a turning point, reflecting the outcome of a process of public deliberation of the worth of obstetric ultrasound investigation as it was offered through the screening program. As soon as the minister publicly criticised the research project, the matter was in the hands of others: the scientists, politicians, parents and commentators who were affecting or being affected by the socio-technical system of obstetric ultrasound. These people came to be engaged in a debate anticipating the future of obstetric ultrasound that in turn had an effect on further course of events. There was something there waiting to be sparked. The heated public debate and the fact that

[10] See for instance Hans Olav Tungesvik. "Skal teknikken ein gong gå føre etikken?" *Aftenposten* (12.11.1999); "To av barna kunne ha vært valgt bort." *Adresseavisen* (13.11.1999); Inge Johansen. "Ultralydprosjektet ved regionsykehuset." *Adresseavisen* (17.11.1999); "Ultralyd: Frigjørende – eller undertrykkende?" *Universitetsavisa* (18.11.1999); Torvid Kiserud. "Ultralyd i tidlig svangerskap." *Aftenposten* (23.11.1999); "Ultralyd, medier og en mor." *Adresseavisen* (25.11.1999).

[11] "Ap-folk stør oppgjør med ultralyd-prosjekt." *Vårt Land* (25.10.99).

[12] "Knuser ultralydprosjektet." *Adresseavisen* (24.11.1999).

[13] "Ventet på rapport I 20 år.", "Som en ripe i lakken." *Adresseavisen* (4.12.1999).

[14] "Avviser kritikken." *Adressavisen* (24.11.1999); "Skuffet Eik-Nes." *Adresseavisen* (18.12.1999).

the researchers suddenly found themselves without general support, I suggest, needs to be understood in light of how the research project came to be linked with, and seen as a challenge to, the established ethos of the screening program.

## 19.4 Two Normative Concerns

The intervention by the Minister sparked two normative concerns on the ethos of obstetric ultrasound screening. The first, which included the Minister's own personal concern, may be understood in light of what has been referred to as the second abortion debate. Given powerful diagnostic tools, abortion might no longer concern only the social conditions of the parents, but would also need to be discussed in terms of the eugenic notion of a "well born" child threatening the unconditional value of humans.

The problem, it seems, was that the research project in question suggested a possible world that people such as the cabinet minister did not under any circumstances want to see stabilised. The research project aimed at learning more about how parents might actually respond to a situation where more knowledge about the foetus was obtained at an earlier stage of pregnancy. But such research questions presupposed the legitimacy of turning the abortion question about a woman's ability to take care of, or want, any child into a question about her ability to take care of, or want, a specific child. The research project would not pave the way for a totally new situation. Because of the increased risk of Down's syndrome, every pregnant woman over 38 years old had already been given the choice of having an amniocentesis. In spite of the fact that Down's syndrome was not officially accepted as a reason for abortion after the 12th week of pregnancy (social reasons had to be given), it had become routine to accept any application. To a limited extent then, the world the cabinet minister did not want to see realised already existed. But it was not realised to the same degree that might have become the case if the new diagnostic tools were to be part of the life world of every pregnant couple. The ethos of the screening program, one could say, would in the Minister's perspective come to embed a eugenic component.

Allowing the diagnostic tool into the world appeared to challenge the unconditional value of human life. In such a world parents would become more responsible for the well-being of the child to which they would give birth. In cases for instance where, according to the outcome of the diagnosis, the child most probably would experience incredible suffering, parents would be forced to pass judgment on issues they had not faced before. The parents would have to pass judgment on whose life is worth living. The burden of responsibility on the parents would become inescapable; the parents would have to bear the responsibility even if they would rather *not* know, because they *could* have known.

Novel technologies may connect issues that have not been connected before. In this case, the issue of human dignity would connect to the issue of what sufferings are worth bearing. If one wanted to defend the inviolability of human life, the question of whose life is worth living is simply not a question one should ask. But such a stance would simultaneously imply that every kind of suffering (as foreseen by the diagnosis) for parents and children would have to be regarded as a non-issue. The separation between the two set of discourses was at stake and the minister's actions may be understood as an attempt to counteract the formation of such linkages. Such linkages would be sustained and mediated in and through a rearranged socio-technical system of obstetric ultrasound and consequently more difficult to undo at a later stage.

The second strand of the controversy concerned worries of an increased 'medicalization' of the pregnancy of women as early diagnostics would potentially increase the power and number of diagnostics tests. "Medicalization" suggests some form of exaggerated medical intervention into healthy or normal pregnancies. Such exaggerated medical intervention might imprint woman's choices and course of action, their experienced lived pregnancies and even their relation to the children after birth.

In practice, almost every pregnant woman accepts the offer of an ultrasound, so apparently women do want the test. It even appears to have become integrated into the very concept of pregnancy. It has become part of the expectation and "happenings" of giving birth to say hello to the foetus and to have its picture taken for the family album. Still, and possibly because of this integration, the question of whether women *really* wanted the screenings is a reasonable question to ask.

Of course, if the medical authorities offer you something, there must be a reason for it, so you would feel uneasy to turn it down. Once a woman accepts the offer, however, she enters the medical world of uncertainty, where the chance is quite high that she ends up in the large group of people living with a "perhaps" diagnosis. A suspicion of a disorder from a doctor is hard to erase. A suspicion tends to stick, even after the woman has given birth to a normal child; maybe there was something wrong – after all the physicians found a reason to look for something. Such a suspicion may take away the bodily unit of pregnancy; its destiny is handed over to others who investigate it by means of some apparatus.

A medical researcher specialising in foetal medicine has made this point through telling her own story of personal transformation.[15] It all started with an early ultrasound test she was offered while she was abroad. She knew very well about the unreliability of the test, and she was even certain that she would accept any baby no matter what the test results would be. But still, she accepted the early ultrasound examination she was offered – and was disturbed. The ultrasound image suggested a chromosome disorder. To her surprise, she became desperate to know more and made use of all the follow-up tests she could get. None of the subsequent tests could eliminate the suspicion from the ultrasound – she knew that. But still, she had to find a way to get a hold of all the information available. Her pregnancy was transformed. She described it as something in which she didn't take part, as she waited for the outcome of the various tests. She described a process of slow self-transformation that occurred as she constantly found herself in new situations with which she had to deal or accommodate. At the end she was seriously confused as to

[15] Seminar on foetal diagnostics at Stortinget, Norway's national assembly, Oslo (20.01.2000).

what she wanted, and who she was and wanted to be. Her question of whether or not to terminate the pregnancy had turned into a question of when it would be legitimate. She found herself facing the problem of selective abortion because she had been trying to act responsibly to protect the physical well-being of her child.

One of the problems of medicalization is that the negative effects are often first experienced when it is too late. Berit Schei, a medical professor in the field of foetal medicine, suggested during the controversy that the ultrasound tests offered to women should come with a warning. Women should be informed about the personal risks they take in accepting the offer of a test. Their personal life-history might be directed into new, unpredictable, and burdensome pathways.[16]

Both sources of controversy here, discussed under the headings of eugenics and medicalization, appear as responses to features of the socio-technical system of obstretic ultrasound, features that potentially would be enforced by the anticipated changes of the system.

Such questions concerning the overall worth of the screening program drew attention to the interconnected nature of technological and political issues.

## 19.5 Ethos, Socio-technical Systems, and Normative Crossover

The controversy was initially framed along the temporally ordered separation of technical and political activities. The scientist's obligations to research and develop diagnostic technologies were expected to precede the politician's obligations to assess political ramifications. The turning point of the controversy appeared when the debate came to focus on overall worth of the screening program. This created opportunity for discussing the politics of obstetric research that otherwise would probably have escaped notice – at least until a later point in time when the screening program would have become less open to change.

The screening program has been analysed in terms of how it constitutes a socio-technical system of obstetric ultrasound. The notion of the ethos of socio-technical systems, this chapter suggests, allow for ways of analysing how the worth of the system is formed as the technical and social orders are built and rebuilt. The ethos changes, alongside changes in the socio-technical system, are implicit within the system and need to be articulated in order to be subjected to critical scrutiny. The ethos came to be articulated as it was engaged with, which in turn drew attention to questions of the appropriate design of the socio-technological system. The notion of the ethos of socio-technical systems, therefore, suggests we use socio-technological systems (rather than technological artefacts) as the unit of analysis for ethico-political scrutiny. Such a focus is not only sensitive to how epistemic and ethico-political activities crossover; it also suggests the ethos of the socio-technical system

[16] Berit Schei. "Hvor går ultralydforskningen ved NTNU?" Vitenskapsteoretisk forum, MTFS, Trondheim (16.05.2000).

as a normative standard. A measure of the legitimacy of this ethos may be found in the moral robustness of the program, that is, its ability to withstand public scrutiny of the overall evaluations of its worth.

The ethos of the screening program came to be engaged with in this controversy as human concerns of medicalization and eugenics challenged the ethos of the program. The controversy of the research project in question appears out of proportion if we do not take this larger social context into account. The research project aimed at evaluating the European trend of incorporating new diagnostic tools into the screening program, as the researchers were actually mandated to do (having a role at a national competence center). A difficulty appeared if one questioned the moral premises of the question being researched, as did the Minister. The new diagnostic tools appeared to connect moral issues the Minister did not like to see connected. Questions concerning human dignity, of the intrinsic worth of life, now appeared to be linked to questions of whose life is worth living. The Minister's intervention, one could say, could be seen as part of the work of evaluating the obstetric trends in Europe, as it sparked a public debate on the question of eugenics as well as worries of increased medicalization.

To summarise, there are three connected elements of the notion of the ethos of a socio-technological system. First, the ethos refers to the immanent moral character of the program. It expresses the worth of the program and possibly what makes it honourable and praiseworthy. To articulate the ethos of the program is to address human identities, and critique and revisions are likely to follow if this program is acknowledged as undesirable. This happened in the case discussed in this chapter. The debate that followed the Minister's intervention could not have become so vivid if there had been nothing there to be sparked. The research project mobilised human concerns that would not disappear simply through criticism of the Minister for crossing the borderlines of science and politics. One suggestion of this chapter is that we need to find ways to analyse and engage the ethos of socio-technical systems as it allows for crossover of epistemic and ethico-political concerns.

Second, the ethos is temporally shaped. 'Ethos', as used as a technical term in rhetoric, draws attention to the process where the legitimacy of the authority of the speaker is constructed, maintained or deconstructed through the speech act (Andersen 1995: 35). The ethos of a technology should likewise be understood as historically shaped in a process of establishing a stable socio-technical system that people are willing to rely on, put their trust in and live with. Engagement with the ethos should consequently take place alongside the process where socio-technological systems are formed. The controversy described, one could say, evolved around the question of a whether a diagnostic tool should be allowed to modulate the ethos of the screening program.

Third, the notion of ethos, understood as shaped in activities that crossover the technical and the social, allows for the normative scrutiny of issues that crossover epistemic and ethico-political concerns. It was the Minister's intervention that drew attention to the broader socio-technological context of the research project. The point is not to defend the Minister's action as exemplary, but to show how this intervention came to engage matters of concern. If the Minister had not hit a nerve, nothing would have been sparked; and he would be remembered as a politician that did not respect basic norms of the science-politics interface. This happened due to the way the debate came to focus on the overall good, or the ethos of the screening program. The ethical discussions of the ethics committee, in contrast, appeared somewhat off the beam in terms of what was at stake in the research project, as the project initially was regarded as legitimate before it became dubious and even cast as bad quality research. Questioning the ethos allow for moral scrutiny in ways that do not reproduce the distinctions between the realms of the technical and the social.

**Acknowledgments** I would like to thank Berge Solberg, Bengt Molander, Bjørn Myskja and Sophia Efstathiou for valuable discussions and comments.

## References

Andersen, Ø. (1995). *I retorikkens hage*. Oslo: Universitetsforlaget.

Bergsjø, P. (1997). Ultralydundersøkelse som rutine i svangerskapsomsorgen. *Tidsskrift for den Norske legeforening, 117*, 2292.

Gulbrandsen, P. (2000). Hva er galt ved selektiv abort? *Tidsskrift for den Norske legeforening, 120*, 1072–1073.

Helsediraktoratet. (1991). *Bruk av ultralyd i svangerskapet* (pp. 1–91). Oslo: Helsedirektoratets retningslinjer.

Hughes, T. P. (1983). *Networks of power. Electrification in Western Society, 1880–1930*. Baltimore: John Hopkins University Press.

Hughes, T. P. (1986). The seamless web: Technology, science, etcetera, etcetera. *Social Studies of Science, 16*, 281–292.

Jasanoff, S. (2004). *States of knowledge: The co-production of science and the social order*. London: Routledge.

Latour, B. (1999). *Pandora's hope*. Cambridge: Harvard University Press.

Merton, R. K. (1996). The normative structure of science. In H. Nowotny & K. Taschwer (Eds.), *The sociology of the sciences* (Vol. 1, pp. 38–49). Northampton: Edward Elgar. (Original work published 1946)

NFR. (1995). *Bruk av ultralyd i svangerskapet: konsensuskonferanse 28. februar–1. mars 1995*. Oslo: Norges forskningsråd.

Sætnan, A. R. (1995). *Just what the doctor ordered? A study of medical technology innovation processes* (STS-rapport; nr 25). Trondheim : Universitetet i Trondheim, Senter for teknologi og samfunn.

Taylor, C. (1985a). What is human agency. In *Human agency and language* (Philosophical papers, Vol. I). Cambridge: Cambridge University Press.

Taylor, C. (1985b). The concept of a person. In *Human agency and language* (Philosophical papers, Vol. I). Cambridge: Cambridge University Press.

Verbeek, P.-P. (2011). *Moralizing technology: Understanding and designing the morality of things*. Chicago: The University of Chicago Press.

Ziman, J. (1998). Why must scientists become more ethically sensitive than they used to be? *Science, 282*, 1813–1814.

# Part III
# Reflections on Process

# Chapter 20
# Translating Values into Design Requirements

**Ibo van de Poel**

**Abstract** A crucial step in Value Sensitive Design (VSD) is the translation of values into design requirements. However, few research has been done on how this translation can be made. In this contribution, I first consider an example of this translation. I then introduce the notion of *values hierarchy*, a hierarchy structure of values, norms and design requirements. I discuss the relation of *specification*, by which values can be translated into design requirements, and the *for the sake of* relation which connects design requirements to underlying norms and values. I discuss conditions under which a certain specification of values into design requirements is adequate or at least tenable.

**Keywords** Value sensitive design • Value • Design • Specification • Requirements

## 20.1 Introduction

In recent years, various authors have argued for incorporating values of ethical importance into engineering design (Flanagan et al. 2008; Friedman and Kahn 2003; van den Hoven 2007). We want cars that are safe and sustainable. We want internet search engines that are transparent in how they gather information, that have no systematic bias towards certain information, that respect our privacy, et cetera.

In this paper I focus on one specific aspect of what has been called Value Sensitive Design (VSD), i.e. the translation of values into more tangible design requirements. I have several reasons for this focus. First, the translation of values into design requirements is a relatively neglected aspect of VSD. Second, design requirements

---

I. van de Poel (✉)
Philosophy Section, School of Technology, Policy and Management,
Delft University of Technology, Delft, The Netherlands
e-mail: i.r.vandepoel@tudelft.nl

D.P. Michelfelder et al. (eds.), *Philosophy and Engineering: Reflections on Practice, Principles and Process*, Philosophy of Engineering and Technology 15,
DOI 10.1007/978-94-007-7762-0_20, 

specify certain properties, attributes or capabilities that the designed artefact, system or process should possess. If VSD is to be successful, the formulation of design requirements is obviously to be (partly) informed by values. Third, design requirements play an important role in guiding the design process. Again, if Value Sensitive Design is to make values bear on the design process, design requirements seem a prime target.

Translating values into design requirements not only happens in VSD, but also in 'regular' design, albeit often implicitly. I therefore start my enquiry with an example that highlights how the value of animal welfare was translated into design requirements for chicken husbandry systems such as battery cages. This example will highlight some of the general characteristics of the translation of values into requirements in design. After discussing the example I will introduce the notion of *values hierarchy*, i.e. a hierarchical structure of values, general norms and more specific design requirements. A values hierarchy is a coherence structure that is held together by two relations. *Specification* is the relation by which higher level elements are translated into lower level elements in the hierarchy. Pursuit *for the sake of* is the relation by which we can connect lower level elements, like design requirements with higher level elements, such as more general norms and values. I will discuss both relations and end with a brief conclusion about the added value of drawing a values hierarchy for translating values into design requirements.

## 20.2 The Design of Chicken Husbandry Systems as an Example[1]

Currently, battery cages are the most common system in industrialized countries for the housing of laying hens. The system makes it possible to produce eggs in an economically efficient and factory-like way. The system, however, has also been heavily criticized for its neglect of animal welfare by reducing chickens to production machines (e.g. Harrison 1964). A main concern in the design of battery cages – and a main reason for the introduction of the battery cage – is economic efficiency. This value has in the course of time been translated into more specific design requirements in terms of egg production per animal, feed conversion (the ratio between the weight of the food fed to the chickens and the weight of the eggs), egg weight and the mortality of chickens, all of which can be measured in tests. Other relevant design requirements relate to egg quality, manure removal and drying, and the cost price and lifetime of systems.

Important moral values in the design of battery cages include environmental sustainability (battery cages cause environmental emissions, especially ammoniac), the wellbeing of farmers (labor circumstances and profitability of the systems) and animal health and welfare. These values have in the course of time been translated

[1] A more extensive discussion with further references can be found in Van de Poel (1998).

into design requirements for battery cages and for alternative chicken husbandry systems, sometimes through government regulation. Here, I will focus on how the value of animal welfare was translated into more specific design requirements in the context of EU (European Union) regulation.

Translating animal welfare into design requirements first of all requires more insight into the notion of animal welfare, and factors that might enlarge or jeopardize it, than was possessed by the engineers and technicians involved in the design of battery cages. The scientific discipline that came to play a key role in making the notion of animal welfare more tangible was ethology. Ethology is a branch of biology that studies the behavior of animals in their natural environment. This 'natural' behavior gives ethologists a kind of reference point with respect to which they can claim to discern 'abnormality' in the behaviour of, for example, chickens in battery cages. Deviant or absent behavior can then be interpreted as possible failure of the animal to adapt itself to the new environment. This led to the notion that chickens have certain 'ethological needs' that should be respected. So, ethology as a science provided a normative standard by which to judge the suffering of animals in general and chickens in this particular case. Of course, this did not mean that all ethologists agreed on the level of animal welfare in battery cages, or on possible measures that might be taken to improve it. However, ethology offered instruments and concepts with which the general and abstract value of animal welfare could be translated into a set of more concrete norms for chicken husbandry systems. The main norms that have been articulated over the course of time are (e.g. Kuit et al. 1989):

1. Chickens should have adequate living space. As the Brambell Committee, installed by the English government and including the ethologist William Thorpe expressed it in the 1960s: "An animal should at least have sufficient freedom of movement to be able, without difficulty, to turn around, groom itself, get up, lie down and stretch its limbs" (cited in Harrison 1993: 120);
2. Chickens should be able to lay their eggs in laying nests;
3. Chickens should have the freedom to 'scratch' and to take 'dustbaths', which implies that 'litter' should be present in the husbandry systems;
4. Chickens should be able to rest on perches.

These norms have in the course of time been translated by governments into more concrete requirements, which have often been adopted as design requirements in the design of chicken husbandry systems. I focus here on the EU legislation. In the 1980s, EU rules with respect to battery cages for laying hens were laid down in Directive 88/116/EEC. This directive stipulated the minimum requirements for laying hens in battery cages coming into use after 1 January 1988. The requirements were: at least 450 $cm^2$ floor area per hen, 10 cm feeding trough per bird, 40 cm height over at least 65 % of the area and a floor-slope of maximally 14 %. These requirements were a further specification of the first general norm above (enough living space), but did not address the other norms. Addressing these other norms was in fact impossible in conventional battery cages and required the development of alternative systems. The main alternative systems that have been developed over the course of time are enriched battery cages and aviaries. Enriched battery cages are cages with special areas for perches, laying nests and litter. Aviaries are characterized by the presence of several levels on which the chickens can drink, eat and rest.

In 1999 new EU legislation was adopted implying a de facto phase-out of the traditional battery cage by 2012; no new traditional battery cages were to be brought into service after 1 January 2003 (EU Council Directive 1999/74/EC). The new directive also contained requirements for enriched cages and for other alternative systems. For enriched battery cages the main requirements are (EU Council Directive 1999/74/EC, article 6.1):

(a) at least 750 $cm^2$ of cage area per hen, 600 $cm^2$ of which shall be usable; the height of the cage other than that above the usable area shall be at least 20 cm at every point and no cage shall have a total area that is less than 2,000 $cm^2$;
(b) a nest;
(c) litter such that pecking and scratching are possible;
(d) appropriate perches allowing at least 15 cm per hen.

For other alternative systems like the aviary, the main requirements are:

1. The stocking density must not exceed nine laying hens per $m^2$ usable area (i.e. about 1,100 $cm^2$ per hen);
2. At least one nest for every seven hens. If group nests are used, there must be at least 1 $m^2$ of nest space for a maximum of 120 hens;
3. At least 250 $cm^2$ of littered area per hen, the litter occupying at least one third of the ground surface;
4. Adequate perches, without sharp edges and providing at least 15 cm per hen.

This example shows how the general value of animal welfare was translated into more concrete design requirements. It is striking that this translation largely took place outside the design process or other engineering practices. Partly, this is the result of certain particularities of this example. Animal welfare was, and still is, a value that is rather alien to engineering and engineers lacked expertise to specify this value. Moreover, there was little market demand for alternative systems. Still, the example highlights a number of aspects that are more generally illustrative for the translation of values into design requirements.

First, the translation of values into design requirements, especially of new values, may be a lengthy and cumbersome process. This also applies to values that are initially less alien to engineering than animal welfare. A nice illustration is Vincenti's description of how the broad notion of flying qualities for aircraft was translated into more specific requirements (Vincenti 1990: chapter 3). As he argues, flying qualities were initially ill-defined, contained subjective elements and were related to different, but related needs of aircraft designers and pilots. It took a mere 25 years and much effort to translate ill-defined flying qualities into more or less well-defined design requirements.

Second, translation may require specific expertise, sometimes from outside engineering. In the case discussed here, ethology provided such expertise. In cases of environmental values, environmental science or ecology may be relevant. For values such as privacy and trust, philosophical analysis may help to better understand these values and translate them into more concrete norms. Even values like safety and usability, which are more familiar to engineering, may require specialized expertise, as witnessed by the emergence of such disciplines as safety science, safety engineering and ergonomics.

Third, translation will often partly take place outside specific design processes. The chicken husbandry example is extreme in this respect; often the final translation from more general norms into specific design requirements will take place within the design process. Nevertheless, in these cases as well engineers will often rely on specifications that are more generally available. Apart from legislation, a main source of such specifications are technical codes and standards, which are usually drawn up by engineers on standardization committees and which lay down requirements or guidelines for dealing with general values and considerations such as safety and compatibility.

Fourth, the translation of values into design requirements is value laden. It can be done in different ways. Sometimes different (sub)disciplines offer different ways of specifying a value. Sometimes specification is made dependent on what is feasible with current technology or on trade-offs with other relevant values. The reason why Directive 88/116/EEC only addressed one of the four more general ethological norms was that it was deemed economically undesirable to formulate requirements that would de facto forbid the commonly used battery cage. From a philosophical point of view, a main question is when certain specifications are adequate or at least tenable.

Fifth, the translation of values into design requirements is context-dependent. Although animal welfare is a general value, its specification is different in the context of the design of chicken husbandry systems than, for example, in the context of toxicity tests or medical experiments. EU Council Directive 1999/74/EC contained as many as three different specifications of requirements for chicken husbandry systems applying to three different types (layouts) for such systems.

Sixth, the example illustrates that values and design requirements have a hierarchical structure. In this case, the general value of animal welfare was first translated by ethologists into a range of norms for holding chickens, and then governments translated these norms into very specific requirements. In the next section, I will be exploring this hierarchical nature of values and design requirements in more detail and introducing the notion of a values hierarchy.

## 20.3 Values Hierarchies

As we saw in the animal welfare example, values and requirements have a hierarchical nature. Design requirements, as it were, constitute the most concrete layer of a hierarchy of values, norms and design requirements that can be identified or defined for a design project.[2] Figure 20.1 gives an example of a values hierarchy.

[2] In the literature such hierarchies have been called objectives hierarchies, objectives networks or objectives trees (e.g. Keeney 1992: chapter 3; Keeney and Raiffa 1993: chapter 2; Cross 2008: chapter 6). What I call a values hierarchy below resembles what Keeney and Raiffa (1993) call an objectives hierarchy and what Cross (2008) calls an objectives tree. Keeney (1992) distinguishes

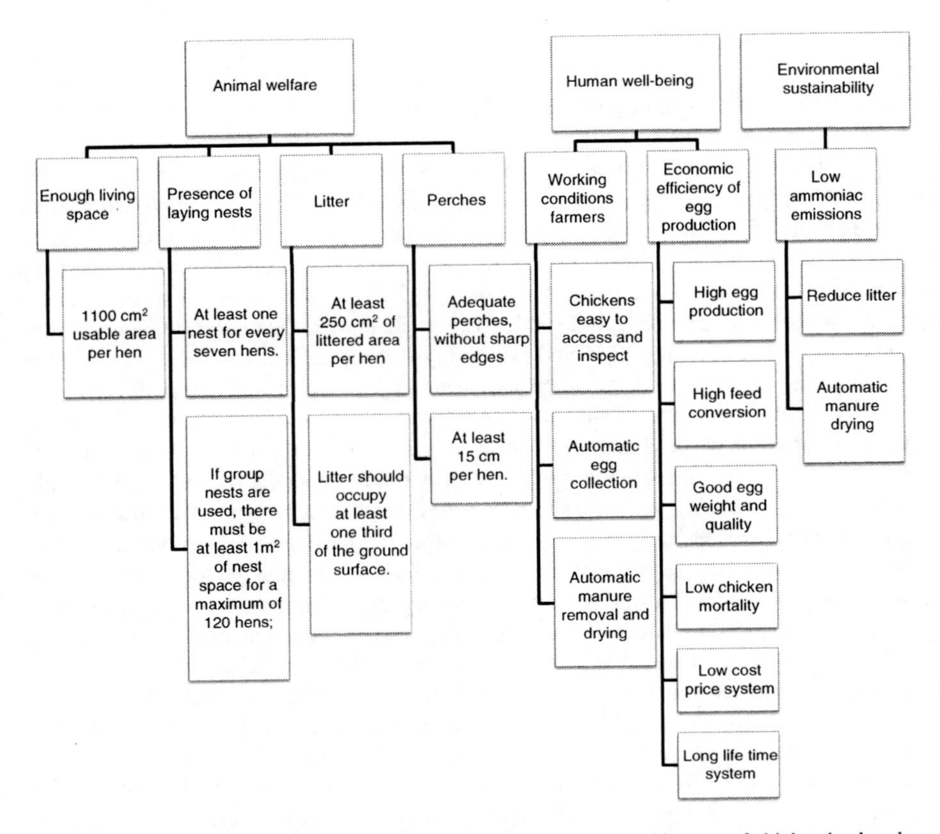


**Fig. 20.1** A partial values hierarchy for the design of aviaries, a specific type of chicken husbandry systems. The design requirements for animal welfare are based on EU Council Directive 1999/74/EC

Whereas the upper layer of a values hierarchy consists of values, and the most concrete layer of design requirements, value hierarchies will usually, as in the example in Fig. 20.1, contain an intermediate layer of norms. I use the notion 'norm' here for all kinds of prescriptions for, and restrictions on, action. One kind of norms that are especially important in design are end-norms. An end-norm is a norm referring to an end to be achieved or strived for (cf. Richardson 1997: 50). The end can be a state-of-affairs but also a capability ('being able to play the piano') or even an activity ('to sing an opera'). End-norms are particularly important in design because design is aimed at the creation of technical artefacts or at least at blueprints for them. End-norms in design then may refer to properties, attributes or capabilities that the designed artefact should possess. Such end-norms may include what sometimes are called *objectives* (strivings like 'maximize safety' or 'minimize costs'

between fundamental objectives hierarchies and (means-end) objectives networks. My values hierarchies come closest to the latter but allow a larger heterogeneity of relations between the elements.

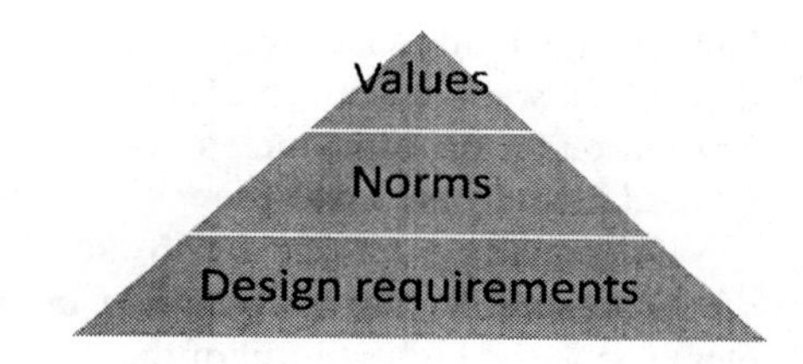


**Fig. 20.2** The three basic layers of a values hierarchy. Note that each of the layers may itself be hierarchically layered

without a specific target), *goals* (that specify a target such as 'this car should have a maximum speed op 150 km/h') and *constraints* (that set boundary or minimum conditions). Figure 20.2 depicts the three basic layers of a values hierarchy.

Figure 20.2 suggests that the formulation of design requirements is based on certain values. Although that is basically what I am claiming here, a range of clarifications is in place to make clear what this claim entails in my view and what it does not. First, it should be noted that the relation between the different layers of a values hierarchy is not deductive. Elements at the lower levels cannot be logically deduced from higher level elements. One reason for this is that the lower levels are more concrete or specific and that formulating them requires taking into account the specific context or design project for which the values hierarchy is constructed. The point is, however, not just that we should take into account contextual information; the point is also that there is usually a certain degree of 'latitude' or 'discretion' in translating higher-level elements into lower-level elements. Such translations are sometimes called *specifications*, a term I will also use.[3] Specification involves (value) judgment and usually more than one specification is possible. This is not to deny that we can formulate criteria for when a certain specification is adequate or tenable (I will be doing so in the next section), but these criteria will usually not narrow down the range of possible specifications to one specification that is the only one allowable.

Second, values hierarchies can be constructed top-down as well as bottom-up. In the latter case, one starts with more specific design requirements and looks for more general norms and values on which these requirements may be based or to which they may contribute. Often constructing a values hierarchy will require working in both directions. We have already seen that working top-down requires specification, but what is involved in constructing a values hierarchy bottom-up? One suggestion is that the elements higher in the hierarchy give an answer to the question *why* we aim for or adhere to certain elements lower in the hierarchy (Cross 2008: 81). This suggests that the higher-level elements have a motivating and justifying role with respect to lower-level elements. I will take up this suggestion by saying that the lower level elements are done *for the sake of* the higher-level elements.

The *for the sake of* relation is antisymmetrical (Richardson 1997: 54–57). If A is done for the sake of B, B is not done for the sake of A (unless A = B). It can easily be seen that values hierarchies are antisymmetrical in this sense. Chickens should

[3] Cf. Richardson (1997). In the engineering literature, specification is also used in a number of different meanings which I do not intend to imply here.

have enough living space for the sake of animal welfare, but it is nonsensical to say that animal welfare is a value for the sake of chickens having enough living space.[4] The reason for the antisymmetry of *for the sake of* is that the elements higher in the values hierarchy are more general and abstract than the lower elements. While you can do something specific for the sake of something more general; the opposite seems impossible. The antisymmetry of the *for the sake of* relation suggests that the elements at the highest level of the values hierarchies are to be done for their own sake. The most obvious candidates for the highest level in the values hierarchy are therefore intrinsic or final values, which are defined as values that are strived for for their own sake (Zimmerman 2004).

A number of things can be done for the sake of something else. The relation of A being done *for the sake of* B can therefore be seen as the placeholder for a number of more specific relations. One possibility is that A is a means to B. Another possibility is that A is a subordinate goal or end, the achievement of which contributes to (the achievement of) B. A third possibility is that A enables the achievement of B, without itself contributing to that achievement. If A takes away an obstacle to B, A may be done for the sake of B.

The *for the sake of* relation is normative. It can neither be reduced to a means-end or causal relation nor to a purely conceptual relation. The best way to capture the normativity of this relation is, I think, to say that the higher elements provide reasons for the lower level elements. The notion of reasons refers here both to a motivational and to a justificatory element. The normativity of the *for the sake of* relation suggests that the higher levels elements justify, or give (moral) authority to, the lower level elements. However, since, as argued earlier, lower levels cannot be deduced from higher levels, justification at a higher level is not automatically transferred to the lower levels. The degree of justification, or normative support, which is transferred from higher to lower levels depends on the plausibility or adequacy of the specifications made.

## 20.4 Specification

I will now further explore the relation or activity of specification by which values are translated into design requirements. Although specification proceeds top-down in a values hierarchy, what I am going to say about whether a certain specification of a value into design requirements is adequate or at least tenable can also be applied as a critical assessment for values hierarchies that are constructed bottom-up. It might then be used to assess whether the design requirements

[4] Note that it does make sense, however, to say that animal welfare is a value (partly) *because* chickens should have enough living space. This suggests two things. First, the relation *for the sake of* is not exhausted by its justificatory part that may be expressed by *because* and, second, the justificatory relation that is expressed by *because* may be bidirectional.

sufficiently cover the value on which they are based and may potentially lead to new design requirements or the reformulation of existing design requirements (or the reformulation of the value).

The specification of values is to be distinguished from an activity that is somewhat related but different in scope and aim: the conceptualization of values. Conceptualization of value is the providing of a definition, analysis or description of a value that clarifies its meaning and often its applicability. Ethologists, for example, conceptualized animal welfare as the fulfillment of certain ethological needs that animals like chickens have in 'natural' circumstances. Usually different conceptualizations of a value are possible. The value of individual human freedom may, for example, be conceptualized as 'the absence of external constraints on individual actions' or as 'the ability to make one's own choices in life.' The second conceptualization strikes me as more adequate because it seems better to capture why we consider 'individual human freedom' a value. Most people do not strive for a life without any external constraints. They have friends and family; make commitments and promises, all of which usually introduce additional constraints, without necessarily experiencing a loss of freedom. What seems more important or essential to freedom is the ability to make such choices yourself, without being forced or manipulated to make them. As this example suggests, some conceptualizations may be more adequate than others. An important criterion for the adequacy of a conceptualization, as suggested by this example, is that the conceptualization does justice to, or at least coheres with, the reasons we have to consider the value valuable in the first place. In many cases different conceptualizations of a value meeting this criterion may be possible.

Conceptualization is largely a philosophical activity that does often not require detailed knowledge of the domain in which the value is applied.[5] This is so because conceptualization does not add content to the value but merely tries to clarify what is already contained in the value. Specification, on the other hand, adds content, and this content is context or domain specific. Specification therefore requires context- or domain-specific knowledge. For example, it might be known that – on the basis of experience and engineering analysis – the main safety risk of a certain type of technical installation is that it explodes. In that case, safety may be specified into the norm 'minimize the probability of the installation exploding.' In other cases, a

[5] It is worth noting that the general conceptualization of animal welfare by ethologists in terms of the fulfillment of certain ethological needs that animals like chickens have in 'natural' circumstances does require very limited domain-specific knowledge. The conceptualization does not require any detailed knowledge of what these needs or what natural circumstances would be, only that these can be somehow identified. Philosophers might indeed criticize this conceptualization of animal welfare on a number of grounds. They may, for example, doubt whether there exists such a thing as 'natural' circumstances and, even if such circumstances would have existed, they may question why these circumstances would provide a normative yardstick (How convincing would it be to argue that killing or rape is part of human welfare or wellbeing because in 'natural' circumstances humans felt a need for them? Of course, animals are not humans). In fact, other conceptualizations of animal welfare are possible, for example, in terms of how animals 'feel', which might be measured for example in terms of stress.

technical installation may be very unlikely to explode but toxic substances may possibly escape from it. Safety may then be specified as 'minimize the probability and amount of toxic releases from the installation' or 'try to replace the toxic substance with a functionally equivalent non-toxic substance.' As these examples illustrate, the adequacy (or least tenability) of a specification is usually highly context-specific. What is an adequate specification of the value of safety for the first type of installation is not an adequate specification for the second type of installation and vice versa.

Although specifying values requires more than philosophical analysis, a philosophical analysis of the activity of specification may be helpful to judge the adequacy, or tenability, of certain specifications that are made in engineering design. For our current purpose, specification may be defined as the translation of a general value into one or more specific design requirements. This translation may be broken down in two steps[6]:

1. The translation of a general *value* into one or more general *norms*;
2. The translation of these *general* norms into more *specific* design requirements.

The first translation implies a transition from the evaluative to the deontic (or prescriptive) domain. Values are relevant for evaluating the worth or goodness of certain options or objects. However, they do not directly imply certain prescriptions or restrictions for action. Norms on the other hand are deontic because they articulate certain prescriptions for or restrictions on action.

For the transition from the evaluative to the deontic domain that is required in the first translation, the relation between values and reasons is relevant. There is no agreement in the philosophical literature on how values and reasons are related. One category of theories, often called 'consequentialism', holds that we have reason to do what has or brings about value, that we should increase the amount of value in the world or even should maximize it. Other theories hold that reasons are prior to values. Elisabeth Anderson, for example, defends what she calls an expressive theory of rational choice (Anderson 1993). According to her statements like 'x is good' or 'x is valuable' can be reduced to 'it is rational to adopt a certain favorable attitude towards x.'

I will not take a position in the theoretical debate about the exact relation between reasons and values. It is, however, worth noting that the positions just briefly mentioned seem to suppose a certain correspondence between values and reasons of the following kind:

> (V) If x is valuable (in a certain respect) or is a value one has reasons (of a certain kind) for a positive response (a pro-attitude or a pro-behavior) towards x.

This statement is intended to be neutral with respect to the question of whether values ground reasons or reasons ground values or that neither can be reduced to the other. As Dancy (2005) notes, whatever position one takes in this debate, something

[6] In practice, the translation may be made in one step, but even then it may be analyzed as involving these two steps.

like (V) seems to be true. The notion of positive response in (V) is meant to capture a range of pro-attitudes and pro-behaviors, such as desiring, promoting, increasing, maximizing, caring for, admiring, protecting, respecting, enjoying, loving et cetera.

Here we are interested in the case where x is a value and (V) tells us that x then corresponds with certain reasons that express a positive response to x. In the design process these may often be reasons to increase or even maximize x if x is a positive value like safety. However, increasing or maximizing a value may not always be a proper response; for some values it may be more appropriate to cherish them, to admire them, to protect them or to respect them. Moreover, although in context of design the proper response to a value may often be to take it into account in the design process and to try to embody it in the design, this is certainly not always the only or even the most appropriate response. Values like freedom and democracy might be appropriately translated into design requirements for a designed product (cf. Sclove 1995), but they may also be translated into requirements for the design process rather than the product designed. My focus is here on the translation of values into design requirements, but a proper response to values in design may be broader than this specific focus.

Two criteria might be formulated for the adequacy or tenability of a certain translation of a value into general norms. The first is that the norm should count as an appropriate response to the value. The second is that the norm, or set of norms, is sufficient to properly respond to or engage with the value. The first criterion tries to avoid inappropriate responses to a value, the second tries to avoid the problem that one response could be selectively chosen which in isolation does not do justice to the value. Applying both criteria requires a judgment that is context-specific. In the context of a beautiful sunset, a proper response to the value of aesthetic beauty is to enjoy it; in the context of architectural design a proper response might be to respect the value of aesthetic beauty and to try to embody it in the design. In the first context, bothering about how the sunset can be made more beautiful would be an odd and inappropriate response, while in the second context admiring the beauty of the building would be odd as long as it has not been designed and built.

The second step in specification is the translation of general norms into more specific design requirements. The requirement can be more specific with respect to the (a) scope of applicability of the norm, (b) goals or aims strived for and (c) actions or means to achieve these aims (cf. Richardson 1997: 73). An example is the specification of the general norm 'maximize the operational safety of a chemical plant' into the following design requirement: 'minimize the probability of fatal accidents (specification of the goal) when the chemical plant is operated appropriately (specification of the scope) by adding redundant safety valves (specification of the means)'. In this case, the design requirement specifies the general norm in three dimensions, but specification may also be restricted to one or two dimensions.

A specification substantively qualifies the initial norm by adding information 'describing what the action or end is or where, when, why, how, by what means, by whom, or to whom the action is to be done or the end is to be pursued' (Richardson 1997: 73). Obviously, different pieces of information may be added so that a general norm can be specified in a large multiplicity of ways. Not all specifications are

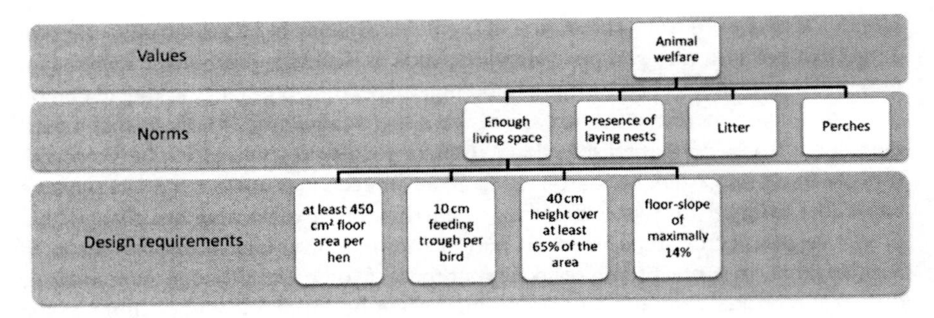


**Fig. 20.3** The specification of animal welfare in EU Council Directive 88/116/EEC

adequate or tenable, however. In general one would want to require that actions – or in our case: designs – that count as satisfying the specific design requirements also count as satisfying the general norm (cf. Richardson 1997: 72–73). In the above example 'maximizing operational safety' is specified as 'minimizing the probability of fatal accidents.' This specification is adequate if in all cases in which the probability of fatal accidents is minimized operational safety is maximized. Now arguably operational safety encompasses not only avoiding or at least minimizing fatal accidents but also avoiding or minimizing accidents in which people get hurt but do not die. This does not make the specification necessarily inadequate, however. Perhaps it is known on the basis of statistical evidence, for example, that in this type of installation there is a strict correlation between the probability of fatal accidents and the probability of accidents only leading to injuries, so that minimizing the one implies minimizing the other. In that case, the specification may still be adequate. In other situations, it may be inadequate and it might be necessary to add a design requirement related to minimizing non-fatal accidents.

We can now also see why the specification of animal welfare in the EU Council Directive 88/116/EEC in the example with which I started may strike us as inadequate (see Fig. 20.3). It translates only one of the more general norms for animal welfare into specific design requirements and neglects the others. Therefore meeting the formulated design requirements hardly seems to amount to a sufficient response to the value of animal welfare in the design of chicken husbandry systems.

## 20.5 Conclusions

In this paper I have discussed the values hierarchy and the relations of specification and *for the sake of* as ways to relate general and abstract values to specific design requirements that can guide the design process. These conceptual tools can be used to translate values into more specific design requirements. They may also be used to reconstruct for the sake of which values certain design requirements are pursued. Usually values hierarchies will be constructed by a combination or iteration of bottom-up and top-down moves, so adding an element of reflection and critical discussion to the formulation of both values and design requirements in the design process.

As we have seen, the specification relation is non-deductive and context-dependent. It implies certain value judgments. Although I have proposed certain criteria to judge the adequacy of a specification, often more than one specification will be reasonably defensible. Given that in design usually one specification has eventually to be chosen, one might wonder how to choose between competing reasonable specifications or how to deal with disagreements between the different parties involved in design about the specification to be used in the actual design process. For the moment I only want to point out that the approach proposed in this paper at least helps to trace more precisely the value judgments and possible disagreements about them, even it does not offer a way to solve these conflicts.

More precisely, the reconstruction of a values hierarchy makes the translation of values into design requirements not only more systematic, it makes the value judgments involved also explicit, debatable and transparent. They become explicit in the specific translations that are made between the different levels of a values hierarchy. This explication creates room for critical reflection on the translations made and makes these debatable among the parties involved. Moreover a values hierarchy may be helpful in pinpointing exactly where there is disagreement about the specification of values in design. Finally, a values hierarchy may, once the designers have chosen a specific specification, make those choices, and especially the implied value judgments, more transparent to outsiders. This is important because design usually impacts on others besides the designers. Although transparent choices are not necessarily better or more acceptable, transparency seems a minimal condition in a democratic society that tries to protect or enhance the moral autonomy of its citizens, especially in cases that design impacts the lives of others besides the designers, as is often the case.

**Acknowledgments** An earlier version of this paper was presented at the Forum on Philosophy, Engineering, and Technology (fPET-2010), Colorado School of Mines, Golden, CO, USA, May 9–10, 2010. I am grateful to NIAS, the Netherlands Institute for Advanced Study, for providing me with the opportunity, as a Fellow-in-Residence, to work on this paper during my stay in the academic year 2009–2010.

## References

Anderson, E. (1993). *Value in ethics and economics*. Cambridge, MA: Harvard University Press.

Cross, N. (2008). *Engineering design methods. Strategies for product design* (4th ed.). Chichester: Wiley.

Dancy, J. (2005). Should we pass the buck? In T. Rønnow-Rasmussen & M. J. Zimmerman (Eds.), *Recent work on intrinsic value* (pp. 33–44). Dordrecht: Springer.

Flanagan, M., Howe, D. C., & Nissenbaum, H. (2008). Embodying values in technology. Theory and practise. In J. Van den Hoven & J. Weckert (Eds.), *Information technology and moral philosophy* (pp. 322–353). Cambridge: Cambridge University Press.

Friedman, B., & Kahn, P. H., Jr. (2003). Human values, ethics and design. In J. Jacko & A. Sears (Eds.), *Handbook of human-computer interaction* (pp. 1177–1201). Mahwah: Lawrence Erlbaum.
Harrison, R. (1964). *Animal machines; The new factory farming industry*. London: Vincent Stuart.
Harrison, R. (1993). Case study: Farm animals. In R. J. Berry (Ed.), *Environmental dilemmas: Ethics and decisions* (pp. 118–135). London: Chapman & Hall.
Keeney, R. L. (1992). *Value-focused thinking: A path to creative decision making*. Cambridge, MA: Harvard University Press.
Keeney, R. L., & Raiffa, H. (1993). *Decisions with multiple objectives: Preferences and value tradeoffs*. Cambridge/New York: Cambridge University Press. (Original edition published in 1976 by Wiley)
Kuit, A. R., Ehlhardt, D. A., & Blokhuis, H. J. (1989). *Alternative improved housing systems for poultry*. Beekbergen: Ministry of Agriculture and Fisheries of the Netherlands, Directorate of Agricultural Research.
Richardson, H. S. (1997). *Practical reasoning about final ends*. Cambridge: Cambridge University Press.
Sclove, R. E. (1995). *Democracy and technology*. New York: The Guilford Press.
Van de Poel, I. (1998). Why are chickens housed in battery cages? In C. Disco & B. van der Meulen (Eds.), *Getting new technologies together. Studies in making sociotechnical order* (pp. 143–178). Berlin: Walter de Gruyter.
Van den Hoven, J. (2007). ICT and value sensitive design. In P. Goujon, S. Lavelle, P. Duquenoy, K. Kimppa, & V. Laurent (Eds.), *The information society: Innovation, legitimacy, ethics and democracy in honor of Professor Jacques Berleur S.J.* (pp. 67–72). Boston: Springer.
Vincenti, W. G. (1990). *What engineers know and how they know it* (Analytical studies from aeronautical history). Baltimore/London: The John Hopkins University Press.
Zimmerman, M. J. (2004). Intrinsic vs. extrinsic value. In E. N. Zalta (Ed.), *The Stanford encyclopedia of philosophy* (Fall 2004 ed.). Stanford: Stanford University. URL: http://plato.stanford.edu/archives/fall2004/entries/value-intrinsic-extrinsic/

# Chapter 21
# Engineering Hubris: Adam Smith and the Quest for the Perfect Machine

**Scott Forschler**

**Abstract** Adam Smith observes, in his *The Theory of Moral Sentiments*, that humans are often driven to perfect a tool or device along certain parameters far beyond the requirements of visible practicality. At times this drive has led to important technological breakthroughs, but at others has led to frustration or even disaster, as other goals are neglected and the perfect becomes the enemy of the good. I illustrate Smith's point with four case studies, three of which led to failure, and one in which the perfectionist drive ended in success: (1) The quest to build a sea-level canal in Panama, continuing the "conquest of nature" theme which was successful in Suez, but which led to ruin in the more hostile geography of the new world; (2) Buckminster Fuller's plan to build inexpensive prefabricated "Dymaxion Houses," scuttled by his endless demands for perfection and complete control over implementation; (3) The quest for a reusable spacecraft, ideally with a "single stage to orbit," which led to the adoption of the expensive and dangerous space shuttle, when refinements of older technologies have proven cheaper and more reliable; and (4) John Harrison's successful creation, after decades of intensive work and the rejection of many good but imperfect prototypes, of a reliable timepiece for solving the "longitude problem."

**Keywords** Adam Smith • Perfectionism • Buckminster Fuller • Panama Canal • Longitude problem

> Neil MacGregor: So what we're looking at in this chopping tool is the moment at which we became distinctly smarter and with an impulse not just to make things, but to imagine how we could make things 'better'.

S. Forschler (✉)
Independent Scholar
e-mail: scottforschler@gmail.com

D.P. Michelfelder et al. (eds.), *Philosophy and Engineering: Reflections on Practice, Principles and Process*, Philosophy of Engineering and Technology 15,
DOI 10.1007/978-94-007-7762-0_21, 

David Attenborough: This object sits at the base of a process which has become almost obsessive amongst human beings. … I think the man or woman who held this, made it just for that particular job and perhaps got some satisfaction from knowing that it was going to do it very effectively, very economically and very neatly.

—"Olduvai stone chopping tool," episode 2 of A History of the World in 100 Objects (podcast).

Adam Smith observes, in *The Theory of Moral Sentiments*, that humans are often infected with a powerful drive to perfect a "system or machine" to a point far beyond what economic or other practical needs would deem prudent. The perfection of the means becomes its own end, one more pleasurable to contemplate, and a stronger motivation to action, than the original end of the invention itself. Indeed, we are often far more impressed by lively descriptions of the means by which a device works than by a description of the human ends it is designed to satisfy, which may themselves be fairly limited in number and comparatively mundane. The flip side of this is that we may be vexed with a means which satisfies some end, but not in what seems the most fitting or perfect manner, which may motivate us to try to improve it even at the risk of interfering with the original end it was intended to satisfy, though with some hope of ultimately satisfying this end more efficiently or effectively.

Smith depicts our interest in perfecting the means to our ends as a kind of deception our imagination can play upon us:

> If we consider the real satisfaction which all these things are capable of affording, by itself and separated from the beauty of that arrangement which is fitted to promote it, it will always appear in the highest degree contemptible and trifling. But we…naturally confound it in our imagination with the order, the regular and harmonious movement of the system, the machine or oeconomy by means of which it is produced. (Smith 1759/1790: §IV.i)

This deception can of course be highly productive, and Smith readily grants that it "rouses and keeps in continual motion the industry of mankind," and is doubtless the source of many tremendous achievements. I would go further and speculate that the ability to analyze the means to some end *as a means*, taking its fitness to the end rather than the end itself as our object of attention, is perhaps one of our most distinct mental capacities, a powerful biological adaptation which sets humans apart from other animals. Furthermore, an overabundance of this attentiveness towards means rather than ends, and the drive to perfect their relationship to some end, may be what sets apart from others some of the most innovative and creative members of our species, including engineers, scientists, designers, and artists.

But for all the benefits this capacity gives us, it can be highly counter-productive, if the perfect becomes the enemy of the good. We often strive to perfect some system or device in ways that provide relatively small advantages compared to the inconveniences we cause ourselves or others in the attempt. Smith considers a relatively trivial example of this, where the owner of a chamber, finding the servants have left all the chairs in the middle, becomes so vexed by the imperfect arrangement that he goes about setting the chairs around the outer walls by himself, so that the room can be traversed more efficiently—even if there is no immediate prospect of any such traffic, and all the owner previously wished to do was to sit in a chair to relax, which he could have done without any trouble.[1]

All of us may occasionally fall into such behavior, but some of us make a career or life out of such obsessions, perhaps leading to more serious harm if they get out of hand. The stereotype of the "mad scientist" who pursues scientific knowledge in itself regardless of the damage caused, or the morbidly perfectionist artist whose genius gets lost in projects that no one else cares about, doubtless have some basis in reality. Engineers, too, may be particularly prone to such obsession, and one substantial concern of professional engineering ethics is to keep the perfectionist drive from getting out of hand. I turn now to three cases where engineering perfectionism derailed potentially useful projects, as well as a final one where it led to much happier results.

## 21.1 Case 1: The Panama Canal

A canal between the Mediterranean and Red Seas at Suez was long a dream of Europeans seeking to access the Far East; by the 1860s, new technology and increasing global trade made it seem feasible, and a French company was incorporated for the task. In this age conquering geographical obstacles by canals and railroads was all the rage, yet this bold undertaking might have been a complete disaster except for the fortuitous invention of dynamite by Alfred Nobel in 1867. This sped up the project considerably, as it was the perfect tool for blasting through the Sinai's hard rock, and by 1870 a complete sea-level link had been cut. The canal was a spectacular success, generating enormous profits and dramatically altering commercial and political history.

The company's founder, Ferdinand De Lesseps, soon turned his attention to Panama, the other great isthmus in the heart of a hemisphere, creating a new company in 1879 to duplicate his success at Suez. Here too he insisted on a sea-level canal, which would not only minimize transit times, but exemplify the uncompromising conquest of natural obstacles. Rather than working with the landscape, the plan was to simply obliterate that part of it which posed a barrier to human goals (McCullough 1977: 237–239). This was a portentous requirement, for the Panamanian Isthmus was the only point in Central America narrow enough to make a sea-level canal feasible, though there were other potential sites for a multi-lock canal, most notably a transit via Lake Nicaragua, which a later study identified as the cheaper route had it been picked from the beginning.

However Panama proved much less tractable than Suez. Instead of dry desert near population centers which could provide labor, it was a thick, wet, malaria-infested

[1] A similar example of obsessive perfectionism in a trivial case might be found in the dishwasher loading scene from the 2008 film *Rachel Getting Married*, based on a real-life dispute between Bob Fosse and the scriptwriter's father, who once hotly contended over the most efficient way to load a dishwasher.

jungle. Workers unfamiliar with local conditions had to be imported, often dying of disease. Instead of hard rock to blast through with clean cuts, the underlying material was largely shale, which slid back into the canal over time, requiring extensive re-blasting, wider cuts, and frequent dredging which continues to this day. The highest point along the route was 330 ft above sea level, almost seven times greater than the 50 ft maximum height at Suez. Unlike Suez, no miracle invention came along to make the construction radically easier, and the builders saw nothing ahead but more hideous difficulties. The company went bankrupt in 1889 without coming close to completing the canal (McCullough 1977: 231–232).

A decade later, the US occupied both Cuba and the Philippines in the Spanish-American War, highlighting its strategic interests in an inter-ocean connection and motivating the resumption of the project in spite of its economic infeasibility. A survey of available canal sites showed that while Nicaragua might have been cheaper from the beginning, the infrastructure and initial digging in Panama now made the latter route somewhat cheaper—but only if a lock canal was built instead of the hopeless sea-level one (Miner 1940: 116). The US took over the operations, aided by some medical discoveries which helped control malaria and other tropical diseases, and opened the locks in 1914—just in time to facilitate military transport during WWI.

De Lesseps's obsession with a sea-level canal epitomizes how a vision of ideal fitness between the means and the goal can interfere with the goal itself; he wanted to bend nature to his will more than he wanted to improve transport or economic efficiency. He brought ruin upon thousands of investors and laborers' lives until a new set of designers accepted a compromise with geographic reality.

## 21.2 Case 2: The Dymaxion House

Buckminster Fuller, an eccentric twentieth century architect and inventor, is best known for his creation of the geodesic dome. But the dome was merely the most famous end-product of his obsession with unconventional solutions to conventional problems which began in the 1920s. He was particularly interested in using tension structures to partly replace the almost exclusive use of compressive structures in buildings. Since many modern materials, especially metal wire, were much stronger in tension than in compression, he proposed a variety of designs leveraging this strength by suspending a structure by wires hung from a central mast, concentrating the compressive element instead of distributing it throughout the structure. This permitted most of the structure to be much lighter and hence less expensive. Utility pipes could also be built into the central mast, and modular rooms suspended around it could be replaced or upgraded over time. The components could be manufactured in one place, shipped in standard compact containers to remote sites, and quickly assembled by following simple instructions (Sieden 1989: 125–129).

Fuller's initial designs considered multi-family and office units, but eventually focused on a single-family dwelling house, in the form of a symmetrical cylinder around the central mast. Inspired by the aluminum grain silos just starting to dominate the rural Midwestern landscape, he switched from a hexagonal to a cylindrical design in his "Wichita House," which was almost put into mass production. But he was also fond of a more general name for his housing designs: "Dymaxion Houses," combining the words dynamic, maximum, and ion.[2]

His unconventional design posed many challenges; a common criticism is that it is hard to hang a picture or mirror on the inside of a curved wall. Both furniture and city lots tend to be rectangular, and rectangular houses accommodate both better than curved structures. While the cost per enclosed volume or floor area could be lower for a Dymaxion House than a conventional one, not all of the extra space can be so easily used, at least not without changing assumptions. Utility and repair contractors also may have balked at working with the radical design of the Dymaxion House. The massive cost savings Fuller promised could only be realized through mass production, forcing significant uniformity in design, which might have proved stifling for consumers seeking to express their individuality. Still, if the houses were cheap enough, many would have accepted the trade offs, and if enough houses could have been produced customers and contractors alike might have gotten used to them. It is hard to imagine aluminum mushroom-shaped houses dominating the market, yet the growth in shoebox-style mobile homes shows that at the right price, such things can sell, aesthetics be damned. Even if the Dymaxion House only satisfied a niche market, a niche out of the post-WWII housing market might have been large, and could have provided many families with a unique alternative to the ticky-tacky options they were otherwise offered. A sober estimate of the final cost of the Wichita House was $6,500, a bargain compared to the $12,000 average home cost of the day, especially given that the components could be shipped anywhere and assembled on site faster than any conventional house could be built from the ground up (Sieden 1989: 282).

But we will never know if the Dymaxion House could have served its time, because Fuller's perfectionism scuttled the plan. In 1944 he made arrangements with a Wichita airplane factory to build a prototype. The factory's demand for airplanes was already shrinking, and it was ready to retool for Dymaxion Houses if the right funding were available. Fuller formed a company for the purpose and found many eager investors. While doing this, he noted his conviction that radical new ideas of this sort have a 25-year gestation time; since his original designs were made in 1927, this foretold a maturation date of 1952 (Sieden 1989: 272).

Anyone who took this as just an estimate or a suggestion was in for a rude awakening. Fuller was obsessive about the date 1952, and insisted that the initial Wichita House was simply not ready. He wasn't going to put his trademark name on a mass-produced product which he knew could be improved, and was hopeful about using the more advanced materials he thought would be available in a few years, feeling the current model was an irksome compromise with imperfect materials. He also insisted upon designing furniture, utilities, and all other house components himself

[2] He rather liked this word, and later applied the adjective to many of his other inventions, including many which were "dynamic" in at best a metaphorical sense.

so as to solve all potential problems in advance, instead of putting out a product that others could add to and improve over time. He constantly put off production by pointing out valid though minor imperfections in the then-current design. The other investors eventually sold their shares back to Fuller in frustration, and the construction plan collapsed (Sieden 1989: 283). Fuller did not restart it in 1952 or any other time. Perhaps this was because by 1952 the window of opportunity had been lost and people had lost faith in his unconventional vision. Perhaps Fuller gave up the idea of a fully custom-designed house as too intractable, to focus instead on his geodesic dome, which was more easily constructed and filled some significant though limited construction niches. In any case, the Dymaxion Houses dotting the American suburban landscape vanished into the realm of what might have been.

As with De Lessep's vision of a sea-level canal, Fuller's perfectionism derailed what might have been a useful product. In both cases, there was a great temptation to cling to a vision of an ideal solution to an outstanding human problem, long after it should have been clear that the vision was infeasible, and stood in the way of a compromise solution which would have satisfied many, if not all, of the envisioned ends at a more acceptable cost.

## 21.3 Case 3: Single-Stage-to-Orbit (SSTO), and the Space Shuttle

It takes a lot of energy to get a payload from the ground into orbit. The main problem is not the height, but the kinetic energy required: objects in Low-Earth Orbit must accelerate to almost 8 km/s in just a few minutes after launch. If the energy is obtained from the most efficient chemical reactions, physics demands that over 90 % of the launch vehicle must consist of fuel, the remainder being the engines, fuel tank, and payload. Furthermore, getting most of this essentially hollow structure into orbit would waste energy, so it is easier just to discard used-up parts along the way. Hence the standard launch design of the space age has been a rocket with two or three stages, stacked on top of each other to minimize air drag. Early stages use up their fuel and associated engines, which are then discarded and fall into the ocean after doing their job of accelerating the remainder of the craft. Staged engines can also each be optimized for the particular speed, air pressure, remaining vehicle weight, and other characteristics of the different launch segments at which they will be used (Bell 2005).

Fuel tanks are not terribly expensive to build, but rocket engines are; and it seems a shame to use up several on each launch. It is tempting to think that if some engine configuration could be built to handle the entire flight profile, and better yet could be reused in future launches, money could be saved and more missions launched. More idealistic dreamers, encouraged by science fiction, lusted for a true space "ship," which like oceanic ships, could be self-contained, plotting their own course with only minimal maintenance between each voyage (Butrica 2003: 3).

This leads to the recurring dream of a Single-Stage-to-Orbit (SSTO) vehicle. Even the most idealistic engineers admit it cannot be done—at least not with chemical rockets. The problem of reaching orbit is just too hard. Indeed, a close parallel here exists between reaching orbit, and crossing the oceans at Panama. It would be nice if we could do it all at once. But given the physical constraints of the situation, we have to settle for doing it piecemeal, with multiple, throw-away stages, or with locks, in the respective cases. Still, the USA's National Aeronautics and Space Administration (NASA) made a momentous decision several decades ago to build something as close to SSTO as could be done with then-current technology: the Space Transportation System (STS), also known as the Space Shuttle.

If we define a rocket "stage" as being a disposable rocket element (engine + fuel tank), then the Shuttle is a "stage and a half" launcher, for it uses and then discards a pair of solid fuel boosters (one stage), as well as an external liquid fuel tank (one-half stage). It retains the engines which burn the liquid fuel intact into orbit for complete re-use on the next mission, along with the sizable payload vehicle, which on re-entry uses aerodynamic braking until it slows down enough to glide to a landing strip. Such a landing is not only more graceful than the old parachute-then-slam-into-the-sea recovery many remember from the Apollo missions, but doubtless is also less jarring to complex equipment that could then be reused on the next mission. With the hope of short turnarounds, some STS promoters thought that each Shuttle could make dozens of flights a year, significantly reducing the costs of space flight through reusability and sheer volume.

There are many criticisms of the Shuttle program; I will focus here only on those relevant to the quest for reusability which it shares with the SSTO dream. It turns out that reusability is not all that it is cracked up to be, and comes with its own costs. The weight and glide-style reentry profile of the shuttle means that it cannot feasibly be set on top of a rocket booster, and must instead be attached to its side. Hence its reentry heat shielding is exposed to both the atmosphere and the booster structure, a vulnerability which doomed the Columbia. In contrast, conventional manned rockets use a much smaller reentry vehicle whose heat shield is entirely contained within the larger craft until long after exiting the atmosphere. In general, the design for a vehicle which satisfies criteria for the radically different profiles of launch and re-entry is enormously complicated, adding to both cost and risk of catastrophic failure. A massive labor team is needed to refurbish the Shuttle between launches, checking not only the heat shields but millions of other components to make sure that no damage from the launch, re-entry, landing, orbital micrometeors, or many other hazards poses a risk to the next mission. The cost of such maintenance dwarfs the cost of both fuel and the structural components for building "cheap, dumb" rockets for each new launch (Taylor 2004: § 12), so that the per-pound payload launch costs of the Shuttle exceed those of most conventional launch systems.

Whether the recently-ended STS program was a vision ahead of its time, or a mistaken attempt at a complete impossibility, is open for debate. But for the time being, barring dramatic new developments in materials science or energy resources, the future of space travel may involve a return to tried-and-tested methods. A multi-stage rocket, like the locks of the Panama Canal, is a concession to reality which we may have to live with for the foreseeable future.

## 21.4 Case 4: The Longitude Problem

Not all perfectionist obsessions lead to a bad end, so I will end with the happier story of John Harrison, the clockmaker who strove in the eighteenth century to build an exact timepiece for oceangoing ships. The demand for precision was great. The British Empire ruled the seas, but its fleet suffered from a long-standing problem. It was easy to calculate one's latitude by noting the angle of fixed stars with the horizon, but with no landmarks in the open ocean, calculating one's longitude was absolute guesswork, unless one knew the exact time and could combine this data with astronomic observations to determine one's distance east or west of known points. After months at sea, navigators often misjudged their location by more than an entire degree of longitude, or 60 nautical miles, often leading to wasted time or serious mishaps.

Some clocks could tell time fairly accurately if fixed in place and undisturbed by the elements; but the roll and pitch of the sea, changes in temperature, and other interrupting factors made them hopeless for ocean travel. The problem was so urgent that in 1714 the British government offered a substantial monetary prize for anyone who could determine a ship's longitude within half a degree. For a transatlantic trip, a ship-borne clock would have to gain or lose no more than 3 s a day, an almost inconceivable accuracy (Sobel 1995: 58–59).

The prize attracted both clever inventors and crackpots, which the public was hard-pressed to tell apart, often seeing the longitude problem the way we look at perpetual motion or cold fusion proposals. But John Harrison fell firmly into the former category. An expert clockmaker, he had already built a surprisingly accurate and durable pendulum clock made almost completely out of wood in 1713, before he turned 20 (Sobel 1995: 64). Obsessed with all things involving time, he was convinced that he could solve many of the problems with existing clocks with careful design of drive mechanisms and selection of materials.

Harrison's first major attempt, H-1, was completed in 1737. Weighing 75 lb, it was accurate within a few seconds per day, enough for a fair chance at the prize. But vividly aware of several correctable flaws in the clock, Harrison refused the trial and asked the prize committee only for some seed money to build a new machine with even greater accuracy.

In 1741 he completed H-2, a clock similar in size but with dramatic innovations including a more uniform drive mechanism and better compensation for temperature changes. The prize committee was impressed and eager to put it to the test; but Harrison was visibly disgusted with the machine, insisting once again that it had substantial flaws demanding correction (Sobel 1995: 85–86). He retired to his workshop and by 1757 finished H-3, whose innovations included a steadier drive mechanism, the world's first bimetallic strip to compensate for temperature variations, and ball bearings to reduce rolling friction (Sobel 1995: 103–104). But apparently inspired by a pocket watch maker who showed that some of Harrison's innovations could be implemented in these more compact devices, Harrison put the test off another 2 years to build H-4, which weighed only 3 lb, yet with an accuracy comparable to the far more massive H-3.

By this time Harrison was getting stiff competition in the form of the "lunar distance method" for calculating the time based on the precise location of the moon with respect to the fixed stars. Attaining the needed accuracy required the compilation of exhaustive observations that took expert astronomers decades to complete, translated into compendious tables carried on board a ship. The required astronomical angles could then be measured precisely with the newly invented sextant, and after some poring over the tables the current time, and thence the longitude, could be determined (Sobel 1995: 88–99). Compiling the necessary data was hardly easier than building a precision clock, yet astronomers were gradually completing the necessary calculations.

Had Harrison eyed only the money, he would have been well advised to make an earlier trial; but he delayed this long enough that his later works were fiercely criticized by champions of the increasingly feasible lunar distance method, delaying further the public recognition of his achievements now that he was finally ready for it. In the end his perfectionism won out, and Harrison's H-4 passed the original criteria for the longitude prize with flying colors in the mid 1760s. Sadly, he never officially won the prize (it was eventually rescinded without having been awarded), although in 1773 Harrison was awarded a significant grant in compensation for his work, in addition to other smaller grants received along the way. But it is safe to conclude that Harrison was no more working for money than Buckminster Fuller was two centuries later; rather, money was simply a means to the end of making the most perfect device which the inventor could imagine and implement by stretching current technology to its limits. Where Fuller failed, Harrison succeeded. Navigators, and anyone else who relied upon the precise measurement of time, owed him much gratitude for years to come.

## 21.5 Conclusion

No single moral or practical lesson can be drawn from the comparison of these cases. Sometimes obsession with perfection pays off; and sometimes it does not. It is both useless and simply false to say that one should only seek perfection if you are good enough to pull it off, for it is doubtful that Buckminster Fuller was any less of a genius or less competent than John Harrison.

Surely an irreducible contingency explains some of the difference; we cannot predict which of many obsessive approaches to innovation will be successful, since after all we are dealing precisely with the unknown, the untried, and the unconventional. We need some obsessive innovators now and then to go against the grain of ordinary thought and practice, breaking our paradigms just when our habits have become most settled. And for every such success, we must tolerate dozens or hundreds of failures or crackpots, as in the case of the Longitude problem. But perhaps we look more kindly on Harrison not merely because he succeeded, but also because his success was not terribly costly, nor would his failure have been. The loss of a few thousand pounds of government venture capital in a risky attempt to solve the problem is more morally affordable than the unnecessary deaths of thousands of workers digging a useless ditch in Panama, or the highly visible loss of trained astronauts or billions of dollars in a reusable rocket whose complexity becomes a money pit.

There is also a difference between putting trust in one's own capacity to solve specific problems, and in the hope of future complementary inventions that will solve your current problems in the nick of time. De Lesseps's foolhardiness was saved once by the unpredictable invention of dynamite, but no similar good fortune would facilitate his second attempt to conquer nature. Fuller looked forward to inserting more advanced materials and technologies into his vision, but they weren't quite there when he needed them. In contrast, Harrison had more concrete ideas about exactly how to solve the problems he saw in his past designs, and he needed only to rely upon his own capacity and determination to innovate, rather than on, we might say, the kindness of strange inventors.

Nor can we conclude anything specific about the dangers of working as a lone inventor as opposed to working on a team. As much as we'd like to believe that many heads can see errors that one may miss, we also know that group-think can lead the latter astray, as it did in the Shuttle program, while Harrison did pretty well on his own recognizance. Still, Fuller clearly would have benefitted from attending more to the practical demand of his investors, and the backers of the STS and De Lesseps would have been well advised to take more seriously existing doubts about the feasibility of the complex proposals they funded. We might learn something here from a brief consideration of James Watt's efficient steam engine, which was furthered by a particular kind of teamwork between an obsessive inventor and a practical businessman.[3] Watt was often obsessive about trying every possible combination of design elements to make his improved engine maximally efficient, while his financier Matthew Boulton pushed for the need to build and sell models that worked, despite their imperfections (Scherer 1965: 176–177). Ultimately Boulton financed Watt's continued design work on the condition that workable models were built and sold at various stages, which of course attracted both interest in and more capital for future improvements.[4]

[3] Thanks to Stephen Goldman for bringing this case to my attention.

[4] After initially submitting this paper, I discovered Matthew Crawford's delightful work on the value of physical craftsmanship, which sheds further light on the necessary and valuable tension between what he calls the craftsman's or engineer's "fiduciary responsibility" to the customer (whether an individual or a whole society) and one's "metaphysical responsibility" to the machine or project under repair or design (Crawford 2009: 117). He contrasts the selfishness of the "idiot" mechanic who doesn't really understand or care about his craft, and does just enough to "get by," with the opposing selfishness of the obsessive craftsman's "tunnel vision," unable to see any goals except his perfectionist ones. While idealists often scoff at the market's tendency to reduce all values to monetizable ones, at its best it provides the constraint of "agreement and convention" guiding us into a saner, middle position, turning our mechanical obsessions into the service of shared human goals (Crawford 2009: 124). The Boulton-Watt partnership clearly exemplifies this kind of effective compromise.

Certainly one lesson which many engineers wisely follow is that it is never enough to optimize a single output variable, like transport time, number of parts, or simplicity; economic, political, and human safety factors must all enter into our final decisions about what to do. We can sometimes forget about the multiplicity of goals engineering projects must satisfy when our eye is caught by one or a few that seem so tantalizingly close to perfection. We must neither entirely succumb to nor disparage our very human propensity to become obsessed with the perfection of the fitness between means and ends, and must hope that we can turn this quest to the service of our common goals rather than become slaves to the perfectionist impulse at the latter's expense.

## References

Bell, J. (2005). The cold equations of spaceflight. *Space Daily*. http://www.spacedaily.com/news/oped-05zy.html. Accessed 26 Oct 2010.

Butrica, A. (2003). *Single stage to orbit: Politics, space technology, and the quest for reusable rocketry*. Baltimore: The Johns Hopkins University Press.

Crawford, M. (2009). *Shop class as soulcraft: An inquiry into the value of work*. New York: The Penguin Press.

McCullough, D. (1977). *The path between the seas: The creation of the Panama Canal, 1870–1914*. New York: Simon and Schuster.

Miner, D. C. (1940). *The fight for the Panama route: The story of the Spooner act and the Hay-Herran treaty*. New York: Morningside Press.

Scherer, F. M. (1965). Invention and innovation in the Watt-Boulton steam engine venture. *Technology and Culture, 6*(2), 165–187.

Sieden, L. S. (1989). *Buckminster fuller's universe: An appreciation*. New York: Plenum Press.

Smith, A. (1759/1790) *The theory of moral sentiments*. http://www.econlib.org/library/Smith/smMS.html. Accessed 25 Jan 2012.

Sobel, D. (1995). *Longitude: The true story of a lone genius who solved the greatest scientific problem of his time*. New York: Penguin.

Taylor, P. (2004). *Why are launch costs so high?* http://home.earthlink.net/~peter.a.taylor/launch.htm. Accessed 26 Oct 2010.

# Chapter 22
# The Technology of Collective Memory and the Normativity of Truth

**Kieron O'Hara**

**Abstract** Neither our evolutionary past, nor our pre-literate culture, has prepared humanity for the use of technology to provide records of the past, records which in many contexts become normative for memory. The demand that memory be true, rather than useful or pleasurable, has changed our social and psychological self-understanding. The current vogue for lifelogging, and the rapid proliferation of digital memory-supporting technologies, may accelerate this change, and create dilemmas for policymakers, designers and social thinkers.

**Keywords** Memory • Lifelogging • Privacy • Social networking • Sensecam

## 22.1 Introduction

The relationship between memory, representation and recollection is highly unusual and counterintuitive. In particular, memories can misrepresent past events in what would seem to be all key respects, and yet still facilitate immediate recognition of veridical representations (e.g. video footage of an event). Many psychologists (Loftus and Palmer 1974; Wells 1993) have been able to show that eyewitnesses can be deeply unreliable in recall, especially if misled by queries or interfering information, yet this does not preclude accuracy in identification. The fact that a person was misremembered as having dark hair and a moustache does not mean that he might not be recalled with the shock of recognition: "yes, that's the fellow!"

Clearly, the 'filing cabinet' metaphor of memory (that it contains a set of representations of the past, organized to facilitate retrieval, such that exposure to a suitable cue

K. O'Hara (✉)
School of Electronics and Computer Science, University of Southampton, Highfield, Southampton SO17 1BJ, UK
e-mail: kmo@ecs.soton.ac.uk

D.P. Michelfelder et al. (eds.), *Philosophy and Engineering: Reflections on Practice, Principles and Process*, Philosophy of Engineering and Technology 15,
DOI 10.1007/978-94-007-7762-0_22, 

will facilitate recall) is as inappropriate as it is naïve (cf. Warnock 1987, 8–9). Memory is constantly changing, in response to conversations with other people about events, constant narration of events by oneself and others, exposure to news reports, photographs, and videos, and inference from the effects of the remembered event. My memory of an event may misrepresent the non-moustachioed man, but once I have seen a photograph of him, I realise that he had no moustache, and my memory adapts accordingly.

As the old Maurice Chevalier song had it,

*We met at nine.*
*- We met at eight.*
*I was on time.*
*- No, you were late.*
*Ah yes, I remember it well.*

The joke here is that the two singers have completely opposite recollections of a significant event in their lives, and yet agree entirely on its identification. As Marcel Proust (still one of the most acute theorists of memory) argued, one's memories are coloured by one's present assumptions and mental models; an apparently insignificant event can appear significant in retrospect because it contained a first encounter with a person whom one later came to love.

In this chapter I shall discuss the use of technology to support recollection. In particular, one often uses representations such as photographs to support recall. I shall make the obviously idealizing assumption that a photograph does not misrepresent the past in the way that a memory can; the camera was pointed and the image captured. Of course images can be Photoshopped, but that requires human intervention to cause the misrepresentation. Further, images can give a false impression, as for example when a trick of perspective makes a distant large object look near and miniature; again, the misrepresentation requires a human interpreter. As a matter of fact I do not think that mechanical reproductions are essentially veridical representations, but it will make the argument simpler and clearer if we pretend that they are, in contrast with human memories which may or may not be veridical.

I will focus on what is normative for memory, and shall argue that the use of technology has increased the prominence of truth in that role. This is not necessarily a bad thing, but it is a newish development which will continue to drive important social and psychological change as technological support proliferates. These considerations should be used to help drive our reactions and regulations in areas such as privacy, deletion, data protection and informational self-determination.

## 22.2 The Technology of Memory

Human memory has always been a rich source of inspiration and metaphor for computer memory (O'Hara et al. 2006a), but our understanding of human, machine and social memory is converging in ways that are more than metaphorical

(O'Hara et al. 2006b). Memory-supporting technology, which at least initially was conceived as a medical resort, has branched out into the areas of leisure, social networking and self-improvement (Garde-Hansen et al. 2009).

Moore's Law has taken such technology out of the medical arena and into the social. The fact that one can more or less store anything one likes means that recording requires a very low cognitive overhead – one needn't worry about the extremely tedious tasks of choosing what information to store, or deciding what to delete when the memory gets full. Meanwhile, improved search and retrieval techniques mean that one can find what one needs relatively straightforwardly. One can, in short, use memory technology indiscriminately – which makes it usable (O'Hara et al. 2009).

Furthermore, the indiscriminate use of such technology chimes in with the associative ways that human memory works. We store all sorts of pieces of 'useless' information, precisely because we do not know at storage time what will be useful in the future. The guesses we make about what memories are likely to be important in the future are unlikely to be right all the time, so the more raw material that is present in our records of the past, the more likely we are to have everything that is useful (Bell and Gemmell 2009).

It has been calculated that it would be straightforward to store 70 years of high quality video taken from a lifetime (Dix 2002); this has prompted the United Kingdom Computing Research Committee[1] to propose 'Memories for Life' as a Grand Challenge for computing research (Shadbolt 2003; O'Hara et al. 2006b) – in other words, a potentially epoch-making area for research where breakthroughs would promote not only computer science, but also social well-being in a wide population (http://www.ukcrc.org.uk/grand-challenge/current.cfm). As a Grand Challenge, research groups have been coalescing in this area, looking for examples of the use of machines to act as companions for humans (Wilks 2010; O'Hara 2010a), or the difficulties for archivists in curating the digital records of noteworthy people.[2] Elsewhere, special-purpose tools have been helping communities use websites as collective memory resources.[3]

Prosthetic memory has been a major area of research. For instance, one device, the SenseCam developed by Microsoft,[4] is a small digital camera designed to take photographs passively, without user intervention, while it is being worn around the neck. It has no viewfinder or display to frame photos, but instead is fitted with a wide-angle lens that maximizes its field-of-view, ensuring that nearly everything in the wearer's view is captured (Hodges et al. 2006). To review the SenseCam output, it is remarkably effective to run the resulting set of pictures as a speeded-up movie (De Bruijn and Spence 2002).

---

[1]An expert panel of the British Computer Society, the Council of Professors and Heads of Computing, and the Institution of Engineering and Technology to promote computing research in the UK (http://www.ukcrc.org.uk/about/index.cfm).

[2]http://www.bl.uk/digital-lives/

[3]See e.g. http://www.bbc.co.uk/dna/memoryshare/ or http://www.livememories.org/Home.aspx

[4]http://research.microsoft.com/en-us/um/cambridge/projects/sensecam/

SenseCams have been shown to have remarkable positive effects on the memories of at least some sufferers of severe memory impairment (Berry et al. 2007). However, these and similar devices are also used more and more frequently to record the behaviour of those with non-impaired memories, either to achieve an objective picture of real-life behaviour (of great value, for example, in market research – cf. Byrne et al. 2008), or simply to record the quotidian details of daily life (Lee et al. 2008; Doherty et al. 2009).

The practice of using such devices to record daily life in an indiscriminate way is called *lifelogging*. The lifelogger simply uses devices that amass information, and then stores the results. The SenseCam is a special-purpose recording device, but one can also use devices with other functions that generate records as by-products; mobile phones, Web browsers, e-mail programs, social networking sites and medical sensors all generate information that is of potential interest to the lifelogger (especially among younger people with their greater tendency to integrate digital and connected technology into their daily lives – O'Hara et al. 2009).

There are many important pioneers in this space, including Steve Mann who has for many years worn devices to record his daily life,[5] and Jennifer Ringley, who achieved notoriety in 1996 for broadcasting the output of a camera in her bedroom across the Web (the so-called JenniCam – Jimroglou 1999). Perhaps the most committed is Microsoft executive Gordon Bell, who has developed a suite of technologies and practices to deal with the giant quantities of information one can generate in a normal life, and who has written about the potentially transformative effects of such technologies for work, health, and learning, as well as in everyday life (Bell and Gemmell 2009).

If such technologies become more ubiquitous, then they will have social effects with which we all will have to deal. A lifelogging world would be characterised by *universality*, both in terms of a high proportion of people owning extensive records of their lives, and of those digital records covering a high proportion of people's activities, so that more people would have access to more of their past lives. Such records are likely to be relatively *durable*; even though there is always a danger of file formats becoming outdated and unsupported by present-day machines, the greater awareness of this problem in the computing industry means that more adaptable general-purpose standards for representational formats are likely to emerge. There is a strong likelihood that lifelogging records would be *shared*, not only because of the relative ease of copying and transfer compared to non-digital formats, but also because of a greater willingness to use the World Wide Web as a sharing format, for instance on social networking sites (O'Hara et al. 2009). The power of a great deal of information amalgamated from several of one's own devices, the lifelogging stores of others, information from social networks (e.g. Facebook or Flickr) and publicly-available information (e.g. using Google or Wikipedia) could be immense in the provision of a rich picture of one's own life (and, as a by-product, of other people's too).

---

[5] http://www.eecg.toronto.edu/~mann/

## 22.3 The Normativity of Truth for Memory

We (and other animals) have memories because they help the organism survive. Our bodies have mechanisms that allow the world outside to change some of their states, allowing adaptation to, and ultimately recall of, significant episodes. There is no need for those episodes to be represented exactly or accurately; it may be that the value of a fear reflex is greater if it is triggered more often than need be (in other words, that the 'memory' of an organism is more effective if it tends to generate falsely positive identifications of threats). Forgetting also has its own adaptive value when the past event was traumatic. Memory's utility stems from the smooth functioning of the self rather than the veridicality of its representations, as the novelist Sebastian Barry suggests:

> It wasn't so much the question of whether she had written the truth about herself, or told the truth, or believed what she wrote and said were true, or even whether they were true things in themselves. The important thing seemed to me that the person who wrote and spoke was admirable, living, and complete. (Barry 2008, 309)

The use of external objects and constructed aspects of the environment to support memory is relatively recent and has tended to colour our perceptions of what is important about memory. Studies of oral cultures, which lack recourse to permanent representations, show that memory and the reconstruction of the past can have very different properties than we are used to in our technological world (Goody 1998; Ong 1982, esp. 57–67, 95–99, 136–152).

In such cultures, verbatim recall of lists or words is rare – unsurprisingly, as it has very little obvious function in such a society. Early anthropologists occasionally dismissed the memories of 'primitives' as flawed because they had difficulty in regurgitating lists of words – yet of what use is that ability when one has no examinations to pass? Recollection becomes a performance, a creative act. History, for instance, becomes indistinguishable from politics, so that when an elder recites the ancestors of a chief through an implausibly large number of generations, what he is really doing is placing the chief in a political context which makes sense. The 'ancestors' that are mentioned allow connections to be made between important dynasties, and so the elder is not performing an impressive feat of memory, but rather reflecting current power structures. Memory of past events, or of a complex ceremony, is distributed across the participants of the discourse. The aim of mnemonics is to stimulate, not to aid recall. All communication is face-to-face, and so there is no need to leave records for others to use in the future, or to 'speak' to people remotely.

In an oral culture, the whole notion of 'misrepresentation' is up for grabs. What is the truth here, when there is no permanent certified 'truth' or record available for comparison? The 'fact' that the chief's great-great-great-great-great-grandmother is such-and-such will be a matter of the completest indifference to him, and so there will be no attempt to keep any kind of record of it; hence when the elder announces a family connection that everyone accepts, what counts is that it is *acceptable*.

The development of literacy gradually provided a certified record against which individual memories could be compared for accuracy. Written words supported recall, but also furnished an independent standard. Adjustment to the literate world took time. In Plato's *Phaedrus*, Socrates took issue with those who relied on the written word; 'You have not discovered a potion for remembering, but for reminding; you provide your students with the appearance of wisdom, not with its reality' (Plato 1997, 552). This attitude remained for centuries; when Montaigne wrote phrases all over the beams in his tower, this was not to remind him of their content but rather to provoke new and interesting thoughts of his own.

With the assistance of technology, writing and later photography evolved from being simply supports of memory. The inheritances of Gutenberg and Daguerre were the fixed objective records that were widely understood and shared through all levels of society. In such an environment, a new aspect of memory became possible. Memory could be held to account against the public record, and could be held as 'wrong' if it contradicted it. Truth became normative for memory.

This, of course, is a caricature of a number of complex psychological, social, technological and philosophical developments; it is not meant to be a potted history of memory. The point is to argue that the spread of the use of technologies as memory supports has created a situation in which truth is normative for memory in ways that it was not, and could not have been, before those technologies existed; and that to treat truth as normative is to downplay other aspects of memory that could have been and no doubt were important in the evolution of the faculty in both non-human animals and human societies.

## 22.4 Worries About Memory-Supporting Technology and Lifelogging

The recent literature has thrown up some particular persistent worries about lifelogging, related to the perception that a person's lifelog contains truths that the human memory does not have, and that it is therefore reliable in a way that the unenhanced human is not. In particular, these are focused around the development of unbalanced, or psychologically disturbing, images, particularly self-images, and around the privacy of the individual.

As an example of the first idea, legal scholar Anita Allen argues that an 'unredacted lifelog could turn into a bigger burden on balance' because 'electronic memory enables destructive reminding and remembrance' (Allen 2008, 56–57). We would be more prone to dredging up horrible memories from the past. 'The lifelogging concept is insensitive to the therapeutic value of forgetting the details of experience' (Allen 2008, 64). 'The technology will enable excessive rumination by persons experiencing unipolar or bipolar depression' (Allen 2008, 64–65). Political scientist Viktor Mayer-Schönberger agrees that the consequences of this technology are that stupid adolescent mistakes can take on disproportionate significance in later life (2009).

On privacy, Mayer-Schönberger also argues that 'comprehensive digital memory represents an even more pernicious version of the digital panopticon' so that 'the future has a chilling effect on what we do in the present' (Mayer-Schönberger 2009, 11–12). Allen sets out in some detail the argument that saving information about oneself would leave one open to invasions of privacy. Not only could one find oneself under surveillance (or, as it is sometimes termed, 'sousveillance') from lifelogger friends and acquaintances (Dodge and Kitchin 2007, 434–437), but also 'a government that has traditionally enjoyed access to communications and correspondence will want access to lifelogs' (Allen 2008, 67).

The purpose of this chapter is not to argue that these worries are unfounded. Quite the opposite; I am sympathetic, although I do think that they are often overstated. The danger, broadly, is that we will be confronted with the truth and nothing but the truth – but not necessarily (in fact, probably not) the *whole* truth.

The development of memory-supporting technology will result in a great deal of reliable information swilling around, relatively easy to access, from all sorts of sources including surveillance, sousveillance, social networking and lifelogging. Our social norms seem to be developing too slowly to keep pace; we live in a world of what we might call 'Intimacy 2.0', where rights to privacy are constantly neglected, eschewed, ignored or undervalued by a society that is increasingly exhibitionist and archival (O'Hara 2010b).

One danger of a situation where there is social upheaval while social norms fail to keep pace is that there will be pressure to conform; lifelogging is currently a fringe activity, and if all lifeloggers are volunteers then it may be unproblematic even if they become a majority. Allen anticipates the possibility that we might reach a situation where someone who wishes to retain control of the information about them (the traditional conception of informational privacy) comes to be seen to be abnormal; in that case, the fact that one does not keep a lifelog may itself be seen as suspicious (Allen 2008, 74). In such a world, our reasonable expectations of privacy (an important aspect of common-law protection of privacy) will decline (McArthur 2001; Bailey and Kerr 2007), with potentially deleterious effects across society.

There is an additional danger of seeing this sort of problem as exclusively a technological one. Not only could memory, which as Sellen and Whittaker argue (2010, 77) is a complex, multi-faceted set of concepts, come to be seen in an impoverished way as what Proust called a 'simple cinematographic vision', but also that what may be sociotechnical problems may come to be seen as amenable to technological solutions.

Entirely technical solutions are very unlikely to work. As has been noted in many quarters, the use of complex privacy controls merely confuses users; privacy-enhancing technologies generally suffer severe usability problems (Sasse and Flechais 2005). The point of lifelogging is that one does not have to think too hard about collecting, storing and retrieving information (O'Hara et al. 2009); one of the ways that social networking sites like Facebook can get people to share information in more lucrative ways (for advertisers) is to set privacy defaults at a low level. Security techniques are similarly flawed; of course good security is a fine thing, but in a socio-technical system it is not just the technology but the way it is used that needs to be made secure. There is no point getting someone to create and regularly change a complex password if they end up having to resort to sticking it onto their computer screen with a Post-It (Inglesant and Sasse 2010).

## 22.5 Mechanisms to Subvert the Record

Hence a recent strand of thought has begun to develop the idea that the record itself could be subverted; this would have the effect of undermining the normative claims of truth. Mayer-Schönberger suggests the use of sell-by dates for information, so that stored information has associated with it a deletion date (Mayer-Schönberger 2009, 171–181). One creates one's Word file, say, and as part of the settings it might include a date when the file deletes itself (say, 1 year after the last edit). One could reset this at any time (as one can reset other metadata parameters, such as read and write permissions or filenames).

This idea has severe usability difficulties associated with it. The idea that one's old essays, letters or whatever might disappear because one forgot to set the delete-by date properly, is disturbing. It is hard to see it catching on; it seems a recipe for irritation (another box to tick before I can start editing my file), misunderstanding (particularly in a corporate context when files may have multiple editors with different ideas about this sort of thing), confusion (how does one calculate the time when information will become useless?), neglect (as one more and more often resorts to the default) and finally horror (oh my God my teenage novel/pictures of Grandpa/bookmarks relating to my old research have disappeared!).

Dodge and Kitchin (2007) suggest that we might subvert the aims of those who wish to breach our privacy by a process of randomized falsification. Lifelogs might be programmed to change a small number of pieces of information so that they misrepresent reality. This is an interesting suggestion, as it uses the normativity of truth to undermine threats to privacy or self-perception; because truth is normative, and because it is possible that information retrieved from the lifelog is false, then the information, or what Bell calls the e-memory (Bell and Gemmell 2009), is that much less valuable.

This solution, though clever, is I think too clever by half. The problem is that although the normativity of truth is a problem, the value of the lifelog is its truth. Randomized falsification undoes some of the worries about memory-supporting technologies at the cost of rendering them less useful. In general, making them less useful will address all the worries given above, because if they are less useful they are less likely to be used, and therefore the anticipated problems with them are less likely to occur. The lifelog's creator wants access to information that is true; he is not interested in having false memories (the pro-lifelogging literature harps on at great length about the fallibility of memory – e.g. Bell and Gemmell 2009, 51–56). So a system that serves up potentially false information seems not to fit the bill at all.

In general, philosophies of deletion and manipulation seem to throw the baby out with the bathwater; the advantages of abundant information seem clear and overwhelming, even if there will be associated difficulties. Information is clearly valuable, and is obviously perceived to be so because so many people spend so much time and effort trying to gather it. Storage and retrieval are incredibly cheap, certainly by historical standards, in which case the germane question is not 'why are we doing this?' but rather 'why *not*?'

## 22.6 Conclusion: The Perils of Rich Representations

Given the usefulness of writing, it seems that Socrates' plaints in the *Phaedrus* were overdone; few would advocate a return to an oral culture, even as an Edenic fantasy. However, his point is well-made in so far as the shift from orality to literacy required corresponding shifts in norms regulating our expectations with respect to discourse in general. It may be, if lifelogging and the use of memory-supporting technologies take off, as its advocates such as Bell predict, that an analogous shift will also be required. We have been used to our pasts decaying from scrutiny at predictable rates; no doubt our e-memories will degrade, but not in a smooth way. One might lose last week's photographs while the ones of that embarrassing party 30 years ago remain stubbornly current. This is a new circumstance, where one's past cannot be expected simply to erase itself, and it is one to which we need to adapt. A past lifelog will have a presence, and we will need to understand what it is saying – and what it is not.

The point is not about good and bad technologies, but rather their use and misuse. We need to guard not against information processing and storage power, but rather what comes with them.

First of all, we need to guard against the closed world assumption. In computing and knowledge representation, this is the assumption that whatever cannot be asserted on the basis of a knowledge base is false – in other words, the assumption that the knowledge base is complete. With respect to a lifelog, or all lifelogs put together, or even the whole of the World Wide Web, this is a very dangerous assumption. To assume that 'if I can't find it with Google it can't be important' is extremely worrying in a world which is partially recorded by digital technologies, but where major inequalities of access correlated with age, educational achievement or nationality are evident.

Second, we must guard against the assumption of, or demand for, consistency. If truth is normative for memory, then inconsistency is symptomatic of a false memory somewhere. Yet given the shades of meaning and understanding underlying memories, it is not only plausible but commonplace to find different people with entirely different memories of an event, created and curated in good faith. A future world where one's testimony was automatically assessed as of less worth than, say, the records of one's Web browsing clickstream, or one's email inbox, or one's camera, would be a very worrisome one. Even if truth remains normative for memory, the e-memories of browsers, e-mail programs and cameras are subject to interpretation too.

Third, we must guard against hindsight. Decisions made under uncertainty may seem to be poor, yet it is extremely easy to underestimate the complexity of real-time decision-making when we are in possession not only of the record of how the consequences of a decision unfolded, but also a richer picture of the context of that decision than could possibly have been available at the time.

Fourth, as many commentators have noted, there is an increasing lack of interest in, and respect for, the distinction between public and private space. In part, this is the result of a lack of care in society as a whole, as I have argued elsewhere. One blatant misrepresentation that is often passed around is that privacy is in the interest of the individual, while publicity is in the interest of wider society ('the community'). Nothing could be further from the truth; abundant information and transparency are often in the interests of the individual, while privacy is in many respects a public good (O'Hara 2010b). Its neglect can often be seen as a tragedy of the commons (Anderson and Moore 2006).

Broadly speaking, our autonomy demands informational self-determination. That is not an easy thing to define or protect, and cannot simply be assimilated to our preferences for sacrificing privacy for material gain. In particular, even though the growth of lifelogging and memory-supporting technologies continues, we should be careful that this does not undermine our reasonable expectations of privacy. We should not be seduced by the richness of the lifelog into accepting all its assumptions, assertions and details.

We should, at all costs, retain the right to be a mystery.

**Acknowledgements** The work reported in this chapter was partly supported by the projects LiveMemories – Active Digital Memories of Collective Life, Bando Grandi Progetti 2006, Provincia Autonoma di Trento, and the EU FET project Living Knowledge (http://livingknowledge-project.eu/), contract no. 231126. Thanks also to the audiences at the 2008 Workshop on Philosophy and Engineering at the Royal Academy of Engineering, and at the 2nd Microsoft SenseCam workshop, held in Dublin, 2010, when many of the ideas in this paper were presented in a keynote entitled *Narcissus to a Man*.

## References

Allen, A. L. (2008). Dredging up the past: Lifelogging, memory, and surveillance. *University of Chicago Law Review, 75*, 47–74.

Anderson, R., & Moore, T. (2006). The economics of information security. *Science, 314*, 610–613.

Bailey, J., & Kerr, I. (2007). Seizing control? The experience capture experiments of Ringley and Mann. *Ethics and Information Technology, 9*, 129–139.

Barry, S. (2008). *The secret scripture*. London: Faber & Faber.

Bell, G., & Gemmell, J. (2009). *Total recall: How the E-memory revolution will change everything*. New York: Dutton.

Berry, E., Kapur, N., Williams, L., Hodges, S., Watson, P., Smyth, G., Srinivasan, J., Smith, R., Wilson, B., & Wood, K. (2007). The use of a wearable camera, SenseCam, as a pictorial diary to improve autobiographical memory in a patient with limbic encephalitis: A preliminary report. *Neuropsychological Rehabilitation, 17*, 582–601.

Byrne, D., Doherty, A. R., Jones, G. J. F., Smeaton, A. F., Kumpulainen, S., & Järvelin, K. (2008). The SenseCam as a tool for task observation. In *People and Computers XXII: Culture, creativity, interaction: Proceedings of HCI 2008* (pp. 19–22). Swindon: British Computer Society. http://www.bcs.org/server.php?show=ConWebDoc.21389. Accessed 9 Jan 2011.

De Bruijn, O., & Spence, R. (2002). *Rapid serial visual presentation*. http://www.iis.ee.ic.ac.uk/~o.debruijn/rsvp.pdf. Accessed 9 Jan 2011.

Dix, A. (2002). The ultimate interface and the sums of life? *Interfaces* (50), 16. http://www.bcs.org/upload/pdf/interfaces50.pdf

Dodge, M., & Kitchin, R. (2007). 'Outlines of a world coming into existence': Pervasive computing and the ethics of forgetting. *Environment and Planning B: Planning and Design, 34*, 431–445.

Doherty, A. R., Gurrin, C., & Smeaton, A. F. (2009). An investigation into event decay from large personal media archives. In *1st ACM International Workshop on Events in Multimedia*, Beijing, China. http://doras.dcu.ie/4722/1/eimm32412-doherty.pdf. Accessed 9 Jan 2011.

Garde-Hansen, J., Hoskins, A., & Reading, A. (2009). Introduction. In J. Garde-Hansen, A. Hoskins, & A. Reading (Eds.), *Save as… Digital memories* (pp. 1–26). Basingstoke: Palgrave Macmillan.

Goody, J. (1998). Memory in oral tradition. In P. Fara & K. Patterson (Eds.), *Memory* (pp. 73–94). Cambridge: Cambridge University Press.

Hodges, S., Williams, L., Berry, E., Izadi, S., Srinivasan, J., Butler, A., Smyth, G., Kapur, N., & Wood, K. (2006). SenseCam: A retrospective memory aid. In P. Dourish & A. Friday (Eds.), *UbiComp 2006: Ubiquitous computing* (pp. 177–193). Berlin: Springer.

Inglesant, P., & Sasse, M. A. (2010). The true cost of unusable password policies: Password use in the wild. In *Proceedings of CHI 2010*. http://hornbeam.cs.ucl.ac.uk/hcs/publications/Inglesant+Sasse_The%20True%20Cost%20of%20Unusable%20Password%20Policies_CHI2010.pdf. Accessed 13 Jan 2011.

Jimroglou, K. M. (1999). A camera with a view: JenniCAM, visual representation and cyborg subjectivity. *Information, Communication and Society, 2*, 439–453.

Lee, H., Smeaton, A. F., O'Connor, N., Jones, G., Blighe, M., Byrne, D., Doherty, A., & Gurrin, C. (2008). Constructing a SenseCam visual diary as a media process. *Multimedia Systems, 14*, 341–349. doi:10.1007/s00530-008-0129-x.

Loftus, E. F., & Palmer, J. C. (1974). Reconstruction of automobile destruction: An example of the interaction between language and memory. *Journal of Verbal Learning and Verbal Behaviour, 13*, 585–589.

Mayer-Schönberger, V. (2009). *Delete: The virtue of forgetting in the digital age*. Princeton: Princeton University Press.

McArthur, R. L. (2001). Reasonable expectations of privacy. *Ethics and Information Technology, 3*, 123–128.

O'Hara, K. (2010a). Arius in cyberspace: Digital companions and the limits of the person. In Y. Wilks (Ed.), *Close engagements with artificial companions: Key social, psychological, ethical and design issues* (pp. 35–56). Amsterdam: John Benjamins.

O'Hara, K. (2010b). Intimacy 2.0: Privacy right and privacy responsibilities on the World Wide Web. In *Proceedings of the 2nd Web Science Conference*, Raleigh. http://journal.webscience.org/294/2/websci10_submission_3.pdf. Accessed 13 Jan 2011.

O'Hara, K., Hall, W., van Rijsbergen, K., & Shadbolt, N. (2006a). Memory, reasoning and learning. In R. Morris, L. Tarassenko, & M. Kenward (Eds.), *Cognitive systems: Information processing meets brain science* (pp. 236–260). Amsterdam: Elsevier.

O'Hara, K., Morris, R., Shadbolt, N., Hitch, G. J., Hall, W., & Beagrie, N. (2006b). Memories for life: A review of the science and technology. *Journal of the Royal Society Interface, 3*, 351–365. doi:10.1098/rsif.2006.0125.

O'Hara, K., Tuffield, M. M., & Shadbolt, N. (2009). Lifelogging: Privacy and empowerment with memories for life. *Identity in the Information Society, 1*, 155–172. doi:10.1007/s12394-009-0008-4.

Ong, W. (1982). *Orality and literacy: The technologizing of the word*. London: Methuen.

Plato. (1997). Phaedrus. In J. M. Cooper & D. S. Hutchinson (Eds.), *Complete works* (pp. 507–556). Indianapolis: Hackett.

Sasse, M. A., & Flechais, I. (2005). Usable security: Why do we need it? How do we get it? In L. F. Cranor & S. Garfinkel (Eds.), *Security and usability: Designing secure systems that people can use* (pp. 13–30). Sebastopol: O'Reilly Media.

Sellen, A., & Whittaker, S. (2010). Beyond total capture: A constructive critique of lifelogging. *Communications of the ACM, 53*(5), 70–77.

Shadbolt, N. (2003). In memoriam. *IEEE Intelligent Systems, 18*(6), 2–3. doi:10.1109/MIS.2003.1200718.

Warnock, M. (1987). *Memory*. London: Faber & Faber.

Wells, G. L. (1993). What do we know about eyewitness identification? *American Psychologist, 48*, 553–571.

Wilks, Y. (2010). Introducing artificial companions. In Y. Wilks (Ed.), *Close engagements with artificial companions: Key social, psychological, ethical and design issues* (pp. 11–20). Amsterdam: John Benjamins.

# Chapter 23
# Plans for Modeling Rational Acceptance of Technology

**Wybo Houkes and Auke J.K. Pols**

**Abstract** We argue that the use-plan analysis of artefact use and design can be combined with the Unified Theory of Acceptance and Use of Technology (UTAUT), a well-tested model for predicting the adoption of information systems in organizational contexts. After presenting the outlines of the use-plan analysis and UTAUT, we show how the basic concepts of the accounts can be mapped onto each other. This indicates that it is possible to develop an empirically informed, evaluative model of 'Rational Acceptance of Technology'. We then demonstrate the mutual benefits of the combination. Specifically, we show how the use-plan analysis can improve and extend UTAUT with conditions for the rational adoption of technology, recommendations for 'adoption-sensitive' design, and conditions for the transfer of control over and responsibility for the technology from designer to user.

**Keywords** Control over technology; rational acceptance of technology • Responsibility • Technology adoption • Use plans

## 23.1 Introduction

If technological items and systems were never used, there would be little need (not to mention opportunity) for a philosophy of technology. And to be sure, philosophers who have evaluated the consequences of adopting technologies have made various implicit claims about what constitutes artefact use.[1] Yet explicit, detailed analyses

[1] Throughout this paper, 'artefact' refers to any technological or socio-technical system.

W. Houkes (✉) • A.J.K. Pols
Department of Philosophy and Ethics, School of Innovation Sciences,
Eindhoven University of Technology, Eindhoven, The Netherlands
e-mail: w.n.houkes@tue.nl; a.j.k.pols@tue.nl

D.P. Michelfelder et al. (eds.), *Philosophy and Engineering: Reflections on Practice, Principles and Process*, Philosophy of Engineering and Technology 15,
DOI 10.1007/978-94-007-7762-0_23, © Springer Science+Business Media Dordrecht 2013

of artefact use and the relation of artefact use to design and production are few and far between, both within the philosophy of technology and in philosophy in general.

One exception is the use-plan analysis, which has been developed by one of the authors, in close cooperation with Pieter Vermaas (Houkes and Vermaas 2010). This analysis primarily serves as a background for a definition of artefact functions and, to a lesser extent, for developing a conception of engineering design. Starting from the often voiced idea that the function of an artefact is that "for which designers intended it to be used", the use-plan analysis presents one way of understanding, or reconstructing, artefact use and design as intentional activities.

The result is a conception of artefact use as the execution of a 'use plan', i.e., a goal-directed series of considered actions, including at least one manipulation of an artefact. This is a rational reconstruction: it allows one to show how the standards of practical rationality have a bearing on artefact use. As such, the analysis is not meant for *describing* episodes of actual use but as a means for *evaluating* their practical rationality, i.e., for assessing whether or not one ought to engage in such use. Here, an action or course of action is characterized as practically or instrumentally rational in case it advances an agent's ends. This minimal characterization is compatible with most conceptions of rationality currently debated in analytic philosophy (see overviews in, e.g., Audi (1989) and Hooker and Streumer (2004)) – the research tradition within which the use-plan analysis is embedded.

Despite this embedding, one might wonder whether the use-plan analysis is sufficiently realistic to play the proposed evaluative role. If, for instance, actual artefact use were never the result of executing plans, all use would, on the use-plan analysis, be irrational – turning the analysis into an error theory of use. To avoid this consequence, it has been argued that use plans can capture the importance of routines and situation-dependence in artefact use (Houkes and Vermaas 2006); and that e.g. serendipity in design and re-appropriation by users are compatible with the view that design is first and foremost the construction of use plans (Houkes 2008).

In this paper, we demonstrate the practical applicability of the use-plan analysis in another way. We argue that the use-plan analysis can be combined with the Unified Theory of Acceptance and Use of Technology (UTAUT) (Venkatesh et al. 2003), a model that cognitive psychologists have developed and tested for predicting the adoption of information systems in organizational contexts. After presenting the outlines of the use-plan analysis (Sect. 23.2) and UTAUT (Sect. 23.3), we first show how the basic concepts – or 'constructs', as technology-adoption researchers would call them – of the accounts can be mapped onto each other. This indicates how an empirically informed, evaluative model of artefact use/technology adoption may be developed. We do not present a full model here; instead, we argue in Sect. 23.4 that such a model would present clear benefits. Specifically, we show how the use-plan analysis can improve and extend UTAUT with conditions for the rational adoption of technology, recommendations for 'adoption-sensitive' design and conditions for the transfer of control over and responsibility for the technology from designer to user.

## 23.2 Use Plans

In this section we present the outlines of the use-plan analysis: a philosophical account of use, design and production that focuses on plans and practical rationality.[2]

The use-plan analysis builds on an approach in the philosophical theory of action in which plans rather than individual intentions are considered as the products of practical deliberation (e.g., Bratman 1987). Farther down its roots lie hierarchical models of cognition and planned action (e.g., (Miller et al. 1960) and more recently (Cooper and Shallice 2006)). The focus in this approach is on complex mental items – plans – that consist of a series of considered actions; the actual actions constitute an execution of a plan rather than the plan itself. Plans are taken as the units of evaluation: in assessing practical rationality, one examines (considered) *courses* of intentional action, rather than individual actions. The mental process of constructing a plan results in a more or less enduring mental state, similar to a belief or intention, and different from a fancy. Plans can also be reconstructed, e.g., when assessing the rationality of actions taken by other people or, retrospectively, one's own actions.

If some of the considered actions are manipulations of an artefact, the plan may be called a 'use plan'. Consider a simple example: Sarah wants to inform Barack about the agenda for next week's meeting and about preparations she wants him to make for one item. Realizing such a goal commonly involves planning, i.e., systematic deliberation about a sequence of actions to be undertaken to realize the goal – in this case, to transmit information and tasks. For example, Sarah might walk to Barack's office and inform him about her wishes; she might send him a handwritten memo by internal mail; she might send him an e-mail, attaching an Open Office document with his task detailed in a comment. In these three scenarios, different plans are involved. The first, walking-and-telling, need not include manipulations of artefacts; thus, in our terminology, walking-and-telling is not a use plan. The second plan presumably includes manipulating a pen, paper and an internal-mail envelope. Hence, this series of considered actions, e.g., putting the written memo in an envelope, is a use plan. The third plan involves manipulation of Sarah's computer and assumes manipulation of Barack's computer; it is a use plan for, minimally, one computer. Note that, while there might be several courses of action for Sarah to get Barack to do his assigned task, these are not use plans for Barack: Barack is a human agent, not an artefact.

Using an artefact may be analyzed as executing a use plan that includes at least one manipulation of the artefact. Designing is reconstructed as constructing and communicating a use plan and, subsidiary to that, as describing the types of items manipulated in the plan. In this analysis, designers primarily aim at aiding prospective users to realize their goals.[3] Central to such assistance is developing a sequence of actions to be undertaken by users and communicating it to them via, e.g., user

[2] A more detailed presentation can be found in Chapter 2 of Houkes and Vermaas (2010).

[3] Cf. Herbert Simon's (1981) characterization of design: "Everyone designs who devises courses of action aimed at changing existing situations into preferred ones" (p. 129).

manuals, explicit instructions or features of the artefact, against a background of known habits and cultural patterns. For instance, a designer may intend to help users with arranging meetings and distributing related tasks. She can realize her aim by coming up with a series of actions including, say, the manipulation of a 'meeting tablet' integrated into office desks: some users may be authorized to write memos on the tablet and to highlight items in it, which will automatically show up on the tablets of all attendants, together with a none-too-subtle noise and request to accept. Designing this plan would involve describing actions with the tablet *and* describing the tablet itself, since it is a currently non-existent item. The latter activity is called 'product-designing'. Although often regarded as the paradigm of engineering design, it is subsidiary to 'plan-designing' on the use-plan analysis.

The actions constituting a plan and their ordering can be communicated verbally: if an agent who knows how to realize a certain goal tells another how he went about realizing it, he communicates a series of actions. Communicating this 'procedural' aspect of artefact use does not, of course, immediately give the other agent the capacity to realize the goal. Some or all steps in the plan may require skills or 'operational knowledge', which the other agent may not possess (Houkes 2006).

On the use-plan analysis, 'user' and 'designer' are roles that may be played by different agents, in a division of practical labor. Designing is aimed at a second-order goal, namely contributing to the realization of practical goals. Constructing and communicating a use plan is the designer's way of realizing this central goal. This is a broad notion of designing that includes the activities of therapists and consultants. Only the subsidiary notion of product designing may distinguish the activities of (some) engineers from those of other practical aides.

The use-plan analysis differentiates the agent roles involved in technology, identifies the goals and actions involved in playing these roles, and thereby relates the various roles. It also provides a framework for evaluation, based on the quality of the plan that is executed or constructed, relative to the circumstances in which it is executed or constructed. Effectiveness is taken as the core quality of a plan: a plan is only of value if agents are likely to realize their goals by executing the plan. Typically, this is judged in combination with efficiency – a comparative value where the reference class is determined by available alternatives, circumstances and skills of an agent. A plan is here called '(practically) rational' if it is both effective and comparatively efficient. Other standards for plans that have been proposed in the literature include goal, means-end and belief consistency – which all concern part of the internal structure of plans.

Since plans can be evaluated in terms of their rationality and use can be described as the execution of a use plan, using an artefact is rational if and only if it is the execution of a rational use plan. Arranging a meeting with the futuristic meeting tablets is rational, provided, among other things, that one justifiably believes that manipulating one's tablet leads to the desired effects on other tablets (i.e., that the executed use plan is effective), that one knows how to manipulate the tablet and that there are no easier and otherwise more appropriate ways of communicating agendas and distributing tasks (i.e., that the executed use plan is comparatively efficient). Thus, the beliefs and specific circumstances of individual users, e.g., concerning their own skills and circumstances, determine whether a specific instance of artefact use is rational or irrational. This distinction between rational and irrational use must be distinguished from that between proper and improper use, which refers to the social institutionalization of use and use plans (Houkes and Vermaas 2010). Using the tablets to arrange private meetings or to avoid responsibility by re-allocating tasks may, for instance, be an improper use of the tablet, although the use plan may be identical and equally effective and efficient.

## 23.3 Models of Technology Adoption

In the late 1980s, management researchers started to develop models to explain and predict user acceptance and adoption of information systems. These technology-adoption models are modifications of more general models of behavioral change, such as the Theory of Reasoned Action (Ajzen and Fishbein 1980) and the Theory of Planned Behavior (Ajzen 1991). In this research tradition, various beliefs and attitudes ('constructs') are proposed as determinants of intentions, which are in turn taken to be important – but not the only – predictors for actual behavior. The proposed models are assessed by testing for the presence of the constructs and the strength of intentions via questionnaires, and by observing actual behavioral change. The Theory of Planned Behavior in particular was found to have predictive success for a wide variety of behavior, both desirable (losing weight, participating in elections) and less desirable (cheating, committing traffic violations).[4]

Models for the use of information systems in organizational contexts have, even more than general models, focused on parsimony: each model attempts to explain the largest amount of variance in intentions and actual behavior with the smallest number of constructs and moderating factors. After the pioneering Technology Acceptance Model (TAM) (Davis 1989), a variety of models have been proposed that add, rephrase, recombine and prune constructs. Surveying the constructs and success of previous work, Viswanath Venkatesh and co-authors (Venkatesh et al. 2003) proposed the Unified Theory of Acceptance and Use of Technology (UTAUT). UTAUT employs four constructs and four moderating factors and has obtained a significantly higher predictive accuracy than its predecessors for ten case studies. This model is, in terms of recency and influence, the culmination point of this line of research and we focus on it here.

The four constructs proposed in UTAUT, with definitions and sample questions as given in (Venkatesh et al. 2003), are:

- Performance expectancy: the degree to which an individual believes that using the system will help to attain goals in job performance. Items on the questionnaire to test for this construct include "Using the system would enhance my effectiveness on the job".

[4] Its large and continued scientific impact is indicated by the citations of Ajzen (1991) in Scopus: 5,177 in September 2010, 6,647 in November 2011 and 10,050 in October 2013.

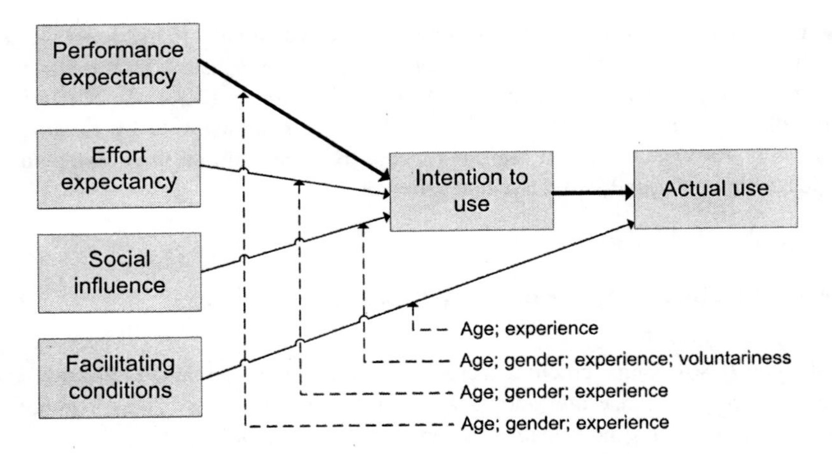


**Fig. 23.1** Structure of the UTAUT model (after Venkatesh et al. (2003)). *Solid arrows* indicate determination relations between basic constructs (*left*) and intention to use and actual use; *dashed arrows* indicate moderating factors

- Effort expectancy: the degree of ease associated with the use of the system (e.g., "Learning to operate the system would be easy for me").
- Social influence: the degree to which an individual believes that important others believe that s/he should use the system (e.g., "People who influence my behavior think that I should use the system").
- Facilitating conditions: the degree to which an individual believes that an organizational and technical infrastructure exists to support use of the system (e.g., "A specific person or group is available for assistance with system difficulties").

This model was tested longitudinally (i.e., at various stages of a 6-month process) for ten real-life cases of information-system adoption in organizations.[5] In some cases, adoption was mandatory, in others voluntary. It was found that, on average, UTAUT explains 70 % of the variance in intention to use and actual use. The structure of the research model tested is depicted in Fig. 23.1. Performance expectancy turned out to be the strongest predictor of intention (hence the larger line width) throughout the time period, but less so for women than for men and for older employees than for younger. Other notable results were that social influence is only significant for mandatory use, although it encompasses effects on social status that could also conceivably affect motivation for voluntary use; that voluntariness only moderates social influence; and that facilitating conditions are non-significant in predicting intention,[6] but significant in predicting actual use.

[5] More precisely, the model was constructed by combining the best predictors of eight existing models for eight cases (e.g., online meeting manager; portfolio analyzer), and the result was tested for two independent cases, which confirmed that UTAUT outperforms its predecessors.

[6] More precisely, this construct is non-significant in models (like UTAUT) that also contain an effort-expectancy construct: beliefs concerning organizational support are apparently and not surprisingly indistinguishable from beliefs about ease of use.

It is safe to say that UTAUT has, at least for the moment, ended the technology-adoption research program it grew out of: after 2003, there have been no published proposals for more parsimonious models that even come close to UTAUT's impact.[7] Four years after publishing on UTAUT, Venkatesh and two of his co-authors (Venkatesh et al. 2007) reported that they were frequently asked whether "technology adoption research is dead".

Still, several shortcomings of UTAUT have been pointed out in the technology-adoption literature. For instance, the model is said not to capture how intentions and behavior typically change over the period of adoption (e.g., initial reluctance to use, followed by gradual acceptance; or sporadic use followed by continuous use), and how individual attitudes with respect to performance and effort expectancy may reflect social norms. From our perspective, we add three more concerns.

Firstly, UTAUT and virtually all its predecessors black-box the design process. The models treat information systems as fixed, finished products; and users interact exclusively with these systems, not with their designers. There is no consideration of the possible effects on user acceptance of beta or prototype testing, customization or other forms of user involvement in design. As such, the technology-adoption literature may be relevant for managers who supervise the implementation of information systems in large organizations, but designers are left in the dark about how they might promote adoption of the systems they design. This is uncomfortably reminiscent of the traditional division of labor in which designers just produce physical systems according to performance characteristics, and users either conform, instigated by marketers and managers, or the product fails.

Secondly, UTAUT is presented as describing user motivation and user behavior – describing the variance in these phenomena is taken as its measure of success. Yet the emphasis on beliefs and intentions conceals evaluative aspects, if only because intentions based on false beliefs may be discredited as irrational. Some formulations in the questionnaire are revealing in this respect. A question such as "Using the system would enhance my effectiveness on the job" implicitly calls for a re-evaluation or rationalization of any previous motivation the respondent might have had to use the system (e.g., one exclusively based on peer pressure), in the light of "effectiveness". This rationalization effect may be inevitable and not only weakens UTAUT's descriptive adequacy, but also puts the need for an evaluative perspective in sharper relief.

Finally, perceptions of control are important in the acceptance of new technologies. Lack of (perceived) control over new technologies might deter intended users or even make them hostile towards those technologies (Baronas and Louis 1988). UTAUT contains only an implicit notion of control that is divided up between two constructs in a problematic way. Specifically, the construct 'perceived behavioral control' in earlier models is re-partitioned into 'self-efficacy' (the user's ability to successfully execute certain actions) and 'controllability' (the degree of belief that performance or nonperformance of the behavior is up to the agent); the former is said to be mediated by effort expectancy, and the latter is included under facilitating

[7] Venkatesh et al. (2003) is cited 3,605 times in Scopus (consulted October 2013).

conditions. This, however, assumes that the self-efficacy/controllability distinction coincides with that between internal and external locus of control (Ajzen 2002). Yet the expected lack of external support may lead to a higher effort expectancy, whereas the lack of willpower may be compensated by external support. Thus, the effort-expectancy and facilitating-conditions constructs reflect both internal and external loci of control, which makes it difficult to find which intuitions concerning control enter the model, and in which constructs. This is evaluatively significant, since it has been argued that having control over actions implies being morally responsible for performing those actions (Fischer and Ravizza 1998). Clarifying the place of control (both perceived and actual) in the UTAUT model thus seems necessary for both discovering to what extent users feel in control of and responsible for their use of the system, and (in line with the call for an evaluative, design-inclusive perspective above) finding whether these intuitions match the desired transfer of responsibility from designers to users.

## 23.4 Rational Acceptance of Technology

In this section, we argue that a combination of UTAUT and the use-plan analysis is possible and profitable. First, we show how the basic constructs of UTAUT can be explicated in terms of elements of the use-plan analysis. The result of fully integrating the accounts would be an explicitly evaluative and empirically informed model of technology adoption. That such an integration is possible is almost trivial, given the common ancestry of UTAUT and the use-plan analysis in hierarchical models of cognition and action. This does not ensure, however, that such a combination is *profitable*. To demonstrate profitability, we close the paper by reviewing three respects in which the resulting model could strengthen and extend UTAUT as well as reveal gaps in the use-plan analysis.

### *23.4.1 Mapping*

The key to mapping the basic constructs of UTAUT onto elements of the use-plan analysis is that the basic constructs are introduced as determinants of Intention-to-use. In the use-plan analysis, a (rational) intention to use an artefact requires a desire to use the artefact and a justified belief that one can execute a use plan for the artefact. This use plan is identified by its goal and the actions involved. Hence, one cannot rationally intend to use something if one does not believe that manipulating it contributes to realizing one's goals.[8] Assuming that there already are ways to realize these goals, and that there are at least some (learning, opportunity, etc.) costs

[8] In other words: that the artefact has a function relative to the use plan.

involved in using the new artefact, a rational intention to use requires the artefact to enhance effectiveness and/or efficiency of goal achievement. Hence, it is no surprise that performance expectancy is the best predictor of intention to use: a rational, or rationalizing, agent needs to have this expectancy, or renounce the intention to use.

The use-plan analysis also makes clear why the other constructs play supporting roles. Besides a belief in the effectiveness of using an artefact and knowledge of a plan, actually using something requires operational knowledge (skills). This means that an intention to use requires beliefs about one's present skills and, if these are found lacking, an intention to acquire the skills necessary to execute the accepted use plan – captured by an assessment of "ease of use". In UTAUT, possibilities for skill acquisition are incorporated in the facilitating-conditions construct: others within the organization may be willing to train the agent; a new artefact may come with instructions or tutorials[9]; or management may make other training facilities available. Other facilitating conditions concern auxiliary items that are needed for the regular execution of a use plan. Using a car, for instance, requires regular refueling; reliable use of a word processor on an office computer may require anti-virus software and regular updates. Still other conditions concern the environment or context of using an artefact: using a gasoline-fueled car is hard in arctic regions; using a particular web browser was once a little too easy given a choice of operating system. These latter two types of facilitating conditions point out that many technologies come in systems: artefacts are rarely used in isolation from other artefacts; they raise or lower the threshold for each other's use, or they are offered as package deals. This means that intentions to use must be evaluated at an appropriate level: agents may have little choice in using certain artefacts in more encompassing systems (or components within artefacts). For example, if all modern cars come outfitted with electric window openers, any car driver who desires to open the window will have to use one, even though some might yearn for the old cranks.

This leaves social influence. Insofar as the use-plan analysis considers this, the construct is related to *proper* use, and the rationality to conform to this, given a division of labor between users and professional designers. Social status and the pressure of peers and superiors within an organization do not, however, have a place in the philosophical analysis. This may reveal that its focus on individual, instrumental rationality is myopic. After all, adopting a practice may not immediately serve personal purposes, but it may make sense in a social context, e.g., given a need to "blend in" or to avoid punishment. If so, it would reveal that the intention-to-use construct is ambiguous and might conceal reluctance to adopt, a finding already suggested by the example of the electric window openers ("Personally, I dislike using this, but abstaining from use would be unwise given social and/or technical constraints"). Otherwise, social influence might indicate forces such as peer pressure, which override instrumental rationality in phenomena such as "group-think".

[9] Manuals provide at most procedural knowledge, i.e., knowledge of a use plan.

### 23.4.2 Rational Acceptance

The considerations above make clear how the combination of UTAUT and the use-plan analysis is an *evaluative* model of technology adoption: it makes it possible to assess to what extent the (justified) beliefs and knowledge are present that contribute to a rational acceptance of a new technology (i.e., a realizable intention to use). Thus, the standards of instrumental rationality enable a principled, but straightforward distinction between constructs that appeal to user perceptions (e.g., perceived ease of use) and constructs that appeal to actual features of usage (e.g., actual ease of use). It leaves open the source of justification of beliefs and intentions: for new technologies, this source may be testimony, i.e., epistemic trust of designers and other agents who recommend use; experienced or knowledgeable users may substitute testimony with previous successes, reliance on their own skills, and in-depth expertise concerning the operation of an artefact. The combined model also allows a distinction between rational and irrational (or less rational) adoption, and between factors that should and should not influence the intention to use. This is relevant for evaluating the role of moderating factors. At first glance, instrumental rationality should not be relative to gender or age; thus, while these factors moderate the determining force of performance expectancy, they should not. The moderating influence of these factors may be brought in indirectly, however, via their effect on elements of planning, such as the desire to excel at certain tasks or the costs in acquiring operational knowledge (via speed of learning).

### 23.4.3 Adoption-Sensitive Design

The use-plan analysis contains an interface between designers and (prospective) users: the former cannot realize the constitutive goal of their activity without making sure that a use plan has been successfully communicated to the latter and that prospective users are in a situation to adopt artefact use on rational grounds. To put this in different terms: it may be in the designers' personal or professional interest to persuade users by any means possible, but being a designer comes with particular (role) responsibilities.

The outline above makes clear which elements are involved in assessing whether adopting a new technology is rational or not: the user needs to know which goal the artefact may serve, how it may be used to realize that goal (i.e., the user must have procedural knowledge), and which skills, auxiliary items and environmental conditions are required. In many cases, much of this information may be presumed available – it is background knowledge in communicating the use plan or distributing the technology. If some of this information is lacking or not self-evident, however, "adoption-sensitive" design requires that it is communicated clearly and effectively; otherwise, designers cannot expect rational users to form intentions to use their technology.

One reason why this might sound too idealistic or too demanding is that, in practice, the responsibility of designers may be shared by agents in other roles, such as marketing staff, instructors and managers. A more complete plan analysis of technological activities should include these agents. However, that there may be a re-allocation or distribution of the responsibility to communicate the mentioned information does not make the responsibility disappear.

Another response might be that these requirements for adoption-sensitive design are familiar from design research. In our opinion, this indicates that our proto-model leads to sensible (albeit unsurprising) recommendations for design – which is certainly an improvement over the lack of any recommendations in UTAUT. Moreover, the need to come up with more specific recommendations, based on literature in design research, might lead to further refinements of the model and thus to a fruitful combination of three lines of research.

### *23.4.4 Being in Control?*

As outlined above, UTAUT does not contain a construct that captures intuitions regarding control of the adopted technology. 'Perceived behavioral control' and 'controllability', which featured in precursor models such as TAM and TPB, have been included under the supposedly more encompassing constructs 'effort expectancy' and 'facilitating conditions'. This muddies the waters by implicitly forging a distinction between internal and external loci of control. A clear notion of control is indispensable in reflecting on technology, however, also on the 'meso'-level of thinking about the interaction between designers and users.[10] The use-plan analysis provides a way to introduce such a notion. In diffusing technological knowledge and artefacts, designers transfer a measure of control over artefacts to users (Pols 2010). Specifically, they enable users to perform certain actions with artefacts, or if that is not directly possible, provide them with instructions on how to acquire the required skills, auxiliary items, etc. necessary for rational artefact use. According to Fischer and Ravizza (1998), if you have the ability to perform certain actions then you are able to exert control over those actions.[11] This, in turn, is what makes people responsible for those actions. This does not mean that users are always solely responsible for rational artefact use: other parties (the designer, a manager) might share responsibility. Similarly, responsibility does not necessarily imply blameworthiness, neither in general nor in the case of technology adoption: if a robber puts a

[10] Most analyses of control in engineering focus on either the 'micro'-level of controlling the output of technological systems or the 'macro'-level of steering (developments within) technological regimes.

[11] This definition of control resembles that of self-efficacy, given in Sect. 23.3. Fischer and Ravizza (1998) contrast this 'guidance control' with 'regulative control', the ability to freely perform one action rather than another, which resembles controllability. This again stresses the close link between (perceived) behavioral control and (perceived) responsibility.

gun to Steve's head, or Sarah's manager threatens to fire her if she does not use the new 'meeting tablets', it is still up to Steve and Sarah to decide what to do. If Steve hands over his wallet and Sarah starts to use the meeting tablets, they exert control over their actions and thus are responsible for them. Given the circumstances, however, it seems rather that the robber and the manager would be to blame for any negative consequences resulting from these actions.

Rational acceptance of technology requires the user to accept part of the responsibility for failed use: if someone claims to possess knowledge of a use plan, together with beliefs that goal realization is compatible with user skills and environmental support, and is not coerced into using the artefact, the person shares the blame or even gets full blame for failure in case the artefact is in working order. This is relevant in many cases of artefact use (e.g., analyses of airplane crashes), but also in the organizational contexts for which UTAUT has been developed. Given, for instance, Sarah's organizational role of coordinating meetings and distributing tasks, her use of the innovative meeting tablets does not only show her adoption of the technology, but also reflects her choice to enact her role responsibility by use of the technology. The central role of use plans in a model of Rational Acceptance of Technology could provide a framework for analyzing the relation between responsibility and technology in organizational contexts.

**Acknowledgments** Research by Wybo Houkes was made possible by the Netherlands Organization for Scientific Research (NWO).

## References

Ajzen, I. (1991). The theory of planned behavior. *Organizational Behavior and Human Decision Processes, 50*, 179–211.

Ajzen, I. (2002). Perceived behavioral control, self-efficacy, locus of control, and the theory of planned behavior. *Journal of Applied Social Psychology, 32*, 665–683.

Ajzen, I., & Fishbein, M. (1980). *Understanding attitudes and predicting social behavior*. Englewood Cliffs: Prentice Hall.

Audi, R. (1989). *Practical reasoning*. London: Routledge.

Baronas, A. M. K., & Louis, M. R. (1988). Restoring a sense of control during implementation. *MIS Quarterly, 12*, 111–124.

Bratman, M. (1987). *Intention, plans and practical reason*. Cambridge, MA: Harvard University Press.

Cooper, R. P., & Shallice, T. (2006). Hierarchical schemas and goals in the control of sequential behavior. *Cognitive Neuropsychology, 23*, 202–221.

Davis, F. D. (1989). Perceived usefulness, perceived ease of use, and end user acceptance of information technology. *MIS Quarterly, 13*, 318–339.

Fischer, J., & Ravizza, M. (1998). *Responsibility and control*. Cambridge: Cambridge University Press.

Hooker, C., & Streumer, B. (2004). Procedural and substantive practical rationality. In A. R. Mele & P. Rawling (Eds.), *The Oxford handbook of rationality* (pp. 57–74). Oxford: Oxford University Press.

Houkes, W. (2006). Knowledge of artefact functions. *Studies in the History and Philosophy of Science, 37*, 102–113.

Houkes, W. (2008). Designing is the construction of use plans. In P. E. Vermaas, P. A. Kroes, A. Light, & S. A. Moore (Eds.), *Philosophy and design* (pp. 37–49). Dordrecht: Springer.
Houkes, W., & Vermaas, P. E. (2006). Use plans and artefact functions. In A. Costall & O. Dreier (Eds.), *Doing things with things* (pp. 29–48). London: Ashgate.
Houkes, W., & Vermaas, P. E. (2010). *Technical functions*. Dordrecht: Springer.
Miller, G. A., Galanter, E., & Pribram, K. H. (1960). *Plans and the structure of behavior*. New York: Holt, Rinehart & Winston.
Pols, A. J. K. (2010). Transferring responsibility through use plans. In I. Van de Poel & D. E. Goldberg (Eds.), *Philosophy and engineering: An emerging agenda* (pp. 189–203). Dordrecht: Springer.
Simon, H. A. (1981). *The sciences of the artificial* (2nd ed.). Cambridge, MA: The MIT Press.
Venkatesh, V., Morris, M. G., Davis, G. B., & Davis, F. D. (2003). User acceptance of information technology. *MIS Quarterly, 27*, 425–478.
Venkatesh, V., Davis, F. D., & Morris, M. G. (2007). Dead or alive? The development, trajectory and future of technology adoption research. *Journal of the Association for Information Systems, 8*, 267–286.

# Chapter 24
# On the Epistemology of Breakthrough Innovation: The Orthogonal and Non-linear Natures of Discovery

**Bruce A. Vojak and Raymond L. Price**

**Abstract** One of the most important roles of researchers in technology-based companies is to develop innovative new products and processes to either increase revenue or decrease cost. However, while some have begun to consider how engineers and scientists "know," most practitioners and researchers of corporate innovation carry unarticulated, less-than-fully-developed assumptions about this topic. In the present work, we gain insight into the orthogonal ("know what") and non-linear ("know how") natures of the epistemology of breakthrough innovation by reflecting on both the characteristics of those who innovate and the characteristics of how breakthrough innovation actually occurs.

**Keywords** Epistemology • Innovation • Breakthrough innovation • Non-linear • Orthogonal

## 24.1 Background

Although rarely, if ever, understood or articulated explicitly, financially-successful companies engage in the practice of epistemology on a daily basis. On the tactical, operational side, businesses seek to establish an environment characterized by what is known and highly predictable. In doing so – by reducing variation, understanding modes of failure and eliminating defects – firms become increasingly certain and rigid as they squeeze cost out of their systems. In contrast, on the strategic, research side, businesses seek to establish an environment characterized by what is unknown and highly unpredictable. In doing so – by creating and discovering innovative new products and processes which either increase revenue or decrease cost – firms

B.A. Vojak (✉) • R.L. Price
College of Engineering, University of Illinois at Urbana-Champaign, Urbana, IL 61801, USA
e-mail: bvojak@illinois.edu; price1@illinois.edu

D.P. Michelfelder et al. (eds.), *Philosophy and Engineering: Reflections on Practice, Principles and Process*, Philosophy of Engineering and Technology 15,
DOI 10.1007/978-94-007-7762-0_24, 

become increasingly skeptical and fluid as they open themselves up to new ways of thinking and new opportunities. Both such activities ensure ongoing company success in terms of the bottom-line financial position of the firm, yet in strikingly different epistemological ways. In some ways is it appropriate to characterize tactical operations as seeking to eliminate the last remnants of what is not known, while characterizing strategic research as seeking to begin to discover and claim vast new territories of what can be known. While both imply and involve a focus on knowing, the difference is between knowing that leads to efficiency and knowing that leads to innovation. For the purposes of the present work, we attend to the latter.

Innovation occurs in various ways in companies, from the incremental (representing minor improvement, such as a new feature being added to an existing design; e.g. increasing the capacity of the hard drive on an iPod) to the truly breakthrough (representing radical change, such as a new product concept; e.g. the initial introduction of the iPod itself along with the associated, compatible iTunes software). An entire host of options exist between these two extremes, such as the transition from the early iPod interface to that of the iPod Touch. This entire spectrum of innovation is critical to company sustainability, with incremental innovation providing low-risk, low-return opportunities and breakthrough innovation providing high-risk, high-return opportunities. Both extremes of innovation deliver important options for future impact. Incremental innovation enables firms to "mine" a product concept, often by either more effectively or more broadly reaching a relatively well-defined customer base. In contrast, breakthrough innovation rejuvenates firms by bringing them into entirely new business domains. For the purposes of the present work, we attend to breakthrough innovation – the extreme situation where the financial impact, and the impact on long-term sustainability, typically is greatest as the firm moves from the currently known through the unknown and then on to the newly known.

While some have begun to consider how engineers and scientists – those who typically contribute breakthrough innovation in firms – "know" from a philosophical perspective (Vincenti 1990; Mitcham 1994; Vojak et al. 2010), most practitioners and researchers of corporate innovation carry unarticulated, less-than-fully developed assumptions about this topic. Our insights into the epistemology of innovation have been developed and refined as part of a larger study of Serial Innovators (SIs), those individuals who have repeatedly conceived and commercialized breakthrough new products in large, mature technology-based firms (see, for example: Vojak et al. 2006, 2010; Griffin et al. 2009). Conducted over the past 11 years and based on over 125 in-depth interviews as well as a large sample survey, this body of research investigates, and has led us to a clearer understanding of, how breakthrough innovation – including the epistemology of innovation – occurs in practice.

## 24.2 Approach

In this chapter, we employ the simple conceptual framework of Fig. 24.1 to guide our reflection. We have observed that SIs come to the process of innovation prepared with a wealth of factual information and are extremely curious, which only serves

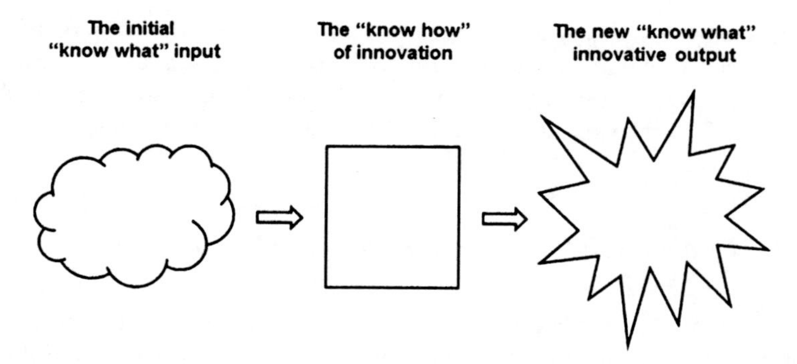


**Fig. 24.1** A simple model of the process by which breakthrough innovators come to know what to do

to add over time to their broad and deep information base, the "know what" input of innovation. Further, these same individuals come to the process of innovation with the "know how" of innovation, the tacit skill of systems thinking. They are expert at connecting the dots of information from both proximate and disparate fields. As depicted in Fig. 24.1, when the "know what" base of factual information serves as the input to the "know how" of innovation in an SI, the result is typically a highly-productive innovative output, a new "know what."

The goal of the present work is to gain new insight into the epistemology of breakthrough innovation. We first consider some of the salient characteristics of each of these three epistemological elements: the initial "know what" input to innovation, the "know how" of innovation, and the new "know what" innovative output. We then turn to consider how these characteristics work together, seeking common themes and trends that may yield additional insight into each, as well as into the whole.

## 24.3 Some Salient Characteristics of the Three Epistemological Elements

### *24.3.1 Characteristics of the Initial "Know What" Input to Innovation*

Many have recognized that breakthrough innovators bring both depth and breadth in their disciplinary knowledge base (Johansson 2004; Brown 2005). While academic researchers typically are characterized by their profound depth of insight in one field of study, industrial innovators are often anecdotally described as "T-shaped" (Guest 1991) in that they know a great deal about their primary discipline (the vertical stem of the "T") and something about many other disciplines (the horizontal bar at the top of the "T"). Further, some have observed that breakthrough innovators are "π-shaped" or even "M-shaped" in that they also exhibit significant depth in multiple other fields.

Additionally, apparently-insignificant insights are observed to disproportionately pave the way to significant breakthrough innovations. Similarly, ever-so-slight differences between two competitors often result in significant differences in ultimate financial performance and success as breakthrough innovation unfolds.

### 24.3.2 *Characteristics of the "Know How" of Innovation*

While seeking to either discern or impose some order in or on it, both industrial practitioners and academic researchers agree that breakthrough innovation is a messy, complex process that does not follow neatly-defined paths. Thus, while a finite set of certain states must be visited as the innovation process unfolds (such as identifying the best problem to address from both company and customer perspectives, understanding the problem deeply, synthesizing what is known into an innovative product concept, and developing insight into how to navigate the internal politics of the firm), these states typically are visited repeatedly, in only a general order initially and with little or no predictability thereafter (see, for example, the "Hourglass Model of Innovation" described in Vojak et al. 2010). Illustrating the iterative, feedback-laden nature of the "know how" of innovation, those describing it at times speak of "chewing on" ideas as they emerge into conscious awareness. The use of analogous language from meteorology also sheds light on the "know how" of innovation; in an effort to stimulate highly-creative, innovative output, no holds barred "brainstorming" is often employed.

### 24.3.3 *Characteristics of the New "Know What" Innovative Output*

Truly innovative output is disruptive, unpredictable and unexpected in its appearance (Schumpeter 1947; Christensen 1997). Further, breakthrough innovation, by the very use of the adjective, implies a rapid transition from non-existence to existence of an innovative insight, not unlike the mental image elicited by considering an object "breaking through" a wall – at one moment it does not exist on the far side of the wall, the next moment it is fully present.

## 24.4 How These Characteristics Work Together to Yield New Insight

While each of these individual observations about the salient characteristics of the three epistemological elements is of interest for its own ability to describe breakthrough innovation, when considered together, a richness of insight emerges.

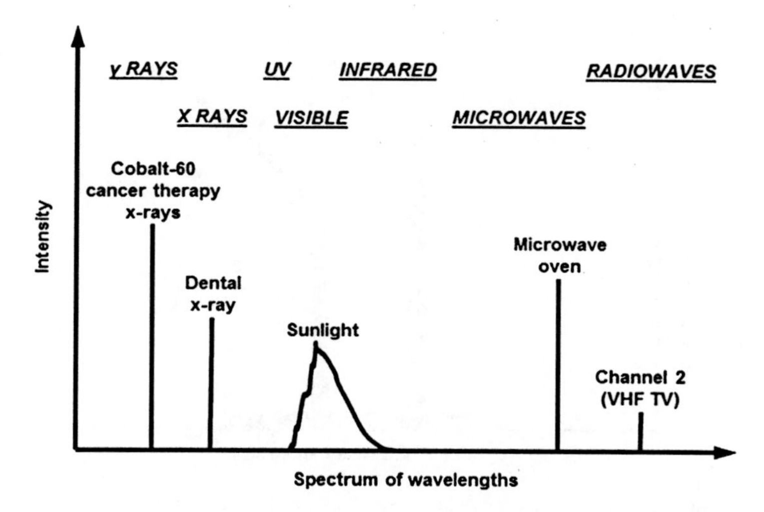


**Fig. 24.2** The electromagnetic spectrum, illustrating orthogonal properties analogous to the spectrum of disciplinary knowledge metaphor applied in this chapter

### *24.4.1 The Orthogonal Nature of Discovery*

The intuitive descriptions of "T-shaped," "π-shaped," or even "M-shaped" people carry with them the familiar appearance of spectra in the physical world, such as the electromagnetic spectrum included as Fig. 24.2, where the horizontal axis represents the range of wavelengths of electromagnetic waves while the vertical axis represents their intensity.

Some familiar types and sources of electromagnetic waves are plotted in this figure to help orient the reader. An important characteristic of the electromagnetic spectrum is that each wavelength on the horizontal axis represents a sinusoidal wave that has the characteristic of being mathematically orthogonal to, and independent of, each and all of the sinusoidal waves represented by every other wavelength. As a result of this characteristic, collectively, the entire electromagnetic spectrum provides one with the ability to construct all possible waves that could ever exist, simply by appropriately weighting and adding the various sinusoidal waves together.

Applying this insight to the intuitive descriptions of the "T-shaped," "π-shaped," or even "M-shaped" people provides us with new insight about the "know what" input to innovation. Consider, for example, the "π-shaped" person illustrated in Fig. 24.3. The horizontal axis represents the span of such a person's disciplinary knowledge and the vertical axis represents its magnitude.

An implication of being able to depict an individual's expertise in this manner is that any arbitrary collection of multidisciplinary knowledge, as might be possessed by an individual, can be represented by the sum of a set of orthogonal, independent functions, one function for each entirely distinct discipline within which something is known – exactly analogous to that which we observed with the electromagnetic spectrum of Fig. 24.2. Thus, the "know what" input of innovation, as well as the

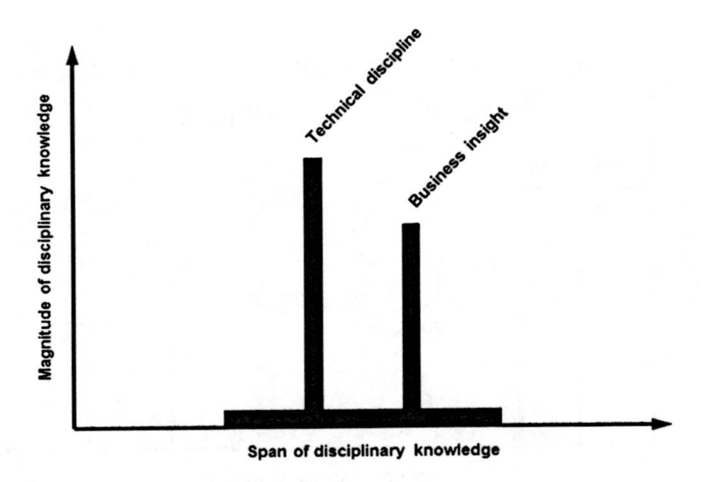


**Fig. 24.3** The spectrum of disciplinary knowledge for a π-shaped person

new "know what" innovative output, can be seen as being comprised of a set of orthogonal and independent pieces of information.

To be clear, disciplinary knowledge is not as purely orthogonal as suggested by the "T-shaped," "π-shaped," or even "M-shaped" metaphors. In fact, some overlap between disciplines is not only expected, but necessary, if only to enable communication between practitioners. Thus, the elements of "know what" that can be understood to be truly orthogonal are likely at a much more lower, more granular level that at that of the discipline. Having said this, however, it is safe to suggest that some pairs of disciplines are often significantly orthogonal (physics and literature) while others are not (physics and music).

### *24.4.2 The Non-linear Nature of Discovery*

Taken both individually and collectively, a number of the characteristics that describe the epistemological elements of innovation (unpredictability, abruptness of change in behavior, feedback, iteration, and extreme sensitivity to slight differences) suggest that some form of non-linear process must be present in the system and, thus, that the underlying nature of innovative discovery can be illustrated mathematically by using chaos theory (Strogatz 2001). Even the use of the analogous language of storms from meteorology supports this observation, as weather is a highly chaotic, non-linear system.

A non-linear system is one whose mathematical description expresses relationships that are not strictly proportional (Gleick 1987). Mathematically, non-linear relationships occur in various ways, such as: a power law relationship (e.g. $y=x^2$, $y=x^{1.7}$, etc.), a trigonometric relationship (e.g. $y=\sin(x)$), or a logarithmic relationship (e.g. $y=\log(x)$). When non-linear terms do not exist, an equation can be broken

**Table 24.1** A comparison of the characteristics of linear and non-linear systems

| Linear systems | Non-linear systems |
|---|---|
| Gradual changes in behavior | Extreme changes in behavior occur abruptly and without warning |
| Modest sensitivity of output to initial conditions | Extreme sensitivity of output to initial conditions |
| Behavior is both deterministic and predictable | Behavior is deterministic, but not predictable |
| Study of linear systems is "Classical" | Study of non-linear systems is known as "Chaos theory" |

down into smaller parts that can be analyzed separately, making an analytical solution possible, something that cannot be accomplished for a non-linear system (Gleick 1987). As a result of these mathematical differences, there exists a striking difference between the behavior and characteristics of the mathematical solutions of linear and non-linear systems. Some of the most salient of these differences are summarized in Table 24.1 (Gleick 1987; Strogatz 2001).

Non-linear systems are observed to abound in nature, ranging from those in weather (e.g. storms), geology (e.g. earthquakes) and sound (e.g. the overtones of a piano). Non-linear systems also play a key role in engineered systems, such as with the up-conversion of an audible signal to a much higher frequency to enable transmission of the original audible signal in a communication system.

Perhaps the most well-known characteristic of non-linear systems is the so-called "Butterfly Effect," which alludes to a system's extreme sensitivity to initial conditions. The name "Butterfly Effect" arose from, and is illustrated by, the observation that "the flap of a butterfly's wings in Brazil can set off a tornado in Texas" – that is, that an ever so slight disturbance in one part of the world can yield extreme consequences for the weather experienced in a distant land. A perhaps more familiar illustration of the "Butterfly Effect" – yet one that carries a much longer history – is the proverb "For the want of a nail," where the lack of just one nail carries with it unfortunate extreme consequences for the kingdom (Gleick 1987):

For want of a nail the shoe was lost.
For want of a shoe the horse was lost.
For want of a horse the rider was lost.
For want of a rider the battle was lost.
For want of a battle the kingdom was lost.
And all for the want of a horseshoe nail.

The "Butterfly Effect" also can be visualized graphically by considering the example of fractals. Most people who are familiar with striking images of fractals, such as those of the well-known Mandelbrot Set, may not realize that the images produced are literally maps of the solutions of a non-linear equation. The color or shading appearing at a given point in a map of a Mandelbrot set (with each point on the map representing a distinct initial condition) represents the rate at which the output of the mapped non-linear function goes to infinity as it is iteratively applied. This is shown graphically in Fig. 24.4 (Burns 2010), with 24.4b representing a 100-fold magnification of the image of Fig. 24.4a. That two adjacent points in an image such

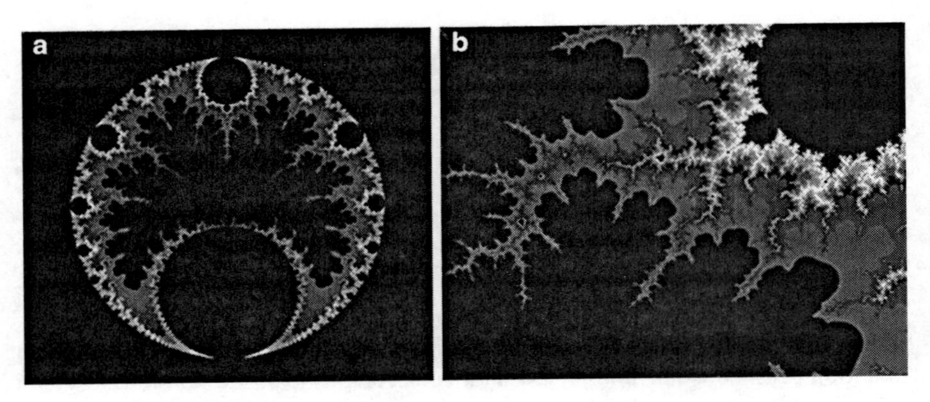


**Fig. 24.4** Mandelbrot set maps, illustrating the sensitivity of non-linear systems to initial conditions

as these can exhibit such a striking contrast of shading is, once again, an illustration of how sensitive non-linear systems are to very slight differences in initial conditions. That the same sensitivity is observed at ever increasing magnifications, such as in Fig. 24.4b, also is characteristic of non-linear systems.

In each of these various illustrations we see vivid examples of how extremely sensitive non-linear systems are to ever so slight variations. That breakthrough innovation exhibits this characteristic, as well as all of the other characteristics of non-linear systems listed in Table 24.1, provides significant substantiation that discovery is, in fact, a non-linear process.

### 24.4.3 *The Non-linear and Orthogonal Natures of Discovery Considered Together*

Bringing the observations of Sects. 24.4.1 and 24.4.2 together, we observe that breakthrough innovators "connect the dots." That is, they gather and synthesize information and insights from many, disparate disciplines and sources in a way that they see a whole that is greater than the sum of its parts. Such transcending and creatively cross-fertilizing or mixing of disciplinary insight has been recognized, as well, by others in the literature (Johansson 2004; Fleming 2007). Further, it has been recognized anecdotally by practitioners who talk about inventing in the "cross terms (the xy terms)" in a polynomial as illustrating where the significant value is in the creation of new ideas. In the present work we take this understanding to a next level.

Perhaps the simplest illustration of simultaneously considering the orthogonal and non-linear of discovery can be found in the multiplication of two orthogonal (i.e. perpendicular) vectors, as illustrated in Fig. 24.5.

Taken alone, these vectors A and B define a plane; any point on this two-dimensional plane can be identified by an appropriately-weighted, linear combination of these two vectors. The multiplication of these two vectors, however, yields an entirely new vector, C, that is simultaneously orthogonal to (i.e. perpendicular to)

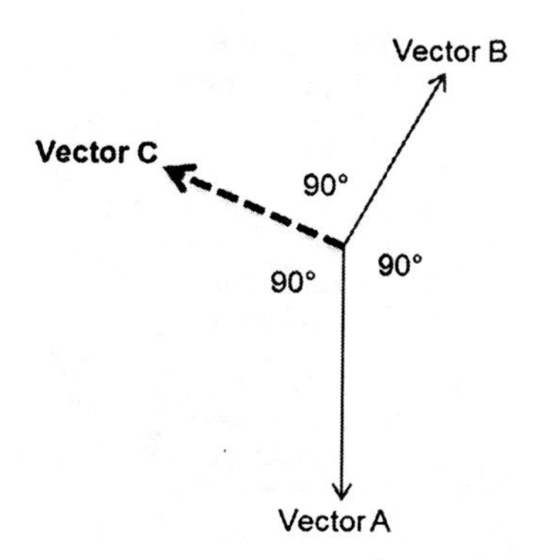


**Fig. 24.5** The vector multiplication of orthogonal vectors A and B, yielding vector C, which is orthogonal to the other two

each of the two original vectors. Further, it is critical to point out that there is no appropriately-weighted, linear combination of the original two vectors that will yield this new vector. If we, very loosely, suggest that the two original vectors represent knowledge ("know what") in two entirely different fields, such as industrial design and electronics technology, then multiplying (non-linear "know how") such knowledge propels one into an entirely new, third dimension of "know what" – in this case, perhaps, a new product concept such as an iPod. This is exactly what we suggest occurs during breakthrough innovation, and it is intriguingly similar to criteria applied in the non-obviousness test used to determine whether an idea is considered an invention in patent law – a simple combination (i.e. addition) of ideas (i.e. orthogonal vectors) is not considered sufficient to pass the test.

Again, by analogy, we note that several systems, ranging from musical instruments (the non-linear mixing of two tones) to wireless communication systems (the non-linear mixing of two signals), display – and critically depend on for their operation – behavior that is mathematically identical to the interaction that we suggest is characteristic of the nature of breakthrough innovation.

## 24.5 Conclusions and Managerial Implications

In summary, we conclude that the "know what" input of innovation can be characterized as being comprised of orthogonal pieces of information and that the "know how" of innovation is non-linear. This is captured schematically in Fig. 24.6, an updated version of Fig. 24.1, which was used to illustrate the conceptual framework for the present consideration of the epistemology of innovation.

That we now can speak of the epistemology of innovation mathematically as a non-linear combination of orthogonal functions opens up the opportunity to enhance managerial insight in support of the development of new products and processes in the commercial realm.

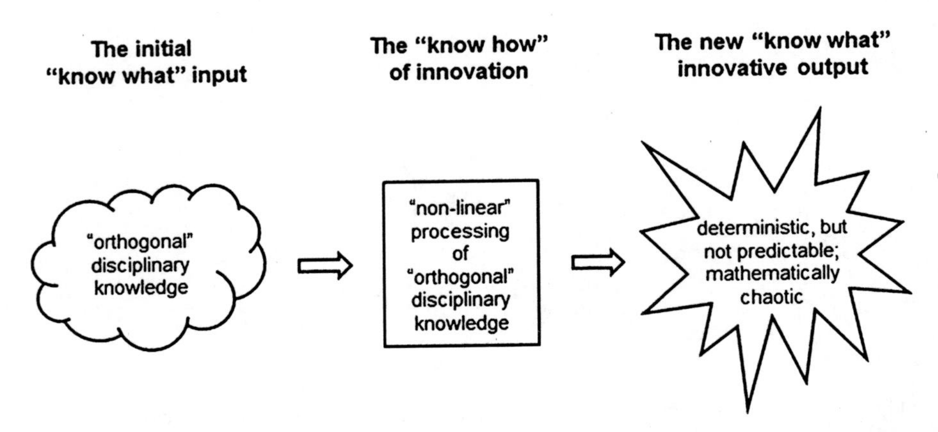


**Fig. 24.6** The simple model of breakthrough innovation first depicted in Fig. 24.1, now including insights developed in this chapter

Regarding these metaphors, they are just that. However, as metaphors they contain a powerful means of illustrating what is occurring as innovators come to know what to do today in order to have impact in the future. The two metaphors work together, not separately. That these two metaphors come together to describe so much of what occurs in innovation is powerful and potentially very useful in application.

As a result, while the epistemology of innovation contains elements of being deterministic, it is not predictable (or at least not predictable with any reliability beyond a forecast horizon, just like the weather) – as such, managerial humility is necessary. Similarly, while the epistemology of innovation displays features associated with non-linear, "chaotic" behavior, it is not random – as such, managerial insight is necessary and useful.

Second, because the non-linear nature of the "know how" of innovation propels the innovator into new, much richer, more filled-in (i.e. orthogonal) dimensions of insight, it represents a critical skill, without which one cannot innovate. Managers are well advised to seek those with extraordinary skill in such non-linear thinking, thinking in which often disparate concepts and insights are combined to yield new and impactful results.

Third, having said this, that the "know what" and "know how" of innovation are complementary indicates that both of the "know what" and "know how" of innovation are necessary. As such, managers should not limit their search for innovators by seeking only those who effectively practice this non-linear "know how" of innovation. The addition and retention of "know what" also is a critical skill. Breakthrough innovators are curious and passionate; they immerse themselves as they add "know what". Breakthrough innovators also have remarkable memories; they persist in recalling facts and details that may be lost on others. Passionate curiosity and strong memory skills, then, represent other skills that should be sought as managers seek to identify potential breakthrough innovators.

Fourth and finally, these metaphors provide direction to managers to "fish where the fish are"; that is, they must work at "the hairy edge of innovation", not necessarily in more comfortable, familiar terrain. They must seek the technical, market, customer, manufacturing and financial equivalent of the boundary of the Mandelbrot set, the places where ever so slight changes and combinations yield remarkably new outcomes.

**Acknowledgements** The authors are indebted to Professor Douglas L. Jones and Professor Alfred W. Hubler (both of the University of Illinois at Urbana-Champaign) for helpful discussions and to Scott Burns for permission to use his Mandelbrot images in Fig. 24.4 (Design by Algorithm; http://www.designbyalgorithm.com).

## References

Brown, T. (2005). Strategy by design. *Fast Company, 95*, 52–54.

Burns, S. (2010). *Design by algorithm*. http://www.designbyalgorithm.com/. Accessed 15 Dec 2010.

Christensen, C. (1997). *The innovator's dilemma*. Cambridge, MA: Harvard Business Press.

Fleming, L. (2007). Breakthroughs and the 'long tail' of innovation. *MIT Sloan Management Review, 49*, 69–74.

Gleick, J. (1987). *Chaos: Making a new science*. New York: Viking Penguin.

Griffin, A., Price, R. L., Maloney, M. M., Vojak, B. A., & Sim, E. W. (2009). Voices from the field: How exceptional electronic industrial innovators innovate. *Journal of Product Innovation Management, 26*, 223–241.

Guest, D. (1991, September 17). The hunt is on for the Renaissance Man of computing. *The Independent* (*London*).

Johansson, F. (2004). *The Medici effect: Breakthrough insights at the intersection of ideas, concepts, and cultures*. Cambridge, MA: Harvard Business Press.

Mitcham, C. (1994). *Thinking through technology*. Chicago: University of Chicago Press.

Schumpeter, J. (1947). *Capitalism, socialism and democracy*. New York: Harper.

Strogatz, S. (2001). *Nonlinear dynamics and chaos: With applications to physics, biology, chemistry, and engineering*. Boulder: Westview Press.

Vincenti, W. (1990). *What engineers know and how they know it*. Baltimore: Johns Hopkins University Press.

Vojak, B. A., Griffin, A., Price, R. L., & Perlov, K. (2006). Characteristics of technical visionaries as perceived by American and British industrial physicists. *R&D Management, 36*, 17–24.

Vojak, B. A., Price, R. L., & Griffin, A. (2010). Corporate innovation. In R. Frodeman, J. T. Klein, & C. Mitcham (Eds.), *Oxford handbook of interdisciplinarity*. Oxford: Oxford University Press.

# Chapter 25
# Uncertainty in the Design of Non-prototypical Engineered Systems

**William M. Bulleit**

**Abstract** Engineering design must be performed under conditions of uncertainty, some of which are obvious and some of which many engineers may never have consciously considered. The level of uncertainty for non-prototypical engineered systems is greater than for systems in which prototype testing is possible. In this paper we consider the uncertainties facing engineers who design non-prototypical engineered systems and some of the ways that those engineers manage uncertainties in a manner that allows design decisions to be made. Uncertainties are dealt with using codes of practice, in order to achieve minimum levels of safety, and quality control measures to minimize human error. The possibility of extreme, unpredictable events can only be dealt with by including engineering details into a system that make it more robust, but are not necessary by minimum standards. These additional engineering details may permit the system to withstand events that far exceed the design capacity.

**Keywords** Uncertainty • Engineering design • Engineered systems • Black swan events

## 25.1 Introduction

Engineering design must be performed under conditions of uncertainty, some of which are obvious and some of which many engineers may never have consciously considered. The level of uncertainty for non-prototypical engineered systems, or "one-off" systems, is greater than for systems in which prototype testing is possible

W.M. Bulleit (✉)
Department of Civil and Environmental Engineering, Michigan Tech, 1400 Townsend Dr, Houghton, MI 49931-1295, USA
e-mail: wmbullei@mtu.edu

D.P. Michelfelder et al. (eds.), *Philosophy and Engineering: Reflections on Practice, Principles and Process*, Philosophy of Engineering and Technology 15,
DOI 10.1007/978-94-007-7762-0_25, 

(Blockley 1992). In this paper we consider the uncertainties facing engineers who design non-prototypical engineered systems and some of the ways that engineers manage those uncertainties in a manner that allows design decisions to be made. Vincenti (1990), writing primarily about aeronautical engineering, describes the problems facing engineers who work with systems where prototypes can be built. Prototype aircraft are always used in the design of new types. All aspects of design discussed by Vincenti, including his variation-selection model for the growth of engineering knowledge, apply to non-prototypical systems, but the engineering of systems where prototypes are not possible presents additional, often significant, difficulties.

Uncertainty can be separated into two broad categories: *aleatory*, related to luck or chance, and *epistemic*, related to knowledge (Der Kiureghian and Ditlevsen 2009). This breakdown has an impact on how engineers cope with the various types of uncertainty and the way we think about each type. We consider five broad sources of uncertainty: time, randomness, statistical limits, modeling, and human error (Bulleit 2008). Some uncertainties are explicitly dealt with using codes of practice (e.g., Ellingwood et al. 1980), some are dealt with through quality control measures, and some are dealt with in implicit ways that we often do not think much about, e.g., heuristics (Koen 2003).

## 25.2 Aleatory Versus Epistemic Uncertainties

In order to think about aleatory versus epistemic uncertainties, consider the flipping of a fair coin. Flipping a fair coin is usually thought of as aleatory uncertainty because the uncertainty appears to be related to chance. But, it might be possible to model the coin flipping process so well, including coin imbalance, air resistance, hand behavior, etc., that the results of the flip would be nearly predictable. If that prediction were possible, then the uncertainty in coin flipping would be primarily epistemic rather than aleatory because an increase in knowledge about coin flipping reduced the uncertainty. In this section we will consider each of the five sources of uncertainty just mentioned in terms of the contribution to each from aleatory and epistemic uncertainties. The manner in which each of these five sources is dealt with in the design of non-prototypical engineered systems is often affected by whether the contributing uncertainties are aleatory or epistemic.

Uncertainties caused by *time* include using past data to predict future occurrences, e.g., using snow load data over the past 50 years or so to predict the snow load over the next 50 years; changes of loadings due to societal change, e.g., bigger and heavier trucks being allowed on the road; changes in material properties, e.g., soil properties can change over time through physical and chemical processes in the ground, and modifications to design standards due to evolving engineering knowledge, e.g., it is possible that a system being designed today is being under-designed because present design knowledge is inadequate. Since we cannot gain enough knowledge about the future to remove all the uncertainties associated with time, most time uncertainties are primarily aleatory in nature.

The uncertainties produced by *randomness* are ubiquitous in engineering. All material properties are random variables, varying about a central value, such as the mean value. All loads, e.g., wind and earthquake, are highly variable. These uncertainties are generally viewed as aleatory, but some portion of them is likely to be epistemic. For example, as we get more knowledge about how the wind behaves around a structure, possibly through wind tunnel tests, we reduce the uncertainty of the wind loads on the structure. In the past, much of that uncertainty would have been attributed to randomness. Thus, although uncertainty from randomness is aleatory, what appears today to be randomness may be limited knowledge; and if that is the case, that portion of the uncertainty is epistemic.

Along with randomness, we also have uncertainties associated with *statistical limits*. One way to help reduce the uncertainty due to randomness is to obtain material property data, such as taking soil borings before designing a foundation, or taking concrete samples to obtain concrete strength properties for the concrete going into a structure. But, in either case, only a small sample of specimens can be tested or a small number of soil borings can be obtained. So, although the data give some information about the future system, there is uncertainty about whether the small sample data set is representative of the soil under the foundation or the concrete in the structure. This uncertainty is statistical in nature. It is primarily epistemic since we can, in theory, reduce the uncertainty by a significant amount by taking a large number of samples. From a practical standpoint, we cannot take enough samples due to the excessive cost of obtaining and testing the number of samples necessary to reduce the uncertainty in that manner. Thus, although much of the uncertainty associated with statistical limits is epistemic, we can never obtain enough data to reduce that variability significantly. As a result, the uncertainties here include a combination of aleatory and epistemic contributions.

The fourth source of uncertainty is associated with the models used in the analysis of the systems being designed. In the design of non-prototypical engineered systems, models are the only way to examine the behavior of the entire system prior to its construction. Thus, *model uncertainty* is a larger contributor to the uncertainty in the built-system behavior of non-prototypical systems than it is for engineered systems where prototypes can be used, e.g., engine design; we can update our models based on the results from the prototype testing, thereby reducing the epistemic uncertainty associated with the system model. There are two types of model uncertainty. The first is related to how well a prediction equation models test data, and the second is how well a system model, e.g., a finite element model, predicts the final behavior of the system. The first type can be dealt with in design by developing a model bias factor that accounts for the bias between the predicted behavior and the test data. Because the bias factor is itself a random variable, the uncertainty in this context seems to be aleatory; but, as in the flipping of the coin, if we can develop better prediction models, then we will reduce the bias between the prediction and the tests. Another way to reduce the uncertainty in the bias factor is to perform more tests for comparison to the prediction model. It is possible, though, that the additional tests may show that the prediction model is worse than believed, which would either falsify the model or increase the uncertainty in the model. Thus, this first type of model error is a combination of aleatory and epistemic uncertainties. The second type of model uncertainty is primarily epistemic. This uncertainty is affected by the engineer's conceptual understanding of the modeling technique being used, how much effort the engineer can afford to refine the model, and the accuracy of the modeling technique when it is used to its fullest. Given enough time and money, a model could be developed that would closely model the built system, but the amount of time and money necessary would be so great that we could not afford to build the system. Thus, model uncertainty is a combination of both aleatory and epistemic uncertainties, because no matter how carefully we model our system we can never account for all behaviors in the final version, except of course the "model" we create when we build the system.

The fifth contributor to uncertainty is *human error*. Human error is a major contributor to the uncertainty in the design and behavior of structures (Petroski 1982, 1994), and many failures are primarily due to it. Model errors, such as conceptual errors in the development of a structural model, could be classified as human error. For instance, the failure of the Tacoma Narrows Bridge was largely a function of a design that assumed that the wind on the bridge only produced lateral forces that deflected the bridge sideways. Other wind effects were considered unimportant to the design, but turned out to be vitally important. This design assumption could be viewed as a conceptual model error or a designer error: a human error. Generally if the engineer acts in good faith and within the state of the art, the failure is viewed as a model error rather than human error, although the distinction can be difficult to see in some cases. The uncertainties produced by human errors are primarily epistemic. Increases in knowledge will help reduce conceptual errors; and good quality control, design checking, and construction inspection can reduce the incidence of design and construction errors.

## 25.3 Designing and Building Under Uncertainty

Whatever the nature of the uncertainty and whatever its cause, the engineer must make design decisions and the system must be built. In this section we will examine some of the ways that engineers of non-prototypical systems make decisions under uncertainty.

Codes of practice are used to help deal with uncertainties caused by randomness, statistical limits, and some aspects of time and modeling. The level of complexity of a code of practice has an impact on the uncertainty in the design (Elms 1999) and must be considered in discussion of design uncertainty because an overly complicated code can lead to errors in interpretation by designers. On the other hand, a code that is too simple increases the probability of conceptual errors, particularly errors in which relevant aspects of the design are ignored or are inadequately considered in the design. In an overly complicated code the uncertainty will be induced by the engineer having difficulty determining what the code writers had in mind; and in a code that is too simple, the engineer will have to make design decisions based on her own interpretation of existing design knowledge (Addis 1990).

Randomness can be dealt with using probability concepts such as exclusion values, extreme values, and return periods. These probability concepts allow us to deal with uncertainties in loads, other environmental effects (e.g., rainfall), and variability in material properties. Much of this type of information is used in codes of practice to help the engineer deal with uncertainties due to randomness. It should be noted here that extreme value distributions and return periods are also ways to manage the effects of time. For instance, a return period is the average period of time between occurrences of the design value related to that return period. A 100-year flood is the flood level that is expected to occur once every 100 years. Of course, that flood level may occur many times over a 100-year period or it may occur only once in 1,000 years. Uncertainties due to randomness are a large part of the uncertainties dealt with in codes of practice by using various types of safety factors. For more detail about safety factors and code design formats see Bulleit (2008).

Human error is managed using quality control methods, such as peer reviews and construction inspection. The majority of designs of engineered systems are checked by the design engineer, other engineers in the design agency, and, in some cases, engineers in other engineering organizations, often referred to as the peer review process. The techniques discussed below for reduction of contingency also act to reduce human error. An example of a self-check arises in the use of complicated and detailed computer modeling of systems. The engineer or engineers performing the computer analyses should use simple 'back-of-the-envelope' calculations to make sure that the results from the computer analyses are reasonable. This type of check helps find severe input errors as well as gross model errors that may be hiding in the analysis. Human errors can also occur during construction. Inspection of the construction of the engineering system is an important part of minimizing the effects of human error on the safety of non-prototypical engineered systems.

Uncertainty is also induced by contingency (Simon 1996). Dealing with contingency is a way to reduce the uncertainty in the final design. All designs are contingent because the object being designed does not yet exist so the design is being done using a visualization of the system. This visualization can lead to model errors caused by differences between the visualized system and the built system. So any technique that can help the engineer visualize the system will help reduce the effects of contingency. Techniques to enhance visualization and understand the built system include blueprints, computer generated three-dimensional models, building information management systems, physical scale models, and even examination of similar systems that have been built in the past. Fabrication of the engineered system can also lead to uncertainties in the built system behavior since the system as built will vary from the design. Large variations would be considered construction errors and should be detected through quality control measures. But small differences between the designed system and the as-built system are inevitable since no design, no matter how detailed, can describe all nuances of what will be built. The goal of the engineer is to reduce the differences between the as-designed system and the as-built system, and reducing contingency is an important part of that effort. Contingency is one of the major differences between science and engineering: science studies objects and systems that exist in nature, whereas engineers must work with objects and systems that do not yet exist. It is contingency that means that truly unique engineered systems, whether designed to a code of practice or not, exhibit more uncertainty than structures that are similar to existing systems (Shapiro 1997). As will be discussed in more detail below, existing systems act as slow feedback prototypes for similar systems built after them.

## 25.4 Time and Again

Time, as discussed above, is one of the causes of uncertainty. But the in-time behavior of engineered systems also acts to reduce the uncertainty involved in the design of future systems. In this section we discuss the influence of feedback on the design and fabrication of engineered systems, both prototypical and non-prototypical.

Deeper examination of the influence of time suggests that we consider the possibility that the difference between non-prototypical and prototypical systems is simply the time scale over which they are built and used. Billington (1983) distinguishes between *machines* and *structures.* Machines have a shorter life and are modified more rapidly than are structures. Aircraft (Vincenti 1990) are an excellent example of machines. Part of the more rapid modification is that prototypes can be and are used, but other aspects are that machines have a shorter life, are generally less costly to replace, and changes to the needs and desires of society can be incorporated into machines more quickly than can be done with structures. Virtually no machines from the nineteenth century are still in regular use, but there are a fairly large number of bridges and buildings from that era still being used. So, machines as prototypical engineered systems can be tested and modified during the design process and, due to their shorter lifespan, can be modified or replaced more often over time. Thus, there are two feedback loops for prototypical systems that allow changes in the design and the built systems, and both feedback loops are shorter than the in-time feedback loop that exists for non-prototypical engineered systems. If we think about it this way, then the controlling factor is the feedback that the engineer gets on the design. If we can build prototypes, then the feedback time is short enough that we can incorporate the feedback into the design itself.

Consider bridges as an example of a non-prototypical engineered system. Engineers have learned a lot about bridges since the nineteenth century, so in a real sense past bridges are prototypes for future bridges. A good example of this learning is the evolution of the design and construction of suspension bridges (Kawada 2010). But, the feedback on bridges has a long cycle that can only be incorporated into later designs. In some cases, the feedback comes from a failure. The failure of the first Tacoma Narrows Bridge only 4 months after it opened gave a significant amount of feedback to designers of suspension bridges. In the case of Tacoma Narrows, the design was a logical extension of a design trend that began in the early twentieth century. Designers began using the Deflection Theory, which accounts for the deformation of the cables, in 1904 for the design of the Manhattan Bridge. The Deflection theory replaced the discredited Elastic Theory, which did not account for the deformation of the cable (Kawada 2010). The problem at Tacoma Narrows was that the Deflection Theory did not account for the dynamics of the bridge. The narrow roadway and its overly flexible stiffening girder acted like a wing, causing the bridge to go into a fatal oscillatory state. The failure of Tacoma Narrows was primarily due to epistemic uncertainties. Depending on your view, the failure could be classified as a human error or a model error. The aerodynamicist Theodore von Karman, who wrote a letter to the editor of Engineering News Record, clearly believed that the failure was due to human error. He showed, with essentially a back-of-the-envelope calculation, that the narrow roadway and flexible stiffening girder would lead to the kind of behavior that caused the collapse (Petroski 1982). But colleagues of Leon Moisseiff, the designer of the Tacoma Narrows Bridge, held him blameless because they believed that he was conscientious in his work and used techniques that were believed to be adequate at the time he designed the bridge, thus, a model error (Kawada 2010; Petroski 1982). The specific aerodynamic behavior that caused the failure is still being discussed today (Delatte 2009).

The feedback cycle for large-scale non-prototypical engineered systems can be significant, at least years and more likely decades. Since the type of feedback that the engineering community receives is often related to failures of the system, information about failures that is suppressed, say by insurance companies, can increase the length of the feedback cycle and even produce more failures. Design evolution continues without feedback from failures, often by increases in theoretical knowledge, as in suspension bridge design, but the design evolution can move in a potentially unsafe direction if information about failures is actively suppressed, or no significant failures occur. In many ways, the manner in which design evolves toward safer systems, particularly for non-prototypical systems, is through failures (Delatte 2009; Kawada 2010; Petroski 1982, 1994).

As a form of internal feedback, engineers who design non-prototypical engineered systems would be wise to take time to reflect on their designs and the thought processes that they used to justify them (Schon 1983). An example of reflection assisting in design is the case of the 59-story Citicorp Building in New York (Morgenstern 1995; Delatte 2009). In that case the engineer, William LeMessurier, realized shortly after the building was completed that a mistake had been made in the design. The building has nine-story columns supporting it at the center of each of the four sides of the building rather than on the corners as would be more traditional. LeMessurier realized, as he was discussing the building design, that the structural analysis used in the design considered only the winds acting on the face of the walls, as would be the case for traditionally located columns. But, with the columns located in the middle of the walls, quartering winds, winds acting at a 45° angle to the building, would control the design. He went back to the design believing that the building was still safe, since the design had called for the steel connections to be welded. Unfortunately the design had been changed, without his knowledge, to bolted connections rather than welded connections. Furthermore, a conceptual design error had caused the bolted connections to be somewhat under designed for some critical connections. The quartering wind error, combined with the design change and the discovery of the additional conceptual error, led LeMessurier to believe that the structure was potentially unsafe. In addition, the high wind season was about to descend on New York. LeMessurier told the owners about the danger, and the building was retrofitted by working at nights to go back in and weld all the critical connections in the building (Morgenstern 1995; Delatte 2009). Although LeMessurier's reflections were driven by discussions with others, the practice of engineers reflecting on their designs can prove useful to assist in ensuring the safety of engineered systems, as it has helped practitioners in other fields (Schon 1983).

## 25.5 Black Swan Events

So far, I have been discussing non-prototypical engineered systems that are complicated but not complex. Bridges and buildings are examples of complicated systems, although as we will see, the effects on these systems may be driven by complex systems interacting with them. The first type of complex system involves human interactions in the operation of the system and typically system responses are tightly coupled. Tight coupling means that small failures in the system can lead to preventative actions, both human and automatic, that cause more small failures and more compensating actions that eventually lead to cascading failures causing system collapse (Perrow 1999). Examples of this type of system include petrochemical plants, the power grid, and nuclear power plants. Complex adaptive systems are the second type. These systems have interactions among agents as well as interactions between the agents and their environment; the agents are also adaptive allowing the system to evolve. Large organizations, human societies, and the human/natural environment system are examples of complex adaptive systems.

Complex, tightly-coupled systems are clearly non-prototypical, but in many cases, such as the power grid, their design evolves such that some parts of the system may be designed using earlier design techniques that have been updated prior to designs for later portions of the system. In complex, tightly-coupled systems, whether designed and built as a single unit, like nuclear power plants, or evolve, such as the power grid, the uncertainties are potentially much greater in both range of possibilities and consequences. A tree falling on a power line in Ohio can lead to a power outage across a large portion of the upper Midwest. The design of complex, tightly-coupled systems generally includes safety devices that are expected to protect human life by preventing or mitigating accidents. These safety devices, both human-operated and automatic, often increase the complexity of the system and may themselves contribute or even initiate system failures. The accident at Three Mile Island was exacerbated by failures of safety devices (Perrow 1999). Another aspect of dealing with complex tightly coupled systems is the tendency for managers and operators of the systems to ignore or actively suppress information about small failures that at the time appear to be a normal part of the operation. The response of managers and engineers to the o-ring problems on the Challenger prior to the fatal day showed a willingness to ignore warnings that should have been heeded (Delatte 2009). Note that the system in this case includes the spacecraft, the launch environment, and managers and engineers at NASA and Morton-Thiokol. Small failures are easy to ignore, particularly in managerial environments where operators and engineers are castigated for reporting problems. The danger in complex, tightly coupled systems is that small failures may not be isolated and may eventually produce cascading failures leading to catastrophe. These cascading failure events are difficult, some would argue impossible, to predict (Perrow 1999).

The uncertainties associated with complex, tightly coupled systems generally encompass all the uncertainties discussed throughout this paper, plus include uncertainties associated with interactions that can produce cascading failures leading to partial or total system collapse. The interactions include human-human interactions, human-system interactions, and interactions between system components, including safety devices. It is these interactions and the uncertain nature of them that produce unintended and unpredictable consequences when the system is affected by internal or external stressors. Consequences that are far outside the realm of experience are sometimes referred to as *black swan* events (Taleb 2007). 'Black swan' refers to the long held belief that all swans were white until black swans were discovered in Australia. Karl Popper (2002) used the black swan example when discussing the problem of induction. Taleb (2007) discussed black swan events from the perspective of his experience in the investment community, although the concept applies broadly. As far back as 1921, Knight (1921/1948) considered the types of uncertainty in the business environment. He divided them into *measurable uncertainties* and *unmeasurable uncertainties.* From a business standpoint, if the probability of an event can be determined, it is a measurable uncertainty and can be managed using insurance. An unmeasurable uncertainty cannot be managed using insurance because it is not possible to determine its probability due to its unpredictability. Knight's unmeasurable uncertainties are black swan events.

Complex adaptive system events can also lead to surprising effects. Two examples are the attack on the Alfred P. Murrah Federal Building in Oklahoma City, Oklahoma in April 1995 and the destruction of the World Trade Center Twin Towers on September 11, 2001. Both of these attacks represent unforeseen and unpredictable, thus highly uncertain, events emerging from the complex adaptive system that is human society. These attacks are also black swan events.

The challenge in dealing with black swan events is that by definition they are not only unpredictable, but also outside the realm of experience of designers and operators. Engineers speak of *robust* systems. Robust systems are able to withstand unusual events due to aspects of the system that increase the capacity of the system beyond what it was designed to do. After the Murrah building bombing, designs for new federal buildings required adequate connection between the columns and the floor system. Since the Murrah building was in a low seismic region, the connections between the columns and floor slabs were minimal. This design, acceptable in Oklahoma City, allowed the blast to lift the slabs up off the columns, which led to the collapse of all floors of the building. If the Murrah building had been designed for seismic conditions, it likely would have responded better to the blast because the columns and floor system would have stayed together, thus exhibiting more robust behavior (Delatte 2009). In the case of the Twin Towers, they withstood the plane crashes, didn't topple over, and stayed up long enough to allow many of the occupants to escape. The original design of the towers included the possibility of a slow-moving Boeing 707 crashing into a tower during landing. This design criterion was driven by the crash of a B-25 into the Empire State building in 1945 while it was flying in fog (Delatte 2009). The high speed crash into the towers of Boeing 767s being flown by terrorists far exceeded the design forces from a slow moving 707, but the building withstood the crash. Even though the towers eventually collapsed, they exhibited robust behavior with respect to the initial crash. The attack would have been much worse if the towers had toppled over under the impact of the aircraft.

By definition it is not possible to design for black swan events, but systems can be designed such that they have details that increase the chances that the system will respond to unusual events in a robust manner. Robustness is difficulty to measure and engineers will often use their own heuristics in attempting to make their systems robust. Whether they are successful or not is only determined when the system experiences a significant unpredicted, and thus not designed for, event.

## 25.6 Conclusion

Uncertainties arising from the effects of time, randomness, statistical limits, modeling errors, and human errors can be separated into two broad categories, aleatory, related to chance, and epistemic, related to knowledge. Uncertainty in non-prototypical engineered systems is greater than in systems in which prototypes can be used because the prototype testing reduces the overall level of uncertainty. Furthermore, prototypes allow feedback to the designer during the design phase of a system, whereas for non-prototypical systems feedback only occurs over relatively long periods of time after the system is built. The often high levels of uncertainty in non-prototypical systems are managed using codes of practice, methods to reduce contingency, inspections during the building of the system, and heuristics, such as ways to increase the robustness of the system, where robustness is the ability of the system to withstand events over and above design levels. Complex systems that have significant component interactions as part of the system can exhibit behavior that is far outside the experience of the designers of the system. Extreme unpredictable events are sometimes referred to as black swan events. In complex systems, efforts to increase robustness are important to the safety of the system and its users, but the efficacy of the efforts can be difficult to measure due to the complexity of the system. Typically, the robustness of the system is only made evident when the system is subjected to an extreme, unpredictable event. Design of non-prototypical systems requires techniques to manage uncertainty that go far beyond the techniques used for systems where prototypes are possible.

## References

Addis, W. (1990). *Structural engineering: The nature of theory and design*. Chichester: Ellis Horwood.

Billington, D. P. (1983). *The tower and the bridge*. Princeton: Princeton University Press.

Blockley, D. I. (1992). Setting the scene. In D. I. Blockley (Ed.), *Engineering safety* (pp. 3–27). London: McGraw-Hill.

Bulleit, W. M. (2008). Uncertainty in structural engineering. *Practice Periodical on Structural Design and Construction, 13*(1), 24–30.

Delatte, N. J., Jr. (2009). *Beyond failure: Forensic case studies for civil engineers*. Reston: American Society of Civil Engineers.

Der Kiureghian, A., & Ditlevsen, O. (2009). Aleatory or epistemic? Does it matter? *Structural Safety, 31*, 105–112.

Ellingwood, B. R., Galambos, T. V., MacGregor, J. G., & Cornell, C. A. (1980). *Development of a probability based load criterion for American national standard A58*. Gaithersburg: U. S. Department of Commerce, National Bureau of Standards.

Elms, D. (1999). Achieving structural safety: Theoretical considerations. *Structural Safety, 21*, 311–333.

Kawada, T. (2010). *History of the modern suspension bridge*. Reston: American Society of Civil Engineers.

Knight, F. H. (1948). *Risk, uncertainty, and profit*. Boston: Houghton Mifflin. Original work published 1921.

Koen, B. V. (2003). *Discussion of the method*. New York: Oxford University Press.

Morgenstern, J. (1995, May 29). The fifty-nine story crisis. *The New Yorker*, 45–53.

Perrow, C. (1999). *Normal accidents: Living with high risk technologies* (2nd ed.). Princeton: Princeton University Press.

Petroski, H. (1982). *To engineer is human*. New York: St. Martin's Press.

Petroski, H. (1994). *Design paradigms: Case histories of error and judgment in engineering*. New York: Cambridge University Press.

Popper, K. R. (2002). *The logic of scientific discovery* (15th ed.). London: Routledge.

Schon, D. (1983). *The reflective practitioner*. London: Temple Smith.

Shapiro, S. (1997). Degrees of freedom: Interaction of standards of practice and engineering judgment. *Science, Technology, and Human Values, 22*(3), 286–316.

Simon, H. A. (1996). *The sciences of the artificial* (3rd ed.). Cambridge, MA: MIT Press.

Taleb, N. N. (2007). *The black swan: The impact of the highly improbable*. New York: Random House.

Vincenti, W. G. (1990). *What engineers know and how they know it: Analytical studies from aeronautical history*. Baltimore: The Johns Hopkins University Press.

# Chapter 26
# Object-Oriented Method and the Relationship Between Structure and Function of Technical Artifacts

**PAN Enrong**

**Abstract** The positive relationship between the structure and function of a technical artifact challenges present-day philosophy of technology and engineering science, since these philosophical approaches cannot support the idea that such a relationship exists. According to the recent Empirical Turn in the philosophy of technology and engineering science, Jeroen de Ridder's 'Functional Decomposition', a reductive design methodology, explains this relationship in the context of rational reconstruction. This explanation does not apply, however, in the context of creative design. In this chapter, I propose a new model to explain the positive relation between structure and function: a holistic methodology which I call the Object-Oriented Method. This model can explain two well-known phenomena associated with the relationship between structure and function, namely underdetermination and realizability constraints.

**Keywords** Design methodology • Functional decomposition • Object-Oriented Method • Structure and function • Philosophy of engineering design

This paper was first presented at the WPE-2008 in London. Chinese Language Rights (Simple and Traditional Chinese) of this paper are with China Social Sciences Press, the text is a translation of parts of the book: *Philosophy of Engineering Design: The Relationship between Structure and Function of Technical artifacts*.

E. PAN (✉)
Teaching and Research Institute of Political Theory,
Zhejiang University, Tianmushan Road 148, 310028 Hangzhou, China

Department of Philosophy, Delft University of Technology,
Jaffalaan, 5, 2600 GA Delft, The Netherlands
e-mail: enrongpan@zju.edu.cn

D.P. Michelfelder et al. (eds.), *Philosophy and Engineering: Reflections on Practice, Principles and Process*, Philosophy of Engineering and Technology 15,
DOI 10.1007/978-94-007-7762-0_26, 

## 26.1 Introduction

The relationship between the structure and function of technical artifacts is one of the most interesting problems in current philosophy of technology and engineering science. Peter Kroes has claimed that an adequate description of a technical artifact will have two elements: (1) a physical/structural description, and (2) a functional description (Kroes 1998). This is known as the dual nature of technical artifacts (DNTA) (Kroes and Meijers 2006). DNTA raises many problems and questions (Mitcham 2002). For example, why do we speak of a dual nature, and not a triple or quadruple one? Is this analogous to Cartesian dualism, and the associated mind-body problem? What is the relationship between these two natures? How is a structural description related to a functional description, and vice versa?

The focus of these questions is on the relationship between the structure and function of technical artifacts. However, in analytical philosophical approaches, there is no positive relationship between structure and function. From a logical perspective, Kroes (1998) argued there was a discontinuity between the two natures, such that a structural description could not be deduced from a functional description, and vice versa. Kroes concluded that a logical gap exists between structure and function (see Fig. 26.1 (Kroes 1998, p. 32)). From an epistemological perspective, it has been argued that functional knowledge is not obtained through structural knowledge (Houkes 2006). From an ontological perspective, Wybo Houkes and Anthonie Meijers argued that any explanation of such a relationship should first successfully explain the phenomena of Underdetermination (UD), i.e. that multiple realizations occur from structure to function, and vice versa; and the Realizability Constraint (RC), i.e. that inferences can be made from statements about function to statements about structure and vice versa (Houkes and Meijers 2006). This has led some to conclude that no present explanation can "offer the conceptual resources needed to describe the relation between these natures" (Houkes and Meijers 2006, p. 118).

Although the relationship between these two natures raises a hard question for recent philosophers of technology, an engineering approach, suggested by the Empirical Turn (ET), offers some insight. ET is "a call to base philosophical analysis concerning technology on reliable and empirically adequate descriptions of technology (and its effects)" (Kroes and Meijers 2000, p. xxiv). Kroes (2002) also suggested that design methodology and the two natures of technical artifacts were so closely related to each other that the former could not be understood without offering insight into the latter. Ridder (2006, 2007) then proposed to describe the relationship between structure and function by using a reductive design methodology known as Functional Decomposition (FD). FD fails to describe, however, the relationship between the two natures in the context of creative design (Ridder 2007, p. 245).

It is a fact that engineers create technical artifacts, and are able to successfully connect structure and function in the context of creative design. Apparently, there is

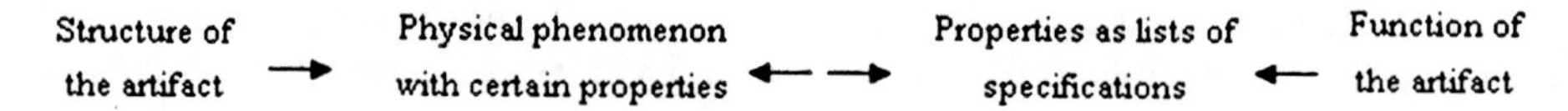


**Fig. 26.1** The schema of a logical gap between structure and function

a certain positive relationship between structure and function that engineers use, but it is unclear what this relationship is.

According to Ridder's solution, there remains a gap between atomic function and atomic structure in the context of creative design (Ridder 2007). However, the failure of FD in Ridder's solution does not mean the design methodology entirely fails to deal with the problem of the relationship between structure and function of technical artifacts. Rather, there is another type of design methodology, termed the Object-Oriented Method (OOM), which offers a holistic design competitor to FD.

The goal of this paper is to highlight the advantages of OOM in explaining: (a) the positive relationship between the structure and function of technical artifacts; and (b) the phenomena of UD and RC, especially within the context of creative design. UD and RC are taken as two criteria for an adequate ontology of artifacts (Houkes and Meijers 2006). This means that any theory about the relationship between structure and function should satisfy UD and RC. In this chapter, I will make two assumptions. One is that natural objects (e.g. rocks), social artifacts (e.g. law), aesthetic artifacts (e.g. paintings), software, and technological byproducts (e.g. engine noise) are not technical artifacts, because technical artifacts are special physical structures constructed by humans for specific purposes (Kroes 2002). The other is that the function of the technical artifact to be designed is known well before the process of design. Therefore, should the technical function be changed, the process of design must also change.

In Sect. 26.2, I will introduce two critical concepts: *Object* and *Class*. My strategy and model will then be presented in Sect. 26.3. In Sect. 26.4, a successful mold design will be introduced in the context of creative design. In Sect. 26.5, through a step by step analysis of this mold design, I will show how a relationship exists between structure and function. Section 26.6 demonstrates how the new model can also serve to explain the phenomena of UD and RC. A summary conclusion is offered in the final section.

## 26.2 Object-Oriented Method: Concepts

The concepts of *Object* and *Class* are foundations of OOM. An *Object* is an entity which can be described in terms of status and behaviors. Similarly, DNTA had claimed that a technical artifact bears two attributes (i.e. structure and function) simultaneously (Kroes 1998). Here, I propose that for a technical artifact considered as an *object*, structure equates to *status* and function equates to *behaviors*.

A *Class* is a concept abstracted from, and representing the common characteristics of, *objects*. It is, therefore, reasonable to regard a *class* as a certain kind or type of *object*. Although an *object* is an instance of one *class*, an *object* can also be an instance of different *classes*. For example, a glass cup is an *object* of the *class* 'cup', as well as a functional name for a *class*, and also an *object* of the *class* 'glass artifact', which is a structural name for a *class*.

The reason that a glass cup could be an object of different *classes* is because technical artifacts closely relate to two contexts of human behaviors: (1) the context of use, and (2) the context of design. Kroes argued that, in the context of use, technical artifacts were "characterized primarily in a functional way" while the structure remained a black box. By contrast, in the context of design, technical artifacts were "described as some kind of physical system" while the function remained a black box (Kroes 2002, p. 292). To simplify, in the context of use the function of an *object* is dominant, while in the context of design the structure of an *object* is dominant.

There is a relationship of Underdetermination between structure and function. For example, the function of a cup is drinking, which can be realized by a glass cup or a paper cup. In the context of use, the functional concept of 'drinking' not only refers to the functional description of a glass cup but also to the functional description of the *class 'cup'*. The *class 'cup'* is the set of cups constructed of glass, paper or plastics.

Similarly, in the context of design, the structural concept of 'cylinder' not only refers to the structural description of a glass cup but also refers to the structural description of the *class 'cylinder'*. The *class 'cylinder'* is a set of artifacts (e.g. a glass cup or a paper cup) whose geometric shapes are cylindrical.

Following in this vein, rather than selecting the *function of object* as dominant in the context of use, I prefer to speak of the *Technical Function of the class*. Similarly, in the context of design, I would prefer assuming the *Structure of the class* as dominant, rather than the *structure of the object*.

## 26.3 Strategy and Model

Jeroen de Ridder's model is based on Functional Decomposition, using a top-down strategy. Following OOM, my strategy is different than Ridder's, in that it adopts a bottom-up approach (see Fig. 26.2).

Since it is difficult to find a positive relationship between structure and function directly within the context of creative design, I intend to explore this relationship indirectly, as did Ridder. I also accept the idea contained in the Function-Behavior-Structure (FBS) model of Rosenman and Gero (1998) that behaviors are the intermediary concept between structure and function. I will, therefore, use the term *behavior* instead of *function* (just as Ridder had done). While exploring the details of behaviors, I will describe them as a mathematical function (MF). MF-S(t) represents the information of behaviors bearing the structure of a technical artifact; whereas MF-TF(t) represents the information of behaviors bearing the function of

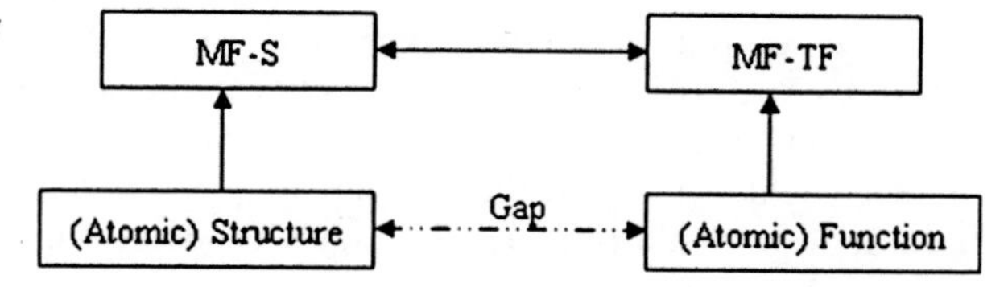


**Fig. 26.2** Bottom-up strategy in the context of creative design

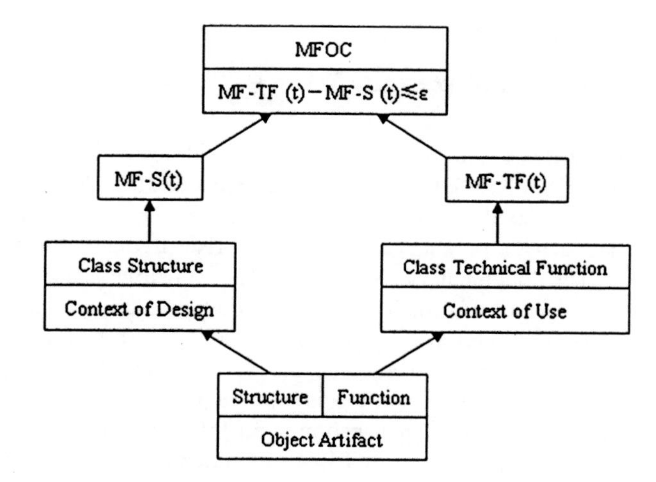


**Fig. 26.3** Method of Mathematical Function of Class (MFOC)

technical artifact. The positive relationship between function and structure emerges when MF-S(t) and MF-TF(t) are represented as equal.

Based on this bottom-up strategy, a method of Mathematical Function of Class (MFOC) was developed (Fig. 26.3). On the bottom level is the artifact as an *object*; which has structural descriptions for the description of status, and functional descriptions for the description of behaviors. The second level is comprised of two contexts for human behaviors. In the context of use, the *class structure* of technical artifacts is the dominant character and, in the context of design, the *class technical function* of technical artifacts is the dominant character. The third level reflects two mathematical functions: MF-TF(t) and MF-S(t). When a special rule is selected to decompose a function to sub-functions, what results is a unique set of spatiotemporal sub-functions (i.e. intended sub-behaviors). Following a similar procedure to decompose structure, a unique set of physical sub-behaviors is found. Theoretically, by applying the same method of mathematical modeling, MF-TF(t) and MF-S(t) are fashioned. MF-TF(t) represents the mathematical formula for the set of spatiotemporal sub-functions (i.e. *class technical function*) in the context of use. MF-S(t) represents the mathematical formula for the set of spatiotemporal sub-functions (i.e. *class structure*) in the context of design. The top level formulation encompasses the MFOC model in its entirety. If the result of the subtraction between MF-S(t) and MF-TF(t) is less than the requirement of error ε, generally, MF- S(t) and MF-TF(t) may be deemed as equal. The positive relationship between the structure and function of technical artifacts then emerges. The error ε comes from the functional requirements of the technical artifacts. If the comparison is more than error ε, a return to the first level to begin the formulaic cycle again is advised (e.g. selecting a new function, then revising the set of spatiotemporal sub-behaviors or applying the rule of decomposition.)

## 26.4 Mold Design: A Case

Let me now introduce a case about mold design which is based on a documentary made by the NHK[1] (the Japanese Broadcasting Corporation) in 2005. This documentary explores the case of mold design, showing how strong the competition is between Chinese and Japanese enterprises in the high-tech mold design sector.

The documentary records a client asking a Japanese factory to design and manufacture a kind of mold that could produce a circular copper annulus from copper plate. The client had been to many mold design factories in China, but no one was prepared to accept such an order. He considered returning to Japan in search of a manufacturer who could satisfy his requirements. In order to obtain this order, the factory had to design a specimen for the client to test.

The engineers clearly listed and stated the requirements of the client. The engineers then imagined a hypothetical context within which the client would use the specimen mold, then attempted to provide details of the mold's intended behaviors. The function of the specimen mold, according to the client, was to bend the copper plate from a flat plate to a circle. Circles, in the engineering community and in the mold design industry, are regarded as the most difficult targets to achieve. As such, they are rarely realized. Time was too limited for the engineers to adequately discuss, research, test, and then perform trial-and-error designs. The engineers, therefore, decided to start by following the traditional method of mold design, as the only method available at that time. If it failed, they hoped they would at least find some clues for a possible solution.

The engineers created a team for the circle project and a young engineer, who had about a decade of experience in mold design, was authorized to be the chief engineer of the project. Before beginning the process of mold design, the engineers decided they would use a method of functional decomposition to describe the process the client engages in when using the specimen mold. They roughly divided the process into four main sub-periods. First of all, the mold must slit a long copper plate into shorter plates, the lengths of which would be the circumference of the circle, then chip regularly placed sawteeth along the two long flanks. Secondly,

[1] http://www.nhk.or.jp/special/onair/051127.html

relying upon decades of experience, the engineers hypothesized that the mold would bend the two longer sides of the rectangle to a right angle while the middle part would remain straight. Thirdly, the mold would bend the middle part from a straight line to a right angle. Finally, the circle copper annulus would be shaped by stamping the mold and pressing it out. The margin of error of diameter in any direction was not to exceed 0.05 mm.

The time for the first test came. Unfortunately, the output did not satisfy the margin of error requirements. Despite attempts to adjust the components of the mold and the procedures, the resulting molds were not circular enough.

Although there were more than 20 procedures involved in creating the mold, the engineers had to test and examine every procedure carefully. After a long period of testing and examining, they still could not locate the cause of the problem. Nevertheless, they had a sense that the components and procedures were correct because they were being manufactured according to the designed blueprint. It was highly probable, therefore, that the method for designing this mold was problematic.

To investigate the design method further, the engineers decided to temporarily halt testing. Their factory had a database which contained all the records of cases and projects the factory had overseen in past several decades. The young chief engineer found some records on the designing and manufacturing of circular shapes. Although these projects had failed,, his predecessor had recorded the whole process: indicating what the project was, where it came from, what problems were encountered, what measures had been taken to compensate for the errors, what the possible reasons for failure had been, and the approaches that had been taken to create a circular mold.

With the help of these records, the engineers were able to return to testing and examining the failed mold. They finally found the problem. In line with their functional decomposition, the mold bent the metal plate in a step by step fashion. Since most parts of the plate were curved, the parting core (see $PB_3$) could then be inserted inside the curved plate. With the stamping of the mold, the parting core and the outer mold shaped the circular copper annulus. However, that was where the problem emerged. Because metal plate has elastic qualities, it was discovered that, at the moment of stamping and shaping, the metal would react to the pressure and the copper annulus would become quite circular. When the pressure of the mold was removed, the annulus would spring back a subtle distance, which led to a error of more than 0.05 mm.

The mold, therefore, required revision. Because there were no successful known theories or tools to reference, such revision depended entirely on the knowledge and skill of the engineers. By accessing his own instincts and professional experience, the younger engineer revised the arc surface of the component by the method of trial-and-error.

Work continued on the project up until the day that the client returned to check the specimen. The outcome was good enough to accept: the average error of the mold's output was approximately 0.03 mm.

## 26.5 Modeling

According to the description of the case, the client hoped that artifact A (the specimen mold), with a function $F_A$ that was used for bending the copper plate to form the copper annulus, would not have an error margin, in terms of diameter, of more than 0.05 mm. Generally, the client had no idea of the black box of structure and procedures behind artifact A. He was concerned that the copper annulus satisfied the error requirements.

The functional description from the client was vague, in terms of how the engineers were to design the specimen mold directly. Thus, before the process of design, the first thing which had to be done was to translate this vague idea into one that was more explicit.

The engineers imagined the context in which the client would use the specimen mold and how the specimen mold would perform its function during the period of work. The engineers chose the method of functional decomposition (FD) to explore the details of how to realize the overall function. As mentioned in the previous section, the engineers presumed that four sub-periods were necessary. In other words, during the working period of time T, there were four sub-periods. At the time of the first sub-period $t_1$, the intended behavior $IB_1$ of the artifact was to cut the long copper plate into shorter plates and chip some sawteeth along the two long flanks. During the period of $t_2$, $IB_2$ bent the two sides of the rectangle to a right angle. At the period of $t_3$, $IB_3$ bent the middle part to a right angle. At the period of $t_4$, $IB_4$ shaped the circular copper annulus by stamping. That is, the functional description from the client was translated to a series of intentional behaviors in the context of use.

Suppose one of the long sides of the copper plate is AOB and O is the middle point. During time T, point O is fixed and is the origin (see Fig. 26.4). At the end of time $t_1$ and after the intentional behavior $IB_1$, the position of point A in spatio-temporal coordinates Time-Position is $(t_1, p_1)$. At the end of time $t_2$ and after the intentional behavior $IB_2$, the position of point A moves to the position of point C, whose coordinate is $(t_2, p_2)$. The coordinates of other points may be similarly derived, for instance, $E(t_3, p_3)$ and $H(t_4, p_4)$. Arc ACEH is the intentional trajectory of point A and its mathematical function is $MF\text{-}TF_A(t)$. That is, the mathematical function of the technical function of the specimen mold is $MF\text{-}TF_A(t)$. Analogously, arc BDFH is the intentional trajectory of point B and its intentional mathematical function is $MF\text{-}TF_B(t)$.

When the process of design was beginning, $MF\text{-}TF_A(t)$ had to be fixed, especially the value of point H. The engineers took $MF\text{-}TF_A(t)$ and the value of point H as the blueprint during the process of design. $MF\text{-}TF_A(t)$ results from the engineers having translated a vague functional description to a explicit one in the context of use.

According to structural tests and examination in the context of design, during time T, point O is still fixed, and it is the point of origin (see Fig. 26.5). At the end of time $t_1$ and after physical behavior $PB_1$, the position of point A in the Time-Position auxiliary plane is $(t_1, p^*_1)$. At the end of time $t_2$ and after physical

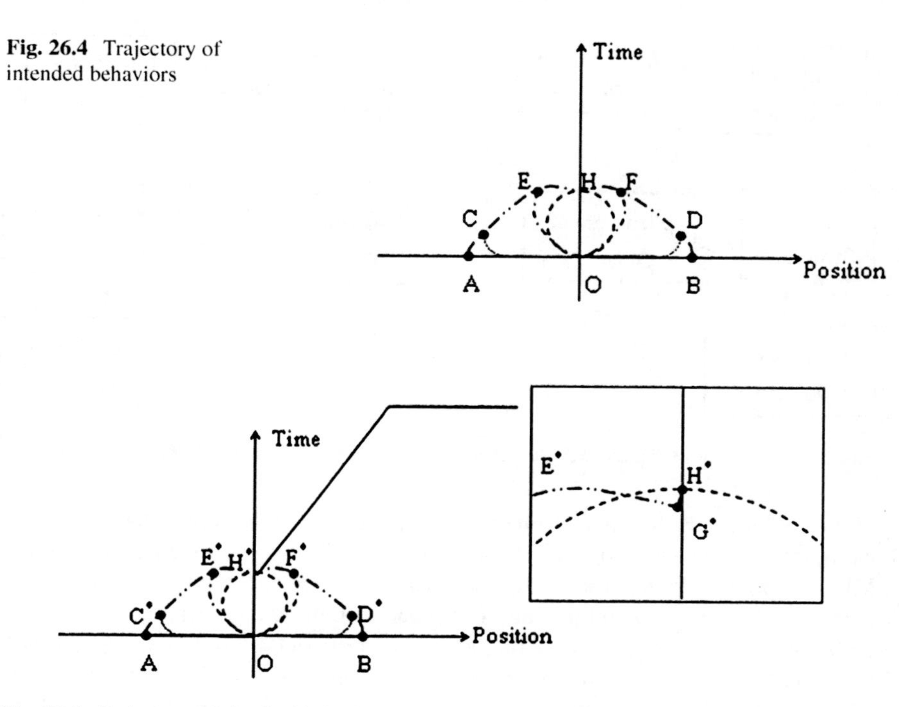


**Fig. 26.4** Trajectory of intended behaviors

**Fig. 26.5** Trajectory of physical behaviors

behavior $PB_2$, the position of point A moves to the position of point $C^*$ whose coordinate was $(t_2, p^*_2)$.

Theoretically, if the design of the mold is successful, the spatial positions, the final effects of IB and PB, will be the same. This means that the points of MF-S(t) and MF-TF(t) in spatiotemporal coordinates Time-Position would be the same, i.e., point $C^*$ will be superimposed on point C. The other points, such as $E^*$, $H^*$, $F^*$, and $D^*$ will, respectively, be superimposed on E, H, F, and D.

Because the metal plate has an elastic character, there is another physical behavior and one more point $G^*$ on Arc $AC^*E^*H^*$ (see the enlargement in the right part of Fig. 26.5). The physical trajectory of point A is $AC^*E^*G^*H^*$ and its mathematical function is MF-$S_A$(t).

On MF-$TF_A$(t) the functional behaviors act from point A to point H, while on MF-$S_A$(t) the structural behaviors act from point A to point $G^*$ then to $H^*$. That is, MF-$TF_A$(t) and MF-$S_A$(t) are not equal, and the functional behaviors and structural behaviors are not the same either. One might conclude that there is not a positive relationship between structure and function.

However, the final proof of the successful outcome of the mold is the fact that the error of any directional diameter is about 0.03 mm, which is less than the 0.05 mm requirement. For the client, it was the goal that he desired. Thus, the factual error being less than the requirement error ε, it may, according to the practical reasoning

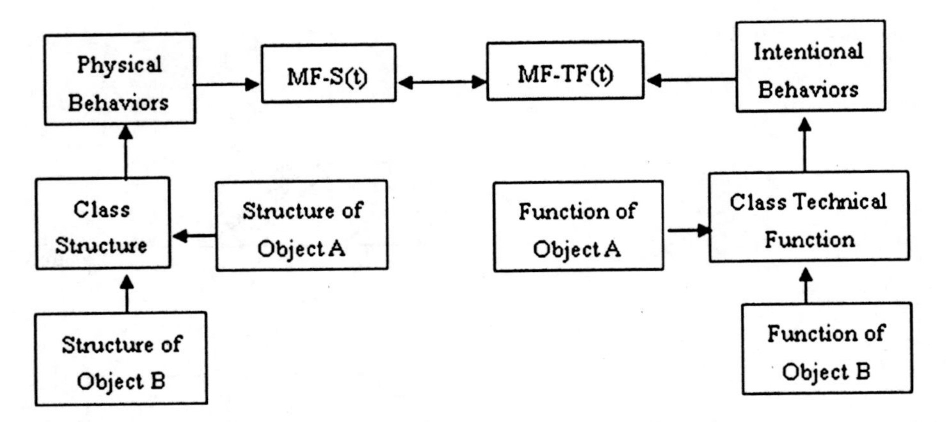


**Fig. 26.6** Model of MFOC between the structure and function of technical artifacts

of the engineers, be ignored. In other words, point $G^*$ and the subtle elasticity would be ignored. Thus, the Arc ACEH and $AC^*E^*G^*H^*$ may be taken as equivalent. $MF\text{-}TF_A(t)$ and $MF\text{-}S_A(t)$ are also taken as equal. Thus, the positive relationship between the structure and function of the specimen mold emerges.

If the error of any directional diameter is more than 0.05 mm, i.e., the elasticity will lead to an error margin which cannot be ignored; point $G^*$ cannot therefore be ignored either. The Arcs ACEH and $AC^*E^*G^*H^*$ would be different arcs. Thus, $MF\text{-}TF_A(t)$ and $MF\text{-}S_A(t)$ could not be taken as equivalent. In this case, there is also no positive relationship between structure and function.

To sum up, I propose a general model of the relationship between physical structure and function (see Fig. 26.6).

## 26.6 Test and Discussion

Houkes and Meijers argued that any theory about the positive relationship between the structure and function of technical artifacts should explain the phenomena of UD and RC.

> Underdetermination
>
> Artifacts should accommodate a two-way underdetermination between artifacts and their material basis: an artifact type, as a functional type, is multiply realizable in material structures or systems, while a given material basis can realize a variety of functions.
>
> Realizability constraints[2]
>
> Artifacts should accommodate and constrain the two-way underdetermination of artifacts and their material basis. There are many kinds of practical inferences from functional to structural statements and vice versa. (Houkes and Meijers 2006, p. 120)

[2] Realizability constraints also relate to malfunction, but malfunction will not be discussed in this paper.

UD and RC imply that there are no logical one-to-one relationships between structures and functions. One physical structure may perform different functions (but not any arbitrary function) and the same function may be performed by different physical structures (but not any arbitrary physical structure).

With regard to the phenomena of UD, there are two situations. One is the multiple realization of function on the basis of one physical structure. Suppose there is a positive relationship between structure of *Object* A and function of *Object* B and the function of *Object* B is another object of the *Class Technical Function*. The Mathematical Function of Class of the function of *Object* B will also be MF-TF(t). The function of *Object* B and structure of *Object* A then create another positive relationship. In other words, the two functions of *Object* A and *Object* B can be realized by one structure of *Object* A. For instance, an object with the structure of an airplane can be used to take people from one place to another or it can be used as a mobile museum.

The other point is that one function can be realized from different physical structures. Suppose there is a positive relationship between the structure of *Object* A and the function of *Object* B, and the structure of *Object* B is another object of the *Class Structure*. The Mathematical Function of Class of the structure of *Object* B will also be MF-S(t). The structure of *Object* B and function of *Object* A then create another positive relationship. In other words, the two structures of *Object* A and *Object* B can perform one function of *Object* A. For instance, a slotted screwdriver and a coin can both be used to tighten screws. In this way, the phenomenon of Underdetermination (UD) can be explained.

With regard to the phenomena of RC, there are also two possible situations. On the one hand, there are multiple possibilities, but not all functions can be realized on the basis of one physical structure. Suppose there is a positive relationship between the structure of *Object* A and the function of *Object* B, and the function of *Object* C is not an object of the *Class Technical Function*. The function of *Object* C would lead to another Mathematical Function of Class MF-$S_C$(t). MF-$S_C$(t) and MF-TF(t) are different mathematical functions and the difference between functional values of these mathematical functions is more than the requirement of error $\varepsilon$. Thus, there is not a positive relationship between the structure of *Object* A and the function of *Object* C. This means that some functions, such as transporting people to outer space, cannot be realized by the structure of a car.

On the other hand, one function is realized by different physical structures, but not by any physical structure whatsoever. Suppose there is a positive relationship between the structure of *Object* A and the function of *Object* B, and the structure of *Object* C is not an object of the *Class Technical Function*. The structure of *Object* C would lead to another Mathematical Function of Class MF-$TF_C$(t). MF-$TF_C$(t) and MF-S(t) are different mathematical functions and the difference value of them are more than the requirement of error $\varepsilon$. Thus, there is not a positive relationship between the structure of *Object* C and the function of *Object* A. This means that some structures, such as a spherical ball, cannot perform the function of fastening slotted screws. Thus, even the phenomenon of the Realizability Constraint (RC) can be explained by the MFOC model of the relationship between structure and function.

## 26.7 Conclusion

The reason that the relationship between the structure and function of technical artifacts poses a difficult problem is because there is no positive relationship between structure and function found in traditional philosophical approaches. This demonstrates the need for an engineering approach via the Empirical Turn, where a positive relationship can be found by looking at case studies of, and the methods of, engineering design. Using the reductive design methodology Functional Decomposition, Ridder successfully explained the positive relationship in the context of rational reconstruction; however, he failed in the other context of creative design

The failure of FD in Ridder's solution did not mean the failure of an engineering approach; we can find the positive relationship between structure and function of technical artifacts by using design methodologies. Thus, I demonstrate here the positive relationship between structure and function of technical artifacts dependent on an Object-oriented Method (OOM), which is different from FD. There are three reasons why I chose OOM. Firstly, Object-oriented Method (OOM) is a holistic design methodology competing with Functional Decomposition. The disadvantages of FD are often the advantages of OOM. Secondly, the definition of the Dual Nature of Technical Artifacts (DNTA) is similar to the definition of *Object* in OOM. Thirdly, the fact that one *Object* can be an instance of different *Classes* satisfies the phenomenon of Underdetermination.

**Acknowledgements** This work was supported in part by the National Natural Science Foundation of China (NSFC) through the Science Fund for Young Scholars under Grant 50905158, by Zhejiang Planning Project of Philosophy and Social Science under Grant 10CGZX06YBQ, by China Scholarship Council (CSC) under Grant 2007U07038. The author would like to thank Peter Kroes, Maarten Franssen, Pieter E. Vermaas, and anonymous reviewers for comments on an earlier version of this paper.

## References

Houkes, W. (2006). Knowledge of artifact functions. *Studies in History and Philosophy of Science Part A, 37*(1), 102–113.

Houkes, W., & Meijers, A. (2006). The ontology of artifacts: The hard problem. *Studies in History and Philosophy of Science Part A, 37*(1), 118–131.

Kroes, P. (1998). Technological explanations: The relation between structure and function of technological objects. *Society for Philosophy and Technology, 3*(3), 18–34. http://scholar.lib.vt.edu/ejournals/SPT/v3n3/KROES.html. Accessed 25 Sept 2010.

Kroes, P. (2002). Design methodology and the nature of technical artifacts. *Design Studies, 23*(3), 287–302.

Kroes, P., & Meijers, A. (2000). *The empirical turn in the philosophy of technology*. Amsterdam: JAI Press.

Kroes, P., & Meijers, A. (2006). The dual nature of technical artifacts. *Studies in History and Philosophy of Science Part A, 37*(1), 1–4.

Mitcham, C. (2002). Do artifacts have dual natures? Two points of commentary on the Delft project. *Techné: Journal of the Society for Philosophy and Technology, 6*(2), 9–12. http://scholar.lib.vt.edu/ejournals/SPT/v6n2/mitcham.html. Accessed 25 Sept 2010.

de Ridder, J. (2006). Mechanistic artifact explanation. *Studies in History and Philosophy of Science Part A, 37*(1), 81–96.

de Ridder, J. (2007). *Reconstructing design, explaining artifacts: Philosophical reflections on the design and explanation of technical artifacts*. Delft: Delft University of Technology.

Rosenman, M. A., & Gero, J. S. (1998). Purpose and function in design: From the socio-cultural to the techno-physical. *Design Studies, 19*(2), 161–186.

# Chapter 27
# The Methodological Ladder of Industrialised Inventions: A Description-Based and Explanation-Enhanced Prescriptive Model

**M.H. Abolkheir**

**Abstract** Are inventions still a mystery? In this short paper, I present a synoptic version of the outcome of a meta-methodological investigation which aims to eradicate any mystery surrounding the possibility of there being a single underlying methodological pattern that successful industrialised inventions share. I identify and clearly define specific *statement-generating phases* through which epistemically (predictively) successful industrialised inventions evolve. Furthermore, I propose an explanation for such phase structure. According to the proposed explanation, the phase structure is an escalating ladder of *individually necessary and jointly sufficient conditions* that steer a given programme from start to finish. In addition, I use the proposed explanation of the phase structure to propose explanations for when epistemic *failure* occurs. The ultimate practical aim of this project is to provide industrial research and development teams with a prescriptive model, which can assist them to minimize their chances of facing epistemic failure and consequently to increase their chances of achieving epistemic success. In this chapter I present the descriptive, explanatory and prescriptive aspects of the Methodological Ladder, alongside a brief analysis of three supporting case studies to illustrate how the Ladder fits actual practice, namely: the microwave oven; the cyclonic vacuum cleaner; and chemotherapeutic penicillin.

**Keywords** Method for inventions • Technological creativity • Inventive design • Ladder of inventions • Penicillin discovery

M.H. Abolkheir (✉)
University of Bristol, Bristol, UK
e-mail: M.H.Abolkheir@Bristol.ac.uk

D.P. Michelfelder et al. (eds.), *Philosophy and Engineering: Reflections on Practice, Principles and Process*, Philosophy of Engineering and Technology 15,
DOI 10.1007/978-94-007-7762-0_27, 

## 27.1 Introduction

Inventions have been thoroughly examined in terms of the social, political, economic and technological challenges that give rise to them and also in terms of the states of affairs that follow their emergence. Such examinations include thousands of detailed historical case studies and anecdotal stories about the intriguing idiosyncrasies and the remarkable engineering ingenuity that was used to see projects through to success. Intensive historical efforts have also been devoted to describing the state of technological knowledge at different stages of human history. Furthermore, some philosophical analysis, especially in the newly emerging literature in the philosophy of technology and engineering sciences, has been undertaken to propose boundaries between "discovery" and "invention" and to examine the nature of engineering design.

However, there still seems to be a mystery surrounding the possibility of getting a firm grip on any underlying methodological pattern, which covers the entire invention process from start to finish, that successful industrialised inventions might have in common, and which can be used as a basis for methodological prescription for future projects.

More specifically, when reviewing the vast literature on creativity and on product design, there seems to be an implied though clear distinction between coming up with an inventive idea (i.e. forming an inventive hypothesis), and developing an inventive idea into a product design. Such a distinction bears some interesting parallels to the distinction that Reichenbach and Popper made in the early part of the twentieth century between the context of *discovery* and the context of *justification*. Popper famously said (Popper 1959, p. 31):

> The question how it happens that a new idea occurs to a man – whether it is a musical theme, a dramatic conflict, or a scientific theory – may be of great interest to empirical psychology; but it is irrelevant to the logical analysis of scientific knowledge. This latter is concerned not with *questions of fact* … but only with questions of *justification* … Accordingly I shall distinguish sharply between the process of conceiving a new idea, and the methods and results of examining it logically.

In principle I argue that such distinctions are ontologically arbitrary as they chop a single continuous phenomenon right in its middle. On closer examination, it turns out that the phenomenon of coming up with technologically inventive ideas actually consists of a ladder of "intermediary justifications". Even the wildest technological ideas or hypotheses which turn out to be successful can clearly be shown to have been grounded from the start in some technologically relevant causal contexts of one sort or another.

Of course the moment the inventive hypothesis actually succeeds in creating a new man-made physical phenomenon is the "big moment". However, that is not to say that the intermediary justifications along the way can be ontologically excluded then taxonomically passed on to a different discipline to worry about, be it empirical psychology or cognitive science.

Due to space limitations, this paper will not address the meta-methodological, ontological and taxonomic foundations of the Methodological Ladder any further, except for one more quick remark regarding yet another discipline: sociology.

In principle I argue that the sociology of technology, important as unquestionably it is, should not be confused either ontologically or taxonomically with an empirical meta-methodology. To avoid engaging in a long debate, there is a single property that the methodological route used by industrial inventors has and which is sufficient to separate it, ontologically, from phenomena that can be the subject of sociological investigations. This property is that whether the methodological route used does or does not ultimately lead to a novel and successful technological prediction (a new man-made physical phenomenon) *is not* at all influenced by human consensus or dissensus (let alone being determined by such factors), but *is* sanctioned by Nature, *alone*. For example, the day the human voice was electrically transmitted for the first time, when Alexander Graham Bell spoke to his assistant while they were in different rooms, was the day a new man-made physical phenomenon came into existence, and the emergence of such a new physical phenomenon was authorised by Nature, not by human society. Such authorisation was neither granted as a result of any prior motives or values (including epistemic values), nor any ultimate noble or evil aims, but was granted because some methodological route or another was used.

Admittedly, the task of the meta-methodology of technology is easier than that of the meta-methodology of pure science, mainly because the history of technology provides the meta-methodologist with the *Archimedean point* of there being new man-made physical phenomena, whose creation has clearly been sanctioned by Nature not human society. With access to such an Archimedean point, the task of the meta-methodologist of technology becomes to reverse-engineer such processes and to establish what methodological pattern was actually used to successfully create such new man-made physical phenomena, so that the description of such a pattern can be used as a basis for methodological prescription for future projects.

In the coming pages I will present the Methodological Ladder, as a description-based and explanation-enhanced single prescriptive model that covers the entire process from start to finish, and which should be understood as a methodological pattern that has been sanctioned by Nature.

## 27.2 The Descriptive Aspects

In all the examined cases, the following ***statement-generating*** phases emerged as a pattern, which rather robustly repeats itself over and over again, irrespective of whether the discipline within which the industrialised invention is located is mechanical engineering, chemistry or biotechnology:

I. ***The Epistemic-Trigger Phase***
II. ***The Novel-Domain Phase***
III. ***The Inventive-Hypothesis Phase***
IV. ***The Technological-Bundle Phase***
V. ***The Industrial-Design Phase***

### 27.2.1 *Preliminary Notes*

#### 27.2.1.1 The Irrelevance of the Singularity/Multiplicity of Personnel

In some cases, the methodological pattern is carried out by a single person: "the inventor," and in other cases it is carried out by a number of persons, who are either working within a single research team or within separate research teams. Such singularity or multiplicity is irrelevant to the Model. What is relevant is that each phase comes into existence with the emergence of one type of statement and is finished with the emergence of a different type of statement.[1]

#### 27.2.1.2 The Irrelevance of the Temporal Duration

In some cases, a phase may start and terminate in a few moments and in other cases a phase may extend over many months or even years. Again, what is relevant is the emergence of different types of statements that mark the start and finish of each phase.

#### 27.2.1.3 The Irrelevance of the Conscious/Subconscious Status

In some cases, moving from one phase to another is conducted by the person or persons involved in a conscious and systematic manner, and in other cases the move takes place with the help of coincidences and the person or persons involved might even be totally oblivious to the process. Again, what is relevant is the sequential emergence of statements.[2]

#### 27.2.1.4 Methodological Copycatting by Epistemic Abstraction

The way in which the descriptive pattern will be presented, as a Methodological Ladder that consists of sequential statement-generating phases, involves a process of *abstracting* the epistemic core from the remainder of the historical data. In some cases, the epistemic core is somewhat obscured in the data and hence requires

---

[1] Kuhn highlighted the role that the multiplicity of personnel may play in troublesome/anomaly-driven discoveries. See Kuhn (1970b, chapters VI, X and XIII), Bird (2000, especially pp. 40–42), and Gillies (2006).

[2] Poincaré argued that "creativity requires the hidden combination of unconscious ideas" and that "ideas are being continuously combined with a freedom denied to waking, rational thought" Boden (1990, especially p. 19).

some epistemic reconstruction.[3] The end-result of such epistemic abstraction is that future technologists can "copycat" the methodological pattern that actually led their predecessors to success, without having to re-live the idiosyncrasies that their predecessors went through.

#### 27.2.1.5 What Is a Confirmed Technological Principle (CTP)?

In the remainder of this paper, I will be using the term Confirmed Technological Principle which I will clarify here. In principle, technological knowledge can be seen as nothing but a network of "hypothetical imperative" (value-neutral) statements of the type: *To achieve (if you wish to achieve) y, then do x*. Such a statement type is clearly a straightforward re-arrangement of the predictive (or inductive) statement: *Doing x leads to y*. In the first instance, this eliminates any confusion with another type of prescriptive statements: the "categorical imperative" of the type: *Do A*, which, for example, is used to communicate value-laden moral requirements to practitioners to abide by, or else get in sociological trouble. However, such a boundary, helpful as it is, still needs further articulation in terms of the different types of hypothetical imperative statements (for example: data-supported; theory-supported etc.).

There is an array of terms that are used in the literature, such as "empirical rule", "technological rule" and "grounded rule".[4] The term Confirmed Technological Principle is used in this paper to connote the following properties:

- **Technological**: refers to it being a value-neutral hypothetical imperative;
- **Confirmed**: refers to it being data-supported – in other words, no matter how much a statement is "supported" by theory, it remains a hypothesis until it is supported by data; but on the other hand, once supported by data a hypothesis becomes a confirmed statement and any further support by theory is a bonus rather than an essential property; and
- **Principle**: refers to it being of a general nature that shows how to achieve a single type of prediction, but lacks the auxiliary details that a final industrial design also needs to predict (mass-producibility, safety, economy etc.).

[3] This may seem to resemble the process of "rational reconstruction" that was advocated by Lakatos and a similar process advocated by Reichenbach, both of which can to some extent be traced back to the Hegelian idea that history has an *underlying logic*. However, there are fundamental differences between these two positions, and between them and the position presented here. Lakatos allows a *value-laden* judgement of "rationality" into his system, whereas I strictly describe the epistemic core without any value-laden contribution. Although Reichenbach uses the term "rational" to mean *epistemic* and consequently I agree with him on this, I find his restriction of the scope of epistemic enquiry to "justification" with the exclusion of "discovery" ontologically arbitrary (as I indicated in the Introduction, above), whereas I describe the entire process (in the context of technological invention) from start to finish. See Worrall and Currie (1978, especially pp. 102ff), Reichenbach (1938, pp. 5ff), and Bird (2008).

[4] See for example, Bunge (1972), Blockley (1980), and Addis (1990).

## 27.3 The Descriptive Phases

### *27.3.1 The Epistemic-Trigger Phase*

This phase comes into existence against a complementary background of the role of technology in fulfilling non-epistemic values: from alleviating suffering and improving the quality of life to achieving social, political, business and financial advantages by creating new man-made phenomena that are mass-producible on an industrial scale. However, it is only an individual epistemic trigger that kick-starts a process that might lead to the successful creation of a new industrialised invention. An epistemic trigger for an industrialised invention is an "intriguing causal relation", which can either be:

- a technological problem (a preferred *effect* for which a *cause* is sought); or
- a technological opportunity (a *cause* for which a preferred *effect* is sought).

A technological problem can be expressed in terms of a preferred empirical result for which there is no established technological means (or no Confirmed Technological Principle) to achieve it. On the other hand, a technological opportunity can emerge from data, scientific models, or Confirmed Technological Principles for none of which there is a technological exploitation *outside* their traditional technological domains. Data triggers emerge from accidental observations, negative experimental data, or previously considered irrelevant data, while the other triggers emerge from newly established or previously considered irrelevant scientific models or previously considered irrelevant Confirmed Technological Principles.

This phase terminates with the emergence of an "Epistemic-Trigger Statement" of one of the following two forms:

- **"There is a technological problem (preferred effect): Problem E"**; or
- **"There is a technological opportunity (available cause): Opportunity C"**.

### *27.3.2 The Novel-Domain Phase*

The statement that emerges from the previous phase kick-starts some sort of *helicopter search* for the most technologically viable domain, which is often helped by historical coincidences and case idiosyncrasies although it can also be carried out using a systematic thought process. The output of such a "helicopter search" is described in the literature by an array of terms, such as "serendipity", "lateral thinking", "thinking outside the box", "flash of light", "sudden mental insight", "creative leap", "cognitive leap" etc. although such terms can be somewhat misleadingly too broad as they can refer to *two* phases (this phase and the next) combined.[5]

An Epistemic-Trigger Statement that is seen by most practitioners in a given technological context as belonging to a given traditional domain is perceived by the "inventor" as belonging to a novel domain. So, a discoloured Petri dish that would have been thrown in the rubbish bin by other scientists is identified by Fleming as belonging to the domain of *pharmaceutical* issues, and consequently opens the door for possible inventive applications. Or, a problem with the vacuum cleaner bag that would have been addressed by designers of domestic appliances as a bag issue is identified by Dyson as belonging to the larger domain of *separation* issues, and consequently opens the door for the possibility of importing inventive solutions. Exposure to a switched-on radar element causing a chocolate bar to melt that would have been dismissed by radar engineers as a hazard is identified by Spencer as belonging to the domain of *food* issues, and consequently opens the door for the possibility of developing inventive applications.

This Phase terminates with the emergence of a "Novel-Domain Statement" of one of the following forms:

- "**Problem E belongs to domain X**"; or
- "**Opportunity C belongs to domain Y**".

### *27.3.3 The Inventive-Hypothesis Phase*

It is all very well establishing "where to look" for an answer, as the output of the previous Phase indicates, but it is quite a different challenge to be able to search then to *zoom in* on a specific inventive hypothesis, which is what happens in this phase. The nature of the Inventive-Hypothesis Phase (this third phase) corresponds to the nature of the Epistemic-Trigger Phase (the first phase).

If the Epistemic-Trigger Phase consists of a problem (an effect for which a cause is sought), the Inventive-Hypothesis Phase consists of identifying a hypothetical cause *within domain X*. If the Epistemic-Trigger Phase consists of an opportunity (a cause for which a preferred effect is sought), the Inventive-Hypothesis Phase consists of identifying a hypothetical effect *within domain Y*. At such an early stage, the hypothetical invention is nothing more than just a vague possibility.

This phase terminates with the emergence of an "Inventive-Hypothesis Statement" of one of the following two forms:

[5]Another term that is also used is "Eureka", which I myself used in an earlier version of the Model but I have now abandoned, as it might confuse this phase – which can involve a high degree of creativity in its own right – with the next phase of zooming in on a specific inventive hypothesis, which again involves a high degree of creativity. "Creativity" is not limited to the early phases of the Model, but is often also present in the subsequent phases, for example in solving experimental obstacles in the Technological-Bundle Phase and/or coming up with supporting innovative ideas at the Industrial-Design Phase.

- **"Within domain X, Problem E might be solved by Cause $C_X$"**; or
- **"Within domain Y, Opportunity C might be exploited to produce Effect $E_Y$"**.

### *27.3.4 The Technological-Bundle Phase*

Experimentation in technology is essentially developmental. Whereas in science experimentation is undertaken to "test" a hypothesis about a phenomenon (natural or man-made) that is already in existence i.e. *after the event*, in technology experimentation is about a hypothetical phenomenon that is still *in the making* (and needless to say might never be proven to be possible to bring into existence). So, the "confirmation" (i.e. epistemic justification) of an inventive-hypothesis happens by trying to bundle it up with Confirmed Technological Principles until the search succeeds in finding one such bundle that makes the hypothetical invention work. This is the phase at which the invention is born, as a Confirmed Technological Principle according to which such and such novel technological result is achievable using such and such bundle of Confirmed Technological Principles. The statement of the "invention" at this phase is not only more precise than that at the Inventive-Hypothesis Phase, but it also almost always stipulates conditions without which the "invention" will either not work at all, or will not achieve a specific level of performance.

This phase terminates with the emergence of a "Technological-Bundle Statement" – which may be supplemented by engineering drawings or any other specific disciplinary format – whose content takes the following prescriptive form:

**"To achieve novel effect E, implement technological bundle: $CTP_1$ … $CTP_n$."**

### *27.3.5 The Industrial-Design Phase*

Following the emergence of the new Confirmed Technological Principle, at this Phase the epistemic focus finally shifts to the search for ways to refine the technological-bundle, by adding more Confirmed Technological Principles, so that it accommodates numerous socio-economic requirements. Such requirements would normally include the choice of materials, mass-producibility, cost, safety, user-friendliness, environmental impact, aesthetics etc., the level of complexity of which varies considerably from simple inventions to complex ones.

This phase terminates with the emergence of an "Industrial-Design Statement" – which may be supplemented by engineering drawings or any other specific disciplinary format – whose content takes the following prescriptive form:

**"To achieve an industrial design that incorporates novel effect E, implement technological bundle: $CTP_1$ … $CTP_{n+p}$."**

## 27.4 The Case of the Microwave Oven: Synopsis and Brief Analysis[6]

Against a background of working as an engineer at the Raytheon Company, developing radar components and also broadly wondering what civil or consumer applications might exist for radar technology, one day in 1945 the inventor (Percy Spencer) walked passed a switched-on magnetron, which is a radar element, and noticed that the chocolate bar he had in his pocket had melted. As he felt no heat, he guessed that it might have been caused by the high frequency radio emissions from the magnetron. The heating effect of microwaves had been noticed by other scientists, including reports of partially burnt birds at the bottom of radar installations, but was dismissed as not deserving of investigation. Following some reasoning and initial experimentation with a bag of popcorn and an egg, the potential for a new way of cooking became a serious possibility, and a more systematic engineering research and development programme was undertaken by Spencer's employers that resulted in the first microwave oven: the "Radarange", which was introduced in 1947. I argue that the following *five* abstracted statements summarise the entire ladder of necessary and sufficient conditions for the development of this invention from start to finish:

I. **The Epistemic-Trigger Statement:** "There is a technological opportunity in the causal effect that exposure to magnetron-generated radio emissions has on a chocolate bar, which is similar to the effect of heat."
II. **The Novel-Domain Statement:** "The technological opportunity offered by magnetron-generated radio emissions belongs to the domain of food issues."
III. **The Inventive-Hypothesis Statement:** "Within the food domain, the technological opportunity might be exploited as a new cooking apparatus."
IV. **The Technological-Bundle Statement:** "To cook foodstuffs by exposure to electromagnetic energy, whose wave lengths fall in the microwave region of the electromagnetic spectrum, use a technological bundle that comprises an evacuated envelope made of a highly conductive material, anode vanes, a cavity resonator, an electron-emissive cathode member, and a magnetic means. Then, every time foodstuffs need to be cooked, expose them to microwave emissions inside the evacuated envelope for predetermined lengths of time."
V. **The Industrial-Design Statement:** [which applies to the first industrial design: the "Radarange"] "To achieve an industrial design that suits volume applications such as restaurants, catering units and hospitals, add the following features to the technological-bundle: cold-water cooling means for the magnetron; a hinged-door with a handle; manual controls for the user; and specify electricity, plumbing and installation instructions and use instructions."

[6] The following are the main references for the industrialised invention of the Microwave Oven: Carlisle (2004), Brown et al. (2002), Hill (1998), Pozar (2005), Scott (1974), Uhlig (2001), Van Dulken (2000), and Patent Specifications No. US 2495429.

## 27.5 The Case of the Cyclonic Vacuum Cleaner: Synopsis and Brief Analysis[7]

Against a background of working as an inventor and designer, one day in 1978 the inventor (James Dyson) dismantled his bag-operated vacuum cleaner at home to see why it loses its extraction power and realised that the bag actually gets clogged soon after first use. By a sheer co-incidence, he was also involved in another "clogging" issue in a different industry: the spray-equipment industry in which gigantic "cyclones" were used to separate powder, drawing on centrifugal force, and he wondered whether this can be used on a miniature scale in a domestic vacuum cleaner. Following initial experimentation using a gaffer tape-sealed cardboard cyclone, a long engineering research and development programme was undertaken that resulted in the first product: the "Cyclone". I argue that the following *five* abstracted statements summarise the entire ladder of necessary and sufficient conditions for the development of this invention from start to finish:

I. **The Epistemic-Trigger Statement:** "There is a technological problem in the permanent clogging of vacuum cleaner bags, which starts soon after first use and leads to a rapid decline in extraction performance."
II. **The Novel-Domain Statement:** "The technological problem of bag-clogging should not be seen as belonging to the domain of filter-separation issues but to the larger domain of separation issues, from which a possible solution might be possible to import."
III. **The Inventive-Hypothesis Statement:** "Within the domain of separation issues, centrifugal force might be exploited to cause the separation instead of the bag."
IV. **The Technological-Bundle Statement:** "To vacuum-clean using a bagless vacuum cleaner that uses centrifugal force to separate dust and debris, use two cyclones: the first being steeply tapered and hence working at a faster speed to pick up dust, while the second cyclone being gently tapered or parallel-walled and hence working at a slower speed and to pick up the larger pieces of debris. Then use the vacuum cleaner in a normal manner."
V. **The Industrial-Design Statement:** [which applies to the first industrial design: the "Cyclone"] "To achieve an industrial design that suits the consumer market, add the following features to the technological-bundle: incorporate the faster (and the smaller) cyclone inside the slower (and the larger and consequently the outer) one. Use clear see-through material for the manufacture of the outer cyclone so that it attracts the visual attention of potential consumer buyers and to show how the machine works, and also so that the user can know when to empty the dirt; use material with a high rubber content for the manufacture of

[7]The following are the main references for the industrialised invention of the Cyclonic Vacuum Cleaner: Dyson (2001, especially pp. 102–114, 121–129), Tidd et al. (2001), Uhlig (2001), Van Dulken (2000), and Patent Specifications No. US 4373228.

the body to increase its durability; and use a telescopic hose with an "instant changeover valve" that allows the user to switch immediately from floor level to overhead cleaning."

## 27.6 Chemotherapeutic Penicillin – Synopsis and Brief Analysis[8]

For many people, Penicillin is known as a discovery and not as a technological invention. Nevertheless, the history of Penicillin clearly indicates the presence of two *epistemically distinct* cases: the first was indeed the discovery of a natural phenomenon, the description of which is analogous to that of the motion of planets, or the behaviour of gases; and the second consisted of the *conversion* of such a natural phenomenon into a man-made one: a technological invention just like the steam engine or the microwave oven.

Against a background of being a physician with first-hand experience in war wounds and antiseptic substances, in 1928 Alexander Fleming (the co-inventor) made his well-known accidental observation of the antibacterial effects that a Penicillium mould seems to have on a strain of bacteria in a culture-plate. Following the observation, Fleming made two main hypotheses: the first was a *descriptive* hypothesis regarding the effects of the mould on bacteria, which he subsequently published in his seminal 1929 paper, and the second hypothesis was a *technological* one: that the antibacterial effects of Penicillin might be convertible into a chemotherapeutic drug. Unfortunately, Fleming failed in developing a technological-bundle that would have converted such a hypothesis into a an industrialised invention; tasks that were not accomplished until more than a decade later at the hands of Howard Florey, Ernest Chain and both the Oxford and Peoria teams.

Following Fleming's failure, the Oxford Team formed the hypothesis that the antibacterial effects of a purified form of Penicillin might work as an intravenously-injectable antiseptic if tested *in vivo*, which neither Fleming nor anybody else had undertaken. Starting with the strain of *Penicillium notatum*, they cultured it in an acidity-and-temperature-controlled medium, then purified it using a novel combination of techniques, namely: "back extraction", "column chromatography" and "freeze-drying". This was followed by the team's important innovation of conducting "animal protection tests", which indicated that Penicillin was active *in vivo* (in mice) against at least three types of pathogenic bacteria. Finally, the team proceeded with testing Penicillin on the human body, achieving unequivocal success. The first Penicillin product was *surface-cultured Penicillin* that was produced by the Oxford

---

[8]The following are the main references for the industrialised invention of Chemotherapeutic Penicillin: Abraham et al. (1941), Bud (2007, especially pp. 33–37, 62–63), Chain et al. (1940), Fleming (1929), Gillies (1993, especially pp. 39–48, 2006), Friedman and Friedland (2000), Hare (1970, especially pp. 169–175), Masters (1946, especially pp. 78, 91–109), Sneader (2005, especially pp. 293–295, 320), and US Patent 2,442,141 and US Patent 2,443,989.

Team using culturing in shallow pans, whose production yield was so low the use of Penicillin on a wide scale was deemed unrealistic. The Oxford Team sought the assistance of the Research Laboratory in Peoria in the USA, which introduced product innovations including the production method of *deep-fermentation*, the use of *corn-steep liquor* in the culture medium and the use of the high-yield strain of *Penicillium chrysogenum*. The result was an increase in the yield from two to eight Oxford units per millilitre to 500 Oxford units per millilitre. The Oxford variant of Penicillin became known as Penicillin F, whereas the American variant became known as Penicillin G and became the dominant variant in clinical use for many years before yet other variants became available.

I argue that the following *five* abstracted statements summarise the entire ladder of necessary and sufficient conditions for the development of this invention from start to finish:

I. **The Epistemic-Trigger Statement**: "There is a technological opportunity in the causal effect that exposure to filtrates of Penicillin have on some pathogenic bacteria and leading to its undergoing lysis."

II. **The Novel-Domain Statement**: "The technological opportunity offered by Penicillin belongs to the domain of pharmaceutical issues."

III. **The Inventive-Hypothesis Statement**: "Within the domain of pharmaceutical issues, Penicillin might be exploited as a chemotherapeutic drug for the treatment of deep-seated infections."

IV. **The Technological-Bundle Statement**: "To use Penicillin as a chemotherapeutic drug, which is injectable into humans to treat deep-seated infections, prepare the Penicillin by culturing it in an acidity-and-temperature-controlled medium, then purify it before giving it to patients in high enough doses to fight infection: and whenever possible give it to patients as early as possible after an injury is inflicted or an infection is developed."

V. **The Industrial-Design Statement**: [This statement applies to the first industrial design (produced in Oxford, UK): the "surface-cultured Penicillin F"] "To achieve an industrial design that can be prepared in advance and is ready for instant use when required, start with a given quantity of *Penicillium notatum*, then culture it in an acidity-and-temperature-controlled medium using surface-culturing under pre-determined conditions; then purify it using "back extraction"; "column chromatography" and "freeze-drying" production techniques; and finally administer it intravenously (using the slow intravenous drip method), according to predetermined dosages."

[This statement applies to the subsequently developed industrial design (produced in Peoria, USA): the "deep-fermented Penicillin G"] "To achieve a mass-producible design that can be prepared in advance in vast quantities ready for instant use when required, start with a given quantity of *Penicillium chrysogenum* and culture it in an acidity-and-temperature-controlled medium of corn-steep liquor, using the production method of deep-fermentation under pre-determined conditions; and finally administer it intravenously, according to predetermined dosages."

## 27.7 The Explanatory Aspects

### 27.7.1 Explaining the Ladder's Structure

I propose that the phase structure is nothing but an escalating ladder of *individually necessary and jointly sufficient conditions* that steer a given programme from start to finish. I will now turn to introduce the Methodological Ladder Diagram, which illustrates how the phase structure works. Following that, I will clarify how the phase structure works in more detail.

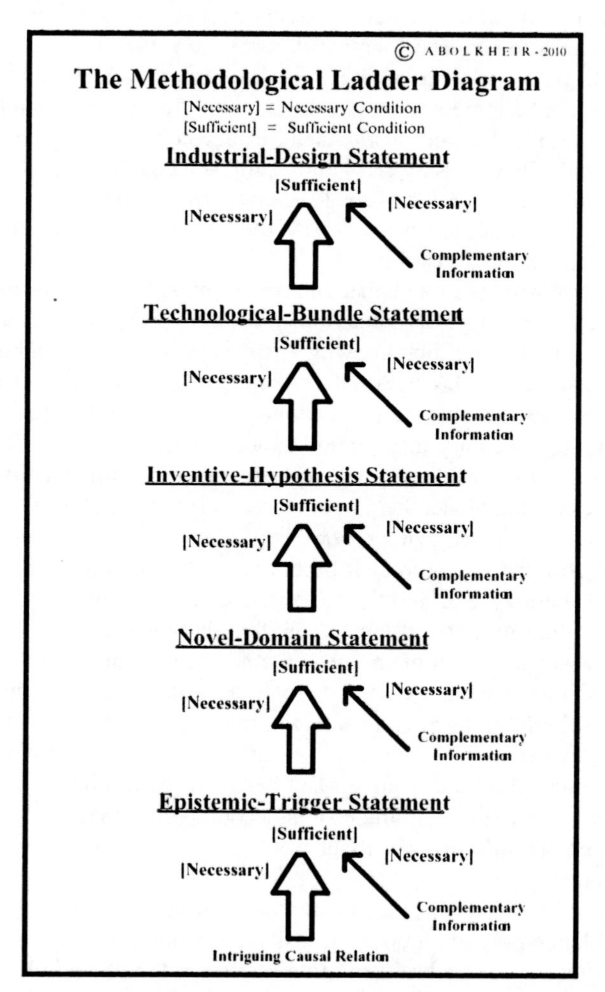


At the very beginning, identifying an intriguing causal relation (i.e. a preferred *effect* for which a *cause* is sought; or a *cause* for which a preferred *effect* is sought)

is a necessary condition for forming an Epistemic-Trigger Statement (i.e. a "mission statement" of what this presumed new programme aims to achieve). But of course it must be a necessary condition, as the intriguing causal relation is all about knowing *one half* of a causal relation before searching for *the other (matching) half*, as how can you know that a "half" that you might come across is or is not a *matching* half, without knowing what the first half is in the first place? However, necessary as it is, it is still insufficient to form the Epistemic-Trigger Statement, as something else is needed before that can happen; after all, just coming across an intriguing causal relation cannot lead to the knower engaging in an inventive programme. The complementary information comes from how industrial development is done to fulfil non-epistemic values, which amateur inventors follow now and again but professional research and development designers do it for a living. Such complementary information is also on its own insufficient for the formation of the Epistemic-Trigger Statement; after all, just having information about how industrial development is done does not always lead to developing new inventions. Finally, with both individually necessary and jointly sufficient conditions in place, sufficiency is established for the formation of the Epistemic-Trigger Statement.

This type of sequence repeats itself four more times until the programme successfully reaches the finish line.

So, having knowledge of an epistemic trigger must be a necessary condition for knowledge of a relevant novel domain, as how can you know whether a domain is or is not "relevant" without first knowing the existence of one member, on the basis of whose properties such relevance can be established? But this on its own is insufficient for the next statement to emerge, as something else is needed before that can happen, and it is the complementary information about candidate domains from which one can be selected. Now, whether your brain scans stored information during your sleep, or a lucky coincidence helps you come across a candidate, or a systematic search at the library guides you, or a brainstorming session at the office lights your way, this search must be completed before you can move forward. So, with both individually necessary and jointly sufficient conditions in place, sufficiency is established for the formation of the Novel-Domain Statement.

Again, knowledge of a domain must be a necessary condition for knowledge of a hypothetical union with a causal counterpart (the other matching half of the intriguing causal relation) *within the domain*, as how can you know of the possibility of a union between two "members" of a given relevant domain, without first knowing what the relevant domain is? But this on its own is insufficient for the next statement to emerge, as something else is needed before that can happen, and it is the complementary information about candidate causal counterparts from which one can be selected.

Once more, whether your brain scans stored information during your sleep, or a lucky coincidence helps you come across a candidate, or a systematic search at the library guides you, or a brainstorming session at the office lights your way this search must be completed before you can move forward. So, with both individually necessary and jointly sufficient conditions in place, sufficiency is established for the formation of the Inventive-Hypothesis Statement.

And again, knowledge of a "hypothesis" (a hypothetical union between two presumed matching halves of an intriguing causal relation) must be a necessary condition for knowledge of a "confirmed" union between the two causal counterparts. As a "confirmed" statement consists of a hypothesis *plus* supporting evidence that elevates the hypothesis to the confirmed status, how can you know a confirmed statement without first knowing the hypothesis on which it is based? But this on its own is insufficient for the next statement to emerge, as something else is needed before that can happen, and it is the complementary information about candidate elements of the technological bundle that might confirm the hypothetical invention and bring it into existence. On the other hand and needless to say, millions and millions of such candidate elements are already in existence, but they do not constitute an invention! If the search for a technological bundle that can bring the hypothetical invention into existence is successful, then both individually necessary and jointly sufficient conditions will be in place and sufficiency is established for the formation of the Technological-Bundle Statement.

Finally, knowledge of a confirmed technological bundle must be a necessary condition for knowledge of an "enlarged" version of the technological bundle, as how can you know how to "enlarge" a confirmed bundle without first knowing the confirmed bundle that is to be enlarged? But again, this on its own is insufficient for the next statement to emerge, as something else is needed before that can happen, which is the complementary information about industrial design. Unless the search that takes place at this phase is successful, the technological bundle will fail to evolve into an industrial design that meets socio-economic requirements. For example, it might not be practically possible to produce the designed object on an industrial scale, and consequently the technological bundle would remain where it was at the end of the fourth phase. However, if successful, then both individually necessary and jointly sufficient conditions will be in place and sufficiency is established for the formation of the Industrial-design Statement. This constitutes the end of *this phase* <u>and</u> the *entire programme*, as any subsequent future development is either a mere incremental improvement of the industrial design or a problem that constitutes an Epistemic-Trigger Statement for yet an entire new programme for the development of a new industrialised invention.

### *27.7.2 Explaining Cases of Failure*

In this section I discuss the "opposite" cases of the three case studies considered above. First, however, it is essentially important to highlight that the explanations I aim to provide here are ***methodological*** explanations: i.e. what methodological steps, had they been taken, would have led to success in creating the new invention? This analysis is done on a purely counterfactual basis. Clearly, such methodological explanation is different from other types, such as sociological or psychological explanations. In other words, a methodological explanation is based on a number of assumptions, such as there being a will to achieve success in creating a new invention and/or the availability of resources to do so. But perhaps that was not the case. Some people are simply content with what they have and are not looking for anything new, others might be totally absorbed by other problems, and others might not perceive any gained value in changing how things are and/or in allocating the required resources. Some people are just lazy! The point here is that this is a thesis in the epistemology of technology, which specifically focuses on methodology. Consequently all other issues – interesting as they may be – are for historians to dig into and investigate, including making judgements of historical blame and credit.

So, here are the three questions, which I attempt to answer using the Methodological Ladder:

1. How exactly did radar engineers – who knew of partially burnt birds at the bottom of radar installations – fail to grab the opportunity and succeed, instead of leaving it to Percy Spencer, years later?

   This failure took place at the very first phase, which was not even completed. Although the data presented radar engineers with an odd causal relation that might have had some potential for a novel technological application, such oddity was not recognised as "technologically intriguing", and consequently no Epistemic-Trigger Statement was generated, which would have created the sufficient condition for the start of the next phase. As I said above, it is for historians to investigate the wisdom of such decisions, as radar engineers were at the time contributing to World War II efforts – a war that was threatening the future of modern civilization – and consequently they might have had more to worry about than partially burnt birds and gooey chocolate bars. However, from the methodological perspective in the context of technological practices, recognising an odd causal relation as technologically intriguing *is* the starting point. No Epistemic-Trigger statement; no invention.
2. How exactly did global vacuum cleaner manufacturers – who were aware of the problem of diminishing extraction – fail to stumble across cyclonic technology, instead of leaving it to James Dyson, years later?

   The failure took place at the second phase: the Novel-Domain phase. The manufacturers in the industry were indeed aware of the diminishing extraction problem, but they all categorised it as a problem that belonged to the traditional (and narrow) domain of bag-issues (filter-separation issues), rather than the novel (and larger) domain of separation issues. Consequently, their efforts would have been directed towards improving the properties of the bag material and increasing the power. In so doing, no Novel-Domain Statement was generated, which would have created the sufficient condition for the start of the next phase of searching for an inventive hypothesis. Again, it is up to historians to establish any perceived wisdom in the decision taken by other manufacturers to avoid the enlargement of the domain, as it could be argued that such an enlargement would almost inevitably have led to changes in production lines and after-sale services, all of which would have required new investments and would have introduced a new element of risk, and which might have suited a newcomer more than established businesses. However, from the methodological perspective in the context of technological practices, searching for a novel technologically viable domain *is* the way to use technological problems as a springboard for developing new inventions. No Novel-Domain statement; no invention.

3. How exactly did Alexander Fleming – who not only formed the Epistemic-Trigger Statement and the Novel-Domain Statement, but also the Inventive-Hypothesis Statement – fail to go the entire distance, instead of leaving it to Howard Florey and the Oxford team, years later?

   The failure took place at the fourth phase: the Technological-Bundle phase. The research environment at St. Mary's where Fleming practiced under the directorship of Almroth Wright was a "scientific" one. Indeed, the first part of Fleming's work was scientific, as it involved the description of a natural phenomenon. However, the second part of his work was not scientific but *technological*. Nevertheless, "scientific values" about how research should and should not be practiced were allowed to deprive the programme of vital pieces of the technological information jigsaw puzzle. Such values included a "scientific" emphasis on the need for explanations before proceeding, and hence deprived Fleming's Penicillin programme of utilizing extremely important information that surfaced from the Prontosil programme: that a therapeutic substance can work *in vivo* even though it does not work *in vitro*, information that was rejected because of lack of theoretical evidence. But what explanations or theoretical evidence did Roman builders have when they succeeded in inventing the masonry arch by trial and error?! Not only that, but a full *theoretical* understanding of Penicillin did not actually become available until 1945, after Penicillin had already gone into mass-production and saved the lives of millions. Technological invention is about the mixing and matching of causal relations while being guided, sometimes by trial and error and sometimes by theory, until a new man-made physical phenomenon comes into existence, following which explanations can be sought, but as an *after-the-event* procedure. Again, it is up to historians to establish any wisdom in the decisions that were made during the earlier Penicillin programme. However, from the methodological perspective in the context of technological practices, searching for a technological bundle (irrespective of the availability of theoretical explanations) *is* the way to convert an inventive hypothesis into a new man-made physical phenomenon. No Technological-Bundle statement; no invention.

## 27.8 The Prescriptive Aspects

So far I have presented the descriptive aspects of the Ladder in the form of a descriptive pattern of the methodological phases through which industrialised inventions proceed, including presenting a brief analysis of three case studies. Furthermore, I have proposed an explanation for the phase structure of the Ladder. For some readers it may be a near-obvious conclusion that if they accept the descriptive aspects, and even more so if they find the proposed explanation plausible then it follows that the Ladder must have some prescriptive aspects which can guide future projects. However, the move from description (and explanation) to prescription still needs a philosophical clarification.

One excellent starting point for such a clarification is the debate that took place in the 1960s between Thomas Kuhn and some of his critics, especially Paul Feyerabend, who said:

> Whenever I read Kuhn, I am troubled by the following question: are we presented with *methodological prescriptions* which tell the scientist how to proceed; or are we given a *description*, void of any evaluative element …?[9]

At one place Kuhn responded by saying:

> The preceding pages present a viewpoint or theory about the nature of science, and like other philosophies of science, the theory has consequences for the way in which scientists should behave if their enterprise is to succeed. … one set of reasons for taking the theory seriously is that scientists, whose methods have been developed and selected for their success, do in fact behave as the theory says they should. My descriptive generalizations are evidence for the theory precisely because they can also be derived from it, whereas on other views of the nature of science they constitute anomalous behaviour.[10]

And at another place Kuhn said:

> Are Kuhn's remarks about scientific development, [Feyerabend] asks, to be read as descriptions or prescriptions? The answer, of course, is that they should be read in both ways at once. If I have a theory of *how* and *why* science works, it must necessarily have implications for the way in which scientists *should* behave if their enterprise is to flourish … Note that nothing in the argument sets the value of science itself … To explain why an enterprise works is not to approve or disapprove it.[11]

Kuhn can be understood to be saying that his theory of scientific development is a "***value-neutral description-based and explanation-enhanced prescription***". In other words, he neither approves nor disapproves of what scientists did to succeed, but he is simply telling us what they did in order to succeed and furthermore he is proposing an explanation for why what they did worked.

Although Kuhn excluded technology from his theory, his empirical meta-methodology and specifically the relationship between description, explanation and prescription provides a fundamental meta-methodological platform for empirical research. The Methodological Ladder of Industrialised inventions follows in the meta-methodological footsteps of Kuhn, precisely as understood above.[12]

---

[9] Feyerabend (1970, p. 198).

[10] Kuhn (1970b, pp. 207–208).

[11] Kuhn (1970a, p. 237); the italics and underlining of the following words are mine: *how*, *why*, and *should*.

[12] See Hoyningen-Huene (1993, p. 6) for a reference to Kuhn's exclusion of applied science and technological invention. As for interpreting Kuhn's meta-methodology, it is of course important to remember that Kuhn's work contained some (at least textual) ambiguities, which have been clarified by subsequent scholars. For example, Bird specifically highlighted the distinction between the descriptive element and the explanatory element of Kuhn's theory, and said that "Kuhn does not

The prescriptive aspects of the Methodological Ladder are presented as five sequential search-and-state prescriptions, as follows:

1. If you have a desire to develop an industrialised invention, search for a technologically intriguing causal relation that you can pinpoint, which can either be a preferred effect for which no technologically recognised cause is known, or a cause for which no technologically recognised preferred effect is known. If successful, then generate an Epistemic-Trigger Statement. In short:

**Search for a technologically intriguing causal relation, then generate an Epistemic-Trigger Statement.**

2. If you have found a technologically intriguing causal relation, then search for a novel technologically viable domain to which the intriguing causal relation can be seen to belong. If successful, then generate a Novel-Domain Statement. In short:

**Search for a novel technologically viable domain, then generate a Novel-Domain Statement.**

3. If you have found a novel technologically viable domain, then search for a hypothetical causal counterpart for the intriguing causal relation. So, if the Epistemic-Trigger Statement is about a preferred effect for which no technologically recognised cause is known, then search for a hypothetical cause that might achieve such a preferred effect: and if the Epistemic-Trigger Statement is about a cause for which no technologically recognised preferred effect is known, then search for a hypothetical preferred effect which might be caused by it. If successful, then generate an Inventive-Hypothesis Statement. In short:

**Search for a hypothetical causal counterpart for the intriguing causal relation, then generate an Inventive-Hypothesis Statement.**

4. If you have found a hypothetical causal counterpart, then, using a combination of trial and error and theory, search for a technological-bundle of causal relations that confirms that the hypothetical causal counterpart can be achieved in reality. If successful, then generate a Technological-Bundle Statement. In short:

**Search for a technological-bundle of causal relations that confirms the inventive hypothesis, then generate a Technological-Bundle Statement.**

5. If you have found a technological-bundle of causal relations that confirms the inventive hypothesis, then search for further socio-economic requirements that the technological-bundle would have to meet, before being accepted as an industrial design. If successful, then generate an Industrial-Design Statement. In short:

**Search for further socio-economic requirements, then generate an Industrial-Design Statement.**

---

himself clearly distinguish these two elements and quite naturally describes the first in terms of the second."; before highlighting the normative/prescriptive element in Kuhn's theory (Bird 2008, p. 4, 2000, p. 67). Due to limitation of space, I will not provide a discussion of the *types* of prescriptions beyond what I mentioned in Sect. 27.2.1.5 above, which is of course nothing but the Kantian distinction between the hypothetical imperative and the categorical imperative. In that respect, Confirmed Technological Principles (as defined in Sect. 27.2.1.5), Kuhn's prescriptions, and the prescriptive aspects of the Ladder which I present in the remainder of this section are all hypothetical imperatives: *To achieve (if you wish to achieve) y, then do x*, which I prefer to call "value-neutral description-based and explanation-enhanced prescription".

## 27.9 A Concluding Remark

I hope that you, the reader, will have found the descriptive, explanatory and prescriptive elements of the Methodological Ladder convincing. There is one final and brief point that I wish to make regarding the ontological nature of methodological practices as a *value-neutral phenomenon*. The reality seems to be that methodological practices are, like logic, a mercenary transporter. As long as you pack your stuff in the correct types of container and tick all the correct boxes on the request form it will be successfully forwarded. This highlights the need for strategic *value-laden* considerations (including rational and indeed moral considerations) to counterbalance the increasingly powerful knowledge that modern societies are amassing. In other words, this paper comes with a warning: the Methodological Ladder can be hazardous, please use it with care and employ it only to good ends.

**Acknowledgments** The Phase Model has evolved over a number of years and has benefited from genuinely helpful feedback from a number of academics. I wish to express my gratitude to the following. First and foremost, I am grateful to Professor A. Bird (Philosophy – University of Bristol) for his general advice over a number of years including raising challenges and making important suggestions and for his meticulous reading of several drafts. I thank Professor D. Gillies (Philosophy – King's College London and Science and Technology Studies – University College London) for his encouraging feedback on an earlier version of the Model back in 2006; and for their brief yet very helpful feedback at different points in time I thank: Professor D. Papineau (Philosophy – King's College London), Dr. J. Booker (Engineering – University of Bristol), Dr. H. Fadil (KPMG), Professor M. Gaber (Engineering – University of Alexandria) and Dr. B. Hicks (Engineering – University of Bath). Following presentations at the Department of Philosophy at University of Bristol, I thank Professor S. Okasha, Dr. A. Pyle, Professor L. Horsten, Professor K. Binmore, Dr. S. Wilkinson, and Miss H. Bradshaw for very helpful feedback. Following a presentation of a shorter version of this paper at the Royal Academy of Engineering in 2008, I received generous and indeed valuable feedback from a number of philosophers and engineers and I thank Professor D. Blockley (Engineering – University of Bristol) for most helpful debates by email; and for their helpful critical feedback I thank Professor P. Kroes (Delft University of Technology), Professor H. Radder (VU University Amsterdam), Mr. W. Grimson (Dublin Institute of Technology); Professor A. Figueiredo (Centre for Informatics and Systems, Technologia da Universidade de Coimbra), and Professor P. Dias (Engineering, University of Moratuwa). Needless to say the responsibility for both the content of this paper and the research methodology used in generating it is solely mine. This manuscript was submitted in 2010 as a summary of the state of the Phase Model project up till 2010. Subsequent developments are not included here.

## References

Abraham, E., Chain, E., et al. (1941). Further observations on penicillin. *Lancet, 2*, 177–189.

Addis, W. (1990). *Structural engineering: The nature of theory and design*. Chichester: Ellis Horwood.

Bird, A. (2000). *Philosophy now: Thomas Kuhn* (J. Shand, Series Ed.). Buckinghamshire: Acumen.

Bird, A. (2008). The historical turn in the philosophy of science. In S. Psillos & M. Curd (Eds.), *Routledge companion to the philosophy of science*. Abingdon: Routledge.

Blockley, D. (1980). *The nature of structural design and safety*. Chichester: Ellis Horwood.

Boden, M. (1990). *The creative mind*. London: Abacus.

Brown, D., Thurow, L., & Burke, J. (2002). *Inventing modern America, from the microwave to the mouse*. Cambridge, MA: MIT Press.
Bud, R. (2007). *Penicillin: Triumph and tragedy*. Oxford: Oxford University Press.
Bunge, M. (1972). Toward a philosophy of technology. In C. Mitcham & R. Mackey (Eds.), *Philosophy and technology; readings in the philosophical problems of technology* (pp. 62–76). New York: Free Press.
Carlisle, R. (2004). *Scientific American inventions and discoveries, all the milestones in ingenuity – from the discovery of fire to the invention of the microwave oven*. Hoboken: Wiley.
Chain, E., et al. (1940). Penicillin as a chemotherapeutic agent. *Lancet, 2*, 226–228.
Dyson, J. (2001). *Against the odds: An autobiography*. London: Texere.
Feyerabend, P. (1970). Consolations for the specialist. In I. Lakatos & A. Musgrave (Eds.), *Criticism and the growth of knowledge* (pp. 197–230). Cambridge: Cambridge University Press.
Fleming, A. (1929). On the antibacterial action of cultures of penicillium, with special reference to their use in silation of *H influenzae. British Journal of Experimental Pathology, 10*, 226–236.
Friedman, M., & Friedland, G. (2000). *Medicine's 10 greatest discoveries*. New Haven: Yale University Press.
Gillies, D. (1993). *Philosophy of science in the twentieth century; four central themes*. Oxford: Blackwell.
Gillies, D. (2006). Kuhn on discovery and the case of penicillin. In W. Gonzalez & J. Alcolea (Eds.), *Contemporary perspectives in philosophy and methodology of science* (pp. 47–63). A Coruña: Netbiblo.
Hare, R. (1970). *The birth of penicillin and the disarming of microbes*. Oxford: George Allen & Unwin.
Hill, A. (1998). *Microwave ovens* (ILSI Europe concise monograph series). Brussels: ILSI.
Hoyningen-Huene, P. (1993). *Reconstructing scientific revolutions: Thomas S. Kuhn's philosophy of science* (A. Levine, Trans.). Chicago: The University of Chicago Press.
Kuhn, T. (1970a). Reflections on my critics. In I. Lakatos & A. Musgrave (Eds.), *Criticism and the growth of knowledge* (pp. 231–278). Cambridge: Cambridge University Press.
Kuhn, T. (1970b). *The structure of scientific revolutions* (2nd ed.). Chicago: The University of Chicago Press.
Masters, D. (1946). *Miracle drug: The inner history of penicillin*. London: Eyre & Spottiswoode.
Popper, K. (1959). *The logic of scientific discovery*. London: Routledge.
Pozar, D. (2005). *Microwave engineering* (3rd ed.). Danvers: Wiley.
Reichenbach, H. (1938). *Experience and prediction*. Notre Dame: University of Notre Dame Press.
Scott, O. (1974). *The creative ordeal: The story of Raytheon*. New York: Atheneum.
Sneader, W. (2005). *Drug discovery: A history*. Chichester: Wiley.
Tidd, J., Bessant, J., & Pavitt, K. (2001). *Managing innovation: Integrating technological, market and organizational change* (2nd ed.). Chichester: Wiley.
Uhlig, R. (2001). *James Dyson's history of great inventions*. London: Constable.
US Patent and Trademark Office. www.uspto.gov; The Microwave Oven: U.S. *Patent No. 2,495,429*; The Cyclonic Vacuum Cleaner: U.S. *Patent No. 4,373,228*; Penicillin: U.S. *Patent No. 2,442,141* and U.S. *Patent No. 2,443,989*.
Van Dulken, S. (2000). *Inventing the 20th century: 100 Inventions that shaped the world*. London: The British Library.
Worrall, J., & Currie, G. (Eds.). (1978). *Imre Lakatos Philosophical Papers Volume 1: The methodology of scientific research programmes*. Cambridge: Cambridge University Press.

# Chapter 28
# On the Feasibility of Nanotechnology: A Chinese Perspective

**WANG Guoyu**

**Abstract** Nanotechnology is growing into a leading technology in emerging strategic industries and opening a huge space of technological possibility. There are, however, different kinds of possibility: real possibility and potential possibility. This chapter gives priority to considering real possibility, namely feasibility. It investigates what feasibility is and critiques the current idea of feasibility. It considers and emphasizes the Chinese concept of feasibility, because it has connotations that the English term does not have, and shows how this concept can make a contribution to feasibility studies dealing with the possibilities (positive and negative) of nanotechnology. From here it summarizes some basic feasibility strategies that can be used for nanotechnology development.

**Keywords** Nanotechnology • Possibility • Feasibility

## 28.1 Introduction

Nanotechnology has been regarded as the core technology that will bring on a new industrial revolution in the twenty-first century. The USA established the National Nanotechnology Initiative in 2000, followed by worldwide developing of nanotechnology by leaps and bounds. Today nanotechnology is gradually being commercialized. Parts of computer chips, trousers that don't wrinkle, DVD players, self-cleaning glass, and opacifiers in sun cream are all concrete examples of nanotechnology. While not comprehensive, this inventory gives an idea of some of the

---

G. WANG (✉)
School of Humanities, Dalian University of Technology, 2 Linggong Road, 116024 Dalian, People's Republic of China
e-mail: w_guoyu@hotmail.com

D.P. Michelfelder et al. (eds.), *Philosophy and Engineering: Reflections on Practice, Principles and Process*, Philosophy of Engineering and Technology 15,
DOI 10.1007/978-94-007-7762-0_28, 

1,000+ manufacturer-identified nanotechnology-based consumer products currently on the market (Woodrow Wilson International Center for Scholars 2010).

There is no doubt that nanotechnology is growing into a leading technology in emerging strategic industries and playing increasingly important roles in bio-pharmacy, genetic engineering, environmental protection, electronic devices, energy technology and space flight and aviation. However, the uncertainty of the results and negative effects produced by nanotechnology worry people and attract widespread attention. The potential risks of artificial nano-materials to humans and the environment, opportunities for attempts at human enhancements, and the convergings of nanotechnology, biotechnology, information technology and cognitive science (Roco and Bainbridge 2010: 9) that threaten privacy, social justice, and related issues all present challenges for traditional ethics.

It is obvious that nanotechnology gives us possibilities that are both positive and negative. The fact also further shows the huge space of technological possibility, and promotes philosophers to further reflect on what are these possibilities, the ontological, epistemological and ethical significance of these possibilities, as well as the possibility of regulating and controlling the possibilities. We need now to distinguish different kinds of possibility: real possibility and potential possibility (Hubig 2006: 150). We need to explore the real possibilities, and seek the conditions that enable the actualization of those possibilities so they can become reality.

This paper aims to systematically and philosophically reflect on the concept of feasibility. First of all, I will investigate what feasibility is and critique the current idea of feasibility. I will then consider the Chinese feasibility concept and emphasize it, because it has connotations that the English term does not, and can make a contribution to feasibility studies dealing with the possibilities (positive and negative) of nanotechnology. Finally, I will consider basic principles that can be used to establish the feasibility and conditions of the feasibility of nanotechnology.

## 28.2 What Is Feasibility?

Feasibility is realistic possibility. The English word "feasibility" and its adjectival form "feasible" has three meanings: (1) capable of being accomplished or brought about; (2) reasonable, logical, likely; (3) used or dealt with successfully, suitable (New English-Chinese Dictionary1986: 448). Feasibility is, first, a capability, meaning capable of being done or carried out; and second, capable of being used successfully; suitable. Sometimes, feasibility is also directly interpreted as possibility. In German, feasibility is named *Ausführbarkeit*, *Durchführbarkeit* and *Machbarkeit*. These words share the suffix "*barkeit*," which means possibility. Judging from word-formation, feasibility has everything to do with possibility. And the explanation of "*faisabilité*" in French is "*possibilité caractère de ce qui est realizable*."

The earliest use of the term "feasibility" in relation to evaluation of technology is to be found in the feasibility study of the Tennessee river basin in America in the 1930s (Encyclopedia of China, volume of chemistry 1985: 378). This study included the following factors: (1) market research, including market demand in both the short and the long term, and the coordination of resources and energy technologies; (2) research on technological advancement, including research into the best work process and equipment required, the arrangement of the factory, and organizational system and personnel training; (3) research into economic rationality, including the prediction of the time taken to complete the project, the calculation of investment costs, the source of funds and the plan for repaying debts, the estimation of production cost and the comprehensive evaluation of investment effects, etc. Generally speaking, a number of plans from which to choose were offered to decision-makers (Cihai 1986: 108–110). After World War II, this research method spread to developed countries and was further developed into the first phase of an engineering project. Since the 1960s, many developing countries also attributed primary importance to the study of feasibility in engineering. In the late 1970s, the research method found its way to China, and in 1983 the Chinese government officially listed it as one of the basic procedures in construction (Encyclopedia of China, volume of economics 1985: 503).

The history of feasibility studies indicates clearly that feasibility refers to a comprehensive and systematic analysis and scientific examination of the technological advancement and economic rationality of an engineering project, in order to maximize the economic results of that project. Feasibility here is firstly a concept of economics or engineering management, including considerations of two major aspects: (1) the feasibility of a technology depends on whether it is a mature technology, and on the possibility for that technology of moving from knowledge and skills to practical products; (2) the economic benefit of a technology, which mainly refers to economic cost and risk assessment. After the 1990s and especially in the twenty-first century, as environmental issues and energy issues become more important, they have been included in the scope of feasibility study. In general, current studies on feasibility include the following dimensions:

1. The necessary preconditions in terms of technology or knowledge, including infrastructure and software;
2. Economic investment and returns, and economic risk;
3. Environmental costs;
4. The supply of resources including energy.

The above elements show that so far, the main concern of a feasibility study is technological and material/economic. There are at least four factors which are ignored:

Firstly, a feasibility study does not consider the purpose of technology. In the traditional technology concept, the purposes of technology are to compensate for a defect or to reduce the burden on humans, and a technology is always good (Gehlen 2003: 3). A feasibility study thus presupposes that the aim of a technology is appropriate; and it only considers the rationality of means in relation to certain purposes, reducing rational analysis to the level of instrumental rationality. It apparently neglects the complexity of modern technology. In traditional feasibility studies the focus is on a system that includes technological and eco-political systems, but overlooks the question of the ultimate purpose of technology and economy.

Secondly, the purpose of technology and technical consequences are regarded as the same. As we believe that technology is intended to provide a good quality of life and the results of technology will be able to improve the quality of life; consequently, the intentions of technology designers, the subjective purpose of the technology and inconsistencies between the outcomes and the objective purpose of technology are ignored.

Thirdly, the inspection of quality and morality of the user is neglected. Technology is the activity of humans, and technical activities must be subject to a code of ethics just as other human activities are. In the face of the complex consequences of the current technology, scientists and engineers should not only be responsible for employers, but also for users, for society, and for future generations. Fourthly, the public's attitude toward the technology is ignored. There are two reasons that may lead to this neglect. On the one hand, we believe that the purpose of technology is consistent with public expectation. If we think that technology can always bring convenience and comfort to life, then who will oppose technology? On the other hand, the use of technology is not taken into the scope of technology. Users and consumers of technology need to be considered alongside designers and implementers of a technology.

"Technology is the realization of the idea, and is based on the purpose-oriented processing of natural objects" (Dessauer 1956: 172–173). In the eyes of Friedrich Dessauer, the German philosopher of technology, we usually use the concept of technology in two senses, that of technical items and of technical process. These items and processes don't exist in nature, but appear as a result of human activity. Therefore, technology is imbued with the values of humans and their pursuit of those values. As a collective collaborative activity, modern technology not only involves multi-subject, but also pluralistic value (Hubig 2005: 70–78). The nonconformity between the subjective wishes of technology designers and the objective realities of the technology itself requires us to carefully reflect on the value orientation of technology. Therefore, a feasibility study of the purpose of technology should be the primary and most important element of a technical feasibility study.

Technology is not only a product; it is also a process and system (Ropohl 1991: 84–85). This dynamic systemic characteristic of technology determines that technology always contains uncertainty and contingency. The uncertainty and contingency can come from external factors of the technology system, such as changes in objective conditions. They also may be due to internal factors of the technology system, such as those arising from the users or designers of technology. These factors give rise to the uncertainty of technology. Uncertainty can be good but can also be disastrous. As a consequence, technical feasibility studies must consider the uncertainties of the consequences of a technology.

Another important feature of modern technology is the way it functions to mediate (German: *Medialität*) (Hubig 2006: 143–148). In particular, nanotechnology, information technology and other emerging technologies are often shown as "enabling technology," which means that the realization of the technology is based on the use of technology, and technology realizes itself in the process of use or even consumption. It is also in this sense that the contemporary German technological philosopher

Günter Ropohl takes "the sum of human actions of using the objective material system" (Ropohl 1991: 84–85) into technical areas. Since technology users and consumers are an element of technical activities, then their attitude to technology, that is their awareness and acceptability of technology, including their cultural norms and their values of community, belong in the scope of a technology feasibility study.

In addition, technical activity is human activity. Its starting point is human, and the end is also human. In order to be safe, people invent technology, and people constantly innovate and create technology to live better. Since it is a human activity, it must be subject to certain moral constraints. The difficulty is that the subject of modern technical activity is not the individual but groups composed of numbers of people. The responsibility of the individual is often overwhelmed by collective action. It is vital for the technical feasibility study to distinguish the responsible subject of technology and especially to pay attention to the virtue of technology's actors.

## 28.3 The Concepts of Action and Feasibility in Chinese Philosophy

It is obvious that the current understanding of the concept of feasibility is not suitable for the systematic analysis of the feasibility of nanotechnology. Strictly speaking, the expression 'nanotechnology' is not accurate. A nanometer is just a scale unit United Nations Educational, Scientific and Cultural Organization 2006). There is no abstract nanotechnology, and there are a number of different 'nanotechnologies', such as nanomaterials technology, nano-biotechnology, nano-catalysis technology, nano-computer technology etc. Different nanotechnologies bring about different problems, and not all nanotechnologies will create ethical issues. Therefore, it is difficult to find a unified ethical norm to regulate nanotechnology. However, we should specifically explore the nature of nanotechnology possibilities and the conditions that turn these into reality. In other words, we should specifically analyze the feasibility of nanotechnology.

In order to extend the feasibility concept in the appropriate way, I introduce the concepts of action and feasibility in Chinese philosophy. Chinese philosophy regards the whole world as a dynamic open process. It does not presuppose a constant and abstract noumenon. Change is seen as the essence of reality. Contrary to Western philosophy, which lays more stress on cognition than on action and emphasizes principles over strategy, Chinese philosophy puts action (i.e. practice) first in the relationship between cognition and action. It considers the ethical principles of action, and pays more attention to strategic principles that judge whether something can be done, how it can be done, and how it can be done feasibly.

Seen from the etymological and semantic point of view, feasible, in Chinese "*xing* (行)" in shell inscriptions is represented as an intersection that looks like [illegible]. According to an ancient book entitled *Words Defined* (说文解字), *xing* as a noun is pronounced "*háng*"; "what *háng* is, is but *dao*" ( 行,道也) (Dictionary of Chinese Characters 1991: 811). The image of *xing* not only has the metaphor of *dao*, but also includes the metaphor of making a choice among numerous paths. As a verb, *xing* has two meanings: If it involves the objective world, especially the heavens, the earth and nature, it refers to the movement of the natural world, like "the movement of the heaven is powerful" (天行健) (Tang 1990: 180); "The motion of nature has its rules. It doesn't continue for Yao, a good king, nor would it stop for Jie, a bad king" ("天行有常,不为尧存,不为桀亡") (Wang 1998: 306). If it involves humans, it means behaviors (行为) and actions of people (行动), such as words and deeds. In *the Analects,* there is a line that goes as follows "Ji Wenzi thinks twice before he does" ("季文子三思而后行") (Liu 1990: 196). Moreover, *xing* could mean approval, meaning "OK", "You are great", "It is done" (行啊!), etc.

As mentioned earlier, the definition of concepts is not a feature of Chinese philosophy. One word often has different meanings in different contexts. As far as the concept of action is concerned, the ancient thinkers were not concerned with what the action is, but how to do it and how it can be feasible. That is the condition of action that we talk about, or the elements of feasibility.

To sum up, the conditions of *xing* (action) include the following aspects:

1. Respect for and compliance with the laws of nature. The so-called "*tian xing you chang*" (天行有常, nature is the true law) (Wang 1998: 306), where "*tian*" (天) refers to nature, and "*chang*"(常) refers to *dao* (道) and laws, that is, the movement of nature follows certain laws. "An actor makes things work by showing respect to the law of nature" (行者,言顺天行气也) (Chen 1994: 166). The "*tian*" here also means the natural law. In the eyes of philosophers in ancient China, neither nature nor humans can go against the natural law: showing respect to the law of nature, everything will be feasible, otherwise nothing is possible. *Dao* is the guiding rule of action, and is the essential condition of feasibility. In addition to *dao*, in the Confucian view, the important thing is humanity (仁), that is: the laws of society and the constant regulations of human relations. As a result, a gentleman must do things in accordance with the *dao* (遵道而行), and show benevolence when doing things (仁以行之).
2. Assess the situation and seize the opportunity. "Is the situation favorable? Do things in accordance with the situation."(坤其道顺乎?承天而时行) (Tang 1990: 178) The way this is done in current society is to grasp the opportunity by considering the situation. One shall do things that comply with the situation, and if one does so, his/her future will be bright. In the *Mencius*, Mencius emphasized the importance of properly grasping the opportunity for farming and sustainable development, when he communicated with King Lianghui: "if you do not miss the farming season, the grain will be enough to eat; if you do not go fishing in the broken pool, fish turtles will be enough to catch; if you cut trees in the best time, trees will be enough to be used. The grains, fish and turtles are enough to eat, and trees are enough to use" (Mengzi 1987: 20).
3. Virtue is a guarantee of the action. Just as the movement of nature has to follow nature's laws, human action has to abide by social law and norms. One who corresponds to the norm is a virtuous person. If the actor has virtue, the action can successfully occur and it will be accepted by other people. Only then is an action feasible. In the *Analects*: Weilinggong 15th, a dialogue between Zizhang

and Confucius about action and feasibility is clearly documented. It runs as follows: "Zizhang asked about *xing*. Confucius answered: Keep your words and show your respect, your *xing* will be successful even among barbarians. Otherwise, you will fail even at home" (Liu 1990: 106). According to Confucius, if a man is virtuous, keeps his word and respects others, he will have no trouble even in foreign lands; on the contrary, if he is not virtuous and cannot keep his word nor respect others, he will not be respected even at home. The first *xing* here refers to action, the second refers to the feasibility of the action.

In accordance with the understanding of Confucius, virtue is the internalization of the *dao* in a person's heart. For virtuous people there are no barriers; they can go everywhere. It means *xing de tong*. *Tong* has two meanings. It can mean there are no barriers. But when used for the action of a person, it means success, or a way that is popular and accepted by the people. On the contrary, if a thing is unpopular and not recognized or accepted by people, then it is not feasible.

## 28.4 Feasibility Strategies for Nanotechnology Development

The Chinese concept of feasibility provides crucial insight for dealing with the challenges associated with nanotechnology, and in many other technologies as well. The Chinese concept of feasibility requires that we pay attention to material and cognitive factors of technology, just as we would in a traditional feasibility study. But it also requires that we consider human factors integral to technical activities; the virtues, not just knowledge and skills of technical actors; the cultural and ethical background of technical action; and people's attitudes to technology. Theory is abstract, while technical activities are always concrete. Abstractly speculating about future ethical issues of technology (Nordmann 2007: 31–46), or abstractly discussing the so-called ethical principles of technology, can only ultimately turn technical ethics into empty preaching. "If the problems faced by technical activities are to judge and weigh, then what is the function of the principles of technology?" (Grunwald 1996: 193).

The current economic feasibility concept needs to be systematically extended, and the Chinese philosophy of feasibility provides guidance for this extension. A feasibility study should include the whole process of technological activity, including the purpose, means and results of technology. Of course elements of a traditional feasibility study would also be included, for instance material elements such as energy, resources, and environment, and elements required for the development of knowledge such as technological tools, equipments, funds and information, etc. Additionally, a feasibility study would include the cultural context and the virtue of the actors and the acceptability of technology to users and consumers. These are included because feasibility has something to do with the perspectives and stances of actors. Agents of technology differ from observers of technology in terms of their respective stances and perspectives. Who thinks this is feasible? Scientists, business people, politicians, or the public? Different interests naturally result in different

**Table 28.1** Matrix of feasibility analysis

| | (1) | (2) | (3) | (4) |
|---|---|---|---|---|
| (1) Who thinks feasible? | Scientist (doer) | Politician (decision-maker) | Entrepreneur (benefitted) | The public |
| (2) Feasible to whom? | Individual (doer) | The State | Society | Natural environment |
| (3) What is feasible? | Target | Method (safety, economy) | Result (safe and acceptable) | – |
| (4) What are the feasible factors? | Natural-law abiding | Materials (resources, energy) | Economy (cost, risk) | Norm (culture, ethic) |
| (5) What is the basis of feasibility | Natural-law abiding | Satisfying human needs | – | Serving the purpose of "good" |
| (6) Feasible when? | Now | Future | – | – |
| (7) Feasible where? | Economically developed area | Economically under-developed area | – | – |

judgments. Seen from the angle of development, the motion, change and development of things in the future constitute an array of possibilities. Conditions of feasibility relate to the diverse factors that influence choices in time and space.

We can analyze the basic content of feasibility from the following seven aspects: Who thinks it is feasible? Feasible to whom? What is feasible? What are the factors that influence feasibility? What is the basis of feasibility? When and where is it feasible? (Table 28.1).

Based on the above elements of feasibility, I think that when we analyze the feasibility of nano-technology, the following strategic principles should be taken into account.

### 28.4.1 Specific Strategy

In Chinese philosophy, there is a general principle that says one should make a concrete analysis of concrete issues, adjust measures to local conditions, and treat different things in different ways. I will refer to this as the "specific strategy." When applied to nanotechnology, this strategy requires that we consider different areas of research, different nanotechnologies, and properties of different nano-materials. Nano-Au, nano-Ag, nano-Sn or nano-semiconductor are very good examples. They evince completely different properties at different nano-measurements.

As mentioned earlier, not all nanotechnology has side effects or may bring potential ethical risks. Not all nanotechnologies are at the same level of maturity. Talking about ethical issues of nanotechnology in a general way will only cause antipathy and non-cooperation of many scientists, and hinder the development of nanotechnology. Such a general discussion of nanotechnology also is not conducive to the development of a mature discussion of nanotechnology in society. While on the one hand, some nanotechnologies can have a negative impact on the environment, on the other hand, nanotechnology also plays an important role in controlling environmental pollution and solving the issue of energy supply. In addition, the attitudes of different people to nanotechnology, even among those in the same cultural circle, are quite different. When suffering from cancer, patients' hopes and expectations of medical nanotechnology far outweigh their concerns about risk. Dieter Birnbacher proved that "Though people stay in the same cultural circle, it is difficult to reach consensus regarding the treatment of some sensitive technical issues" (Birnbacher 2009: 81–88). Culture and values are indispensable when conducting a feasibility study.

### *28.4.2 Real-Time Strategy*

Chinese philosophy considers it important to make real-time assessments and judgments in accordance with the current situation of nanotechnological development. I will refer to this as a "real time strategy" integral to a proper feasibility study. The key to real time is "time." Only by grasping "time" can our actions be reasonable and appropriate. David Guston and Daniel Sarewitz, two U.S. science policy analysts, developed a technical assessment model called "real-time technology assessment" (RTTA) for the risk assessment of nanotechnology. In addition, "tracing evaluation of the process" (Guston and Sarewitz 2001: 98–118) has been developed by researchers of the Institute for Technology Assessment and Systems Research in Karlsruhe, Germany. In both cases researchers emphasize real-time assessment, real-time decision making, and real-time adjustments in the ongoing development of a technology (Fiedeler et al. 2004).

At present, we are in a world of rapid technological development. Nanotechnology keeps pace with nanoscience, and in some cases moves ahead of it. That is to say, even when the mechanisms, toxicity and basic theory of nano-materials are not clear, nanotechnologies have been turned into products that go to market and are in use. In light of the characteristics of cumulative and long-term technical consequences, the traditional method of assessing technical consequences is not suitable for the assessment of nanotechnology. Some consequences of nanotechnology are far from clear, so the ethical and social impact is even more difficult to predict. On this occasion, if nanotechnology is misused, it could lead to unpredictable ethical disasters. Real-time tracking of the forefront of nanotechnology development, starting from the beginning of the study on nanotechnology, synchronously conducting research on the safety and ethical consequences of nanotechnology, timely development of relevant policies and laws and ethical norms, and guiding the healthy development of nano-science are not only vital to the sustainable development of nanotechnology, but also are crucial to the sustainable development of humans.

### 28.4.3 Dynamic Strategy

Chinese philosophy requires an open, dynamic ethical attitude. I will refer to this as the "dynamic strategy" integral to a feasibility study. According to the situation of nanotechnology development and social acceptability of nanotechnology, the relevant ethical and legal norms should be promptly adjusted, in order to "take advantage of an opportunity that comes one's way." In other words, we should take advantage of opportunities and keep abreast with the times. We should respect tradition while facing the future (Hubig 2005: 77).

The main characteristic of the dynamic strategy is to ensure the controllability and modifiability of our actions. Its premise is to recognize that our understanding of technology and science is a gradual process and we must recognize our own ignorance. We shall not advance rashly when conditions are not ripe. On the other hand, scientific truth is also interpretable and relative to the context. When nanotechnology is shown to be safe, then it should be applied and developed. When the safety of nanotechnologies is uncertain, we should be careful. Consequently, we must adjust our development strategy in a timely manner.

### 28.4.4 Holistic Strategy

Chinese philosophy requires that in pursuing the responsible development of nanotechnology we not only proceed from a region or a country, but also from the overall interests of mankind. I will refer to this as the "holist strategy" integral to a proper feasibility study.

Scientists are the subjects of nanotechnology development and research. Due to the fact that nanotechnology involves cognition and transformation of Nature at the level of molecules and atoms, it is hard for the general public to understand, intuitively, the advantages and disadvantages of nanotechnology. Scientists should not only focus on the national interest but, when major issues about sustainable development are involved, they should proceed from the overall interests of mankind, and explicate to the public, without any reservation, the merits and faults, good and bad effects of the technology being developed. The public should get involved in the decision-making about nanotechnology. Participation in nanotechnology is the right and responsibility of the public.

In brief, abstract speculation and assessment of the possibility of nanotechnology – whether talking about potential huge economic and social benefits, or potential threats and risks – are not conducive to the sustainable development of nanotechnology, but are contrary to the pursuit of a better life. What we need is the practical, real-time and comprehensive investigation on the feasibility of concrete nanotechnology.

**Acknowledgments** This work was financially supported by the National Basic Research Program of China (2011CB933401) and National Social Science Foundation (09BZX048). Part of this chapter was previously published in: Fischer, P., Luckner, A., & Ramming, U. (hrsg.). (2012). *Reflexionen des Möglichen. Zu Christoph Hubigs Philosophie der Medialität.* Band 23 der Reihe "Technikphilosophie" hrsg. von Klaus Kornwachs, Münster: Lit-Verlag.

## References

Birnbacher, D. (2009). The convention on human rights and biomedicine of the council of Europe in Germany. In Wang Guoyu & Liu Zeyuan (Eds.), *Cross-cultural dialogue of ethics of science and technology*. Beijing: Science Press.

Chen, L. (1994). *Baihutong Shuzheng*. Beijing: Zhonghua Book Company.

*Cihai*. (1986). Shanghai: Shanghai Dictionary Press.

Dessauer, F. (1956). *Streit um die Technik*. Frankfurt am Main: Josef Knecht.

*Dictionary of Chinese characters*. (1991). Chengdu: Sichuan Dictionary Publishing House.

*Encyclopedia of China, volume of chemistry*. (1985). Beijing: Encyclopedia of China Publishing House.

*Encyclopedia of China, volume of economics*. (1985). Beijing: Encyclopedia of China Publishing House.

Fiedeler, U., Fleischer, T., & Decker, M. (2004). Technikfolgenabschätzungen zur Nanotechnologie: Roadmapping als neues Instrument. *FZK-Nachrichten*, *36*(4), 230–234. http://www.itas.fzk.de/deu/lit/2004/fiua04a.pdf

Gehlen, A. (2003). *Man in the age of technology* (He Zhaowu, He, Bing, Trans.). Shanghai: Verlag Wissenschaftsbildung.

Grunwald, A. (1996). Ethik der Technik Systematisierung und Kritik vorliegender Entwuerfe. *Ethik und Sozialwissenschften EuS*, *7*, 191–204.

Guston, D. H., & Sarewitz, D. (2001). Real-time technology assessment. *Technology in Society*, *23*(4), 98–118.

Hubig, C. (2005). *Ethik der Technik als Provisorische Moral* (Guoyu Wang, Trans.). *World Philosophy*, *4*, 70–77.

Hubig, C. (2006). *Die Kunst des Möglichen Technikphilosophie als Reflexion der Medialität*. Bielefeld: Transcript.

Liu, B. N. (1990). *Lunyü Zhengyi*. Beijing: Zhonghua Book Company.

Mengzi. (1987). *The thirteen classics with annotations and commentary*. Beijing: Zhonghua Book Company.

*New English-Chinese dictionary* (Complementary Ed.). (1986). Shanghai: Shanghai Translation Publishing House.

Nordmann, A. (2007). If and then: A critique of speculative nanoethics. *Nanoethics*, *1*(1), 31–46.

Roco, M., & Bainbridge, W. S. (Ed.). (2010). *Coverging technologies for improving human performance. Nanotechnology, biotechnolgy, information technology and cognitive science* (Cai, Shushan usw, Trans.). Beijing: Tsinghua University Press.

Ropohl, G. (1991). *Technologische Aufklärung*. Frankfurt am Main: Suhrkamp.

Tang, M. (Ed.). (1990). *Zhouyi Pingchuan*. Beijing: Zhonghua Book Company.

United Nations Educational, Scientific and Cultural Organization. (2006). *The ethics and politics of nanotechnology*. http://unesdoc.unesco.org/images/0014/001459/145951e.pdf

Wang, X. Q. (1998). *Xunzi Jijie*. Beijing: Zhonghua Book Company.

Woodrow Wilson International Center for Scholars. (2010). www.nanotechproject.org/inventories/consumer. Accessed 15 Dec 2010.

# Chapter 29
# Engineering Innovation: Energy, Policy, and the Role of Engineering

**Zachary Pirtle**

**Abstract** Efforts to mitigate anthropogenic global warming have led to a new focus on energy innovation. The historical US approach to energy innovation has been too trapped by the so-called linear model of science and innovation, which posits basic research as being central to the process of innovation. While historians and economists of technology have long criticized the linear model, it still frames public innovation programs to negative effect. After surveying the history of the US Department of Energy, I discuss relevant debates in the philosophy of science about the importance of basic scientific understanding and scientific laws. I suggest that a long obsolete approach toward the philosophy of physics is one of many contributors to the lingering power of a science-focused approach toward innovation. To assist in further development of an enriched philosophy of engineering and innovation, I present principles taken from the innovation studies literature.

**Keywords** Innovation policy for climate change • Mario Bunge • Linear model of science • Nancy Cartwright • Energy innovation

The views expressed here are my own and do not necessarily represent those of the National Aeronautics and Space Administration or the United States.

Z. Pirtle (✉)
National Aeronautics and Space Administration, Washington, DC, USA
e-mail: Zachary.pirtle@fulbrightmail.org; zpirtle@nasa.gov

D.P. Michelfelder et al. (eds.), *Philosophy and Engineering: Reflections on Practice, Principles and Process*, Philosophy of Engineering and Technology 15,
DOI 10.1007/978-94-007-7762-0_29, 

## 29.1 Introduction

There are reasons to believe that the dialogue concerning climate change has been too focused on the science of global warming and insufficiently focused on the engineering of cheap, low-carbon emitting energy technologies (Pielke 2010). In the United States, concern about climate change and an inability to pass legislation on climate policy has contributed to a new focus on energy innovation policy (CSPO/CATF 2009; Hayward et al. 2010; American Energy Innovation Council 2010; PCAST 2010). Instead of debating what effects global warming will bring, this new effort is trying to identify the right points in the energy innovation system where engineering efforts can be placed in order to develop the new low carbon energy technologies that will help lower overall carbon dioxide emissions.

The search for scientific and technological fixes for societal problems warrants caution and humility. With climate policy, some have hoped to induce political action based on consensus over the science of climate change. However, as Daniel Sarewitz has observed, some problems need to be resolved through political discourse; overly focusing on the science can preclude important debates about political values from occurring, which may unintentionally delay the conclusion of policy debates (Sarewitz 2004). The lack of action on climate policy may reflect an attempt to improperly scientize the political debate (Pielke 2010). At the same time, technological fixes, which attempt to remove the cause of political disputes with technology, can be incomplete or have negative unintended consequences (Sarewitz and Nelson 2008). Simply put, technology cannot solve many societal problems. But, unlike with science, the creation of new technological options can change political dynamics in that they can create new policy options that can satisfy groups with conflicting values. The recent shift toward energy innovation on the global warming debate may help create technologies that can reduce the cost that society needs to pay to address global warming, which may make political agreement on climate change policy easier to attain.

Engineers and philosophers of engineering have a role to play in encouraging a new approach to innovation, as new approaches to innovation policy can greatly benefit from a more nuanced understanding of the nature of science, engineering and innovation. To illustrate this, I present a caricature of the Department of Energy's efforts to innovate as it relates to the linear model of science. I then connect it to debates about what engineering is, characterizing Mario Bunge and Nancy Cartwright as providing visions of engineering that differ based on the priority given to scientific laws. I see Cartwright's view as creating a vision of engineering that aligns well with principles of innovation that have been put forward by innovation scholars, which I survey at the end. Developing a more refined view of engineering and innovation can be an important part in getting technology to yield helpful benefits to society.

For purposes of this paper, I have not distinguished innovation from engineering more generally. Just as engineering is not merely applied science, so too is innovation not merely engineering. However, the connection between engineering and innovation is far stronger than the connection between applied science and engineering. The best definitions of engineering acknowledge the engineer as businessperson and artisan (Layton 1986). A successful innovation implies a technology that is more than functional but that is desired in the broader economy and society. An engineer's new technology is not fully successful unless she succeeds in getting it used once it is developed, in other words, unless she successfully innovates. The dimensions of innovation systems are broader than what a typical analysis of engineering might entail, but the two are close enough that I at times interchange the two terms here in this paper.

## 29.2 The US Department of Energy and the Linear Model of Science, Engineering and Innovation

Energy innovation is a good domain from which to draw larger lessons about the connections between science, engineering and innovation. The aspects of the broader US innovation system focused on energy have been profoundly influenced by government activities. Since President Richard Nixon's focus on clean energy technology, the United States has encouraged the innovation of low-carbon intensity technologies. The Department of Energy (DOE) was formed in 1977 out of organizations that had previously been part of the Energy Research and Development Administration, the National Science Foundation, and the Atomic Energy Commission (AEC). The AEC was the earliest institutional precursor of the DOE, and many of the AEC's physicists and centers came directly from the Manhattan project. Crudely put, recent criticisms of DOE's innovation approach can be explained by its origin in labs run by scientists, who continue to look to the high age of physics and who approach innovation by putting research first.[1]

The department is still trying to define its core mission, which has evolved significantly over time, as well as how best to pursue it. The AEC was focused on the development of nuclear power and weapons systems, duties which are still part of the DOE's larger mission today. The oil and fuel crisis of the mid-1970s created a national security imperative to advance energy sources independent of the Middle East. In response, the DOE pursued technologies for solar photovoltaics, energy efficiency, breeder nuclear reactors, and synthetic fuels created from coal, among others, with mixed results (Cohen and Noll 1991). Many of those technologies overlap with the DOE's more recent mission, which still promotes national security while more deeply focusing on ways to produce energy without emitting greenhouse gases.[2]

[1] For criticisms of activities of the AEC and how it might have differently directed the development of nuclear technology, see Morone and Woodhouse (1989).

[2] The DOE's previous work on creating synthetic fuel from coals (which would not be a low-carbon intensity technology) helped establishes the resources needed to develop carbon capture and sequestration technologies that can minimize emissions from coal plants.

As the fuel crisis of the mid-1970s receded and the DOE's mission changed, the DOE budget focused on energy innovation decreased dramatically. As shown in Gallagher and Anadon (2010), the DOE budget focused on energy research, development and demonstration (RD&D) dropped considerably from its over six billion dollar high in the 1970s to less than two billion dollars in the 1990s. As attention on climate change has increased in the last decade, the US budget on energy RD&D has increased slightly. The 2009 American Recovery and Reinvestment Act brought a one-time appropriation of over six billion dollars to energy RD&D, but this money will not be repeated on an annual basis. It seems likely that the annual US spending rate will continue to be just over three billion dollars, which is much lower than the spending rates brought on by the fuel crisis of the 1970s.

Within the broader realm of energy RD&D, several programs focus on different energy technologies. These programs are managed by different branches of the DOE, each with their own unique cultures and that loosely lead the more than 20 US national labs that are funded by the DOE. The DOE has separate organizations for science and its other non-security related technical activities: the Office of Science, managed by the Undersecretary for Science, and a set of Offices, collectively under the purview of the Undersecretary of Energy. For low-carbon energy technologies, two important offices within the DOE are the Office of Energy Efficiency and Renewable Energy (EERE) and the Office of Fossil Energy and Power Systems. EERE is itself divided into ten branches covering efficiency and renewable energy technologies.

Many innovation experts have been critical of the DOE's approach to innovation (Weiss and Bonvillian 2009). Some have criticized the DOE's approaches toward engaging with industry, arguing that they are insufficient (CSPO/CATF 2009). The focus seems to be on doing good research, not on connecting with industry. The DOE's efforts at commercialization have also met with challenges. In one of its larger endeavors, the DOE focused on rapidly commercializing a technology without sufficient industry buy-in (see the Clinch River Breeder Reactor chapter in Cohen and Noll 1991). Others criticize the DOE for having a fragmented mission and for its inability to train a workforce for the broader energy industry (Duderstadt et al. 2009). Still others have criticized its loan guarantee program as being too difficult and costly for many companies to profitably engage with (CATF 2009).

The DOE's science-centric innovation approach and its difficulty in connecting with private industry evince an attitude that scholars have called the linear model of science, technology and innovation (CSPO/CATF 2009). The linear model has a long history in US science policy, dating back to before World War One (Kevles 1995, Chapter 4). In 1945, the US president's science and technology czar, Vannevar Bush, published *Science: The Endless Frontier*, which embodied the linear model in an extremely influential way. Bush laid a vision of science as the driver of innovation:

> Basic research is performed without thought of practical ends. It results in general knowledge and an understanding of nature and its laws ... [Basic research] creates the fund from which the practical applications of knowledge must be drawn. New products and new processes do

> not appear full-grown. They are founded on new principles and new conceptions, which are in turn painstakingly developed by research in the purest realms of science. Today it is truer than ever that basic research is the pacemaker of technological progress. In the nineteenth century, Yankee mechanical ingenuity, building largely upon the basic discoveries of European scientists, could greatly advance the technical arts. Now the situation is different. (Bush 1945)

Simply put, the linear model is this: science provides basic research, from which understanding and scientific laws can be used to explore practical options which will underlie new innovations. The mechanism by which research will produce innovation is not known in advance, but there is a religious-like faith that something beneficial will come eventually. Later in the report, Bush cautioned that government and industry must ensure that they sufficiently support basic research, and not be lured into overly focusing on applied or commercial projects.

This description of the innovation process coheres with the characterization of the DOE made by its critics. As mentioned, DOE has an Office for Science that is separated from other technical offices, which is a symbolic embodiment of the linear model. The lack of sufficient success at DOE using this innovation model helps underscore the limits of the linear model, which historians have long criticized. The historical record shows that innovation is a more complex process than what Bush described. For example, in the Manhattan Project, which Bush helped supervise, many theoretical breakthroughs were induced based upon the practical challenges encountered in engineering a workable bomb (Kevles 1995). The reality of innovation is not a linear tale of research inspiring engineering, but a complex and interdependent process where engineering challenges can inspire fundamental insights and vice versa.

While scholars of innovation (such as those surveyed later in this chapter) have long tried to describe this deeper complexity surrounding innovation, the linear model still shapes the national dialogue about energy innovation. In the next section, I'll discuss some conceptual motivations that are part of the reason why the linear model continues to linger.

## 29.3 Behind the Linear Model, an Old Philosophical Debate About Science and Laws

The linear model is connected to notions about ideal science, represented by physics, that were prominent in the earlier parts of the twentieth century. These old debates about the relationship between laws, science and engineering still shape discussion today. While the philosophy of science has moved on to more nuanced views, the lay notion of scientific laws still tacitly supports the continued force of the linear model for innovation.

To help in drawing this connection, I will sketch out the view of Mario Bunge, a philosopher of science who wrote extensively about technology. For Bunge, a "law is a confirmed hypothesis that is supposed to depict an objective pattern.

The centrality of laws in science is recognized upon recalling that the chief goal of scientific research is the discovery of patterns." (Bunge 1967, p. 305). Like many others, Bunge is interested in precise and deep understanding of phenomena. Laws are a powerful way to describe that understanding, but there are limits: "every law has a limited extension or domain of validity – one beyond which it becomes definitely false" (op cit, p. 347). Bunge notes that laws require more than mere empirical generalizations, or claims that describe patterns of behavior in reality. He requires that one be able to "derive them from stronger assumptions belonging to some theory, i.e. to explain them" (op cit, p. 355), where a theory is a larger, more comprehensive collection of laws. Bunge and others saw scientific law as emblematic of the most fundamental understandings that science can make of the world.

From this, Bunge draws some strong distinctions between science and engineering. For him, engineering is a scientifically based subset of the broader realm of technology. Whereas artisans are likewise technologists because they deal with artefacts, engineers deal with quantitatively sophisticated works in design and implementation. Bunge claims that engineers do not deal with the laws of physics; their concern with action places an emphasis upon rules, which guide action (Bunge 1983, p. 68). While artisans use rules but not science, technological rules connect to science through the "grounding" of rules upon scientific law, which means that the rule "is based on a set of law formulas that could be used to rederive them" (op cit, p 68). In later work, Bunge argues that while some engineering disciplines may not have grounded rules, more mature engineering disciplines are based upon a fund of scientific knowledge (Bunge 1985, p. 235). While Bunge does not necessarily embrace the notion that engineering developments are in practice actually derived from laws, the views are connected. Bunge claims that a majority of technological developments arise from investments in "basic science," that is, that many, if not most, technological developments are in fact derived from scientific laws (op cit, p. 238).

The epistemological primacy Bunge gives to science-derived laws in engineering gives at least a cursory primacy to the importance of science in the innovation process. It should not be seen as a coincidence that the simplistic view of scientific laws espoused by Bunge meshes with Bush's linear model, which likewise focuses on "an understanding of nature and its laws" and a "fund" for knowledge. The law-based vision of science (as espoused in different ways by intellectual predecessors of Bunge) likely shaped and led to Bush's articulation of the linear model of science.[3]

[3] Proving this point would likely require an exhaustive study of the connection between the broader science policy dialogue and the history of knowledge, Kevles (1995) surveys some of that, with discussions of the dialogue in the century before Bush. Boon (2011) provides useful context on Aristotle and subsequent debates on the epistemology of technology. Kealey (2008) has a sarcastic bias against the government support of science, but this can be separated from his valuable intellectual history of the linear model from Alexander the Great to Bacon to today.

However, the philosophy of science now characterizes scientific laws and their role in understanding the world in a different way. Published in 1983, Nancy Cartwright's book, *How the Laws of Physics Lie*, describes the limits of laws and attacks their perceived status as the focal point of science (Cartwright 1983). While many disagree with Cartwright's conclusions, discussions of laws in science no longer proceed on the linear terms that underlay Bunge's view.

Cartwright does not describe engineering in detail, though she sees it as central to the scientific process. For Cartwright, science is a process of connecting scientific models to reality by carefully engineered experiment. Cartwright argues that there are no fundamental, law-like truths from which other ideas spring. Instead, different disciplines establish models with which to view and understand the world, and the disciplines relate to one another in a chaotic and uneven way. Cartwright's account may map more closely to the complex and interdependent innovation cycle that actually occurs. In her 1999 book, *The Dappled World,* she presents a metaphor that might better help frame science, engineering and innovation:

> [W]e live in a dappled world, a world rich in different things, with different natures, behaving in different ways. The laws that describe this world are a patchwork, not a pyramid. They do not take after the simple, elegant and abstract structure of a system of axioms and theorems. Rather they look like – and steadfastly stick to looking like – science as we know it: apportioned into disciplines, apparently arbitrarily grown up…For all we know, most of what occurs in nature occurs by hap, subject to no law at all. What happens is more like an outcome of negotiation between domains than the logical consequence of a system of order. The dappled world is what, for the most part, comes naturally: regimented behaviour results from good engineering. (Cartwright 1999, p. 1)

What would Bunge say in response to Cartwright's views? His philosophy did become more complex and moderate over time (see Bunge 2007), and his initial work quoted above acknowledged that every law has exceptions beyond which it is inaccurate. However, the importance he still assigns to scientific laws might encourage the characterization of more mature engineering disciplines as being based on laws, which would continue to reinforce the linear model. If the epistemology that Bunge described were fundamentally right, then a physics-centered approach toward innovation might be most effective. And, crudely put, perhaps future basic research will discover exactly the right, simple and unified laws necessary to induce new technological revolutions.[4]

But as Cartwright points out, history offers good reason for an empiricist to doubt the existence of such thorough and comprehensive laws (Cartwright 1999, p. 12). Preparing to innovate in a dappled world offers a more secure path for using science and engineering to help humanity. The next section highlights some of our best insights about how to do that.

[4] There will always be new cases of some new law-based theoretical discovery leading to an innovation. But would a linear model-driven innovation system be maximally effective? The question is about whether different approaches would be more productive, and to what extent are the successes we already have are a result of more engineering-focused strategies, and not due to the linear model.

## 29.4 Principles to Help Guide Energy Innovation

The innovation studies literature helps illuminate the conceptual accounts just examined. In the spring of 2010, I wrote a set of principles for the Consortium for Science, Policy and Outcomes and the Clean Air Task Force, entitled "Four Policy Principles for Energy Innovation and Climate Change: A Synthesis" (CSPO/CATF 2010). In it, I presented the results of my survey of recent literature by scholars and practitioners of innovation, identifying key principles for successful innovation systems. The experts behind these reports, listed in the following text box, come from a variety of backgrounds, and had experience both inside and outside of government. While some reports were focused on single issues, I found that most agreed with the major principles that I identified.[5]

The following are the principles developed in the CSPO/CATF 2010 study, slightly modified. Please see the full brief for more context, as well as the more comprehensive CSPO/CATF study, *Innovation Policy for Climate Change* (2009).

**Reports Examined in the CSPO/CATF 2010 Synthesis**

America's Energy Problem (and How to Fix it), by Richard Lester, from the Massachusetts Institute of Technology (MIT) and its Industrial Performance Center. This report examines the magnitude of the energy-climate challenge and the current context surrounding energy innovation while advocating for a better "system of innovation institutions" (Lester 2009).

*Structuring an Energy Technology Revolution*, by Charles Weiss and William Bonvillian. Written by experts on innovation policy, this book presents a framework for innovation policy, seeking to create appropriate policies for different technologies and to overcome institutional hurdles (MIT Press, 2009).

*Innovation Policy for Climate Change*, by Arizona State University's Consortium for Science and Policy Outcomes and the Clean Air Task Force (CSPO/CATF). This report relies on expert analysis from three workshops for obstacles to innovation for three different energy technologies. Led by Dan Sarewitz and Armond Cohen (2009).

Technology Policy and Global Warming, by David Mowery, Richard Nelson and Ben Martin. This overview paper examines the best historical analogies for energy innovation. Surveys key historical episodes of innovation in the US and UK, including agriculture and information technology (2009).

(continued)

[5] None of the experts should be seen as endorsing my formulation of the principles. I also did not attribute every principle to every author. Nevertheless, I feel that there was wide consensus among experts about the core principles.

(continued)

An Energy Future Transformed, by Xan Alexander. From the Climate Policy Center of Clean Air Cool Planet. This report provides recommendations and an analytical framework for the new Advanced Research Projects Agency for Energy (ARPA-E). The author is a former manager in the Defense ARPA and provides a 1 year operating plan for ARPA-E (2009).

Coal without Carbon, by the Clean Air Task Force (CATF). Used groups of expert authors to develop research, development and demonstration road maps for critical clean coal and geologic sequestration technologies. Introduces the idea of a First Project Demonstration Fund to support demonstration projects. Led by Joe Chaisson (2009).

Energy Discovery-Innovation Institutes, by J. Duderstadt et al. From the Metropolitan Policy Program at the Brookings Institution. This report advocates creating regionally focused innovation hubs to connect academic and federal researchers with private industry, oriented around particular innovation tracks (2009).

Clean Energy Technology Pathways, by the National Commission on Energy Policy (NCEP), which is part of the Bipartisan Policy Center. Using a system-level framework, this report draws on models of future energy technology 'mixes' and the effect that each technology will have on the others. Examines cross-cutting challenges to energy technology development (2009).

Various publications by the Energy Research, Development, Demonstration and Deployment (ERD3) Policy Project, which is part of the Energy Technology Innovation Program at Harvard University. These reports make recommendations on the energy innovation policy and for management of innovation institutions. Led by Venkatesh Narayanamurti, Laura Diaz Anadon, and Matthew Bunn (Anadon et al. 2009; Narayanamurti et al. 2009).

Accelerating Energy Innovation, by researchers from the National Bureau of Economic Research (NBER). These studies examine the history of innovation in the life sciences, chemistry, agriculture, and information technology industries, highlighting insights for accelerating innovation in energy technologies. These works were released as a book edited by Rebecca Henderson and Richard Newell 2011.

To encourage energy innovation, the US government should:

1. **Recognize that innovation policy is more than research policy**: innovation occurs through a complex set of interactions, most of which occur in the private sector. The best way to sustain innovation is to have technologies deployed in the field, where engineers and scientists can then begin to optimize existing technologies and work to improve them. A focus on policy for research can be useful, but only touches on a small part of the broader energy innovation system.

(a) ***Align front-end R&D with Deployment programs***. Following (1), it is clear that deployment programs can be essential. However, a lack of coordination between research, development and demonstration programs (RD&D) and deployment programs can hinder the effectiveness of both. Harvard's Energy Research Development, Demonstration and Deployment (ERD3) team in particular emphasizes the importance of connecting the work of research agencies with applied programs.

(b) ***Focus on both policy and technical challenges that technologies will face, especially as they enter the market***. Many of the challenges to innovation are non-technical in origin, and result from existing competition and entrenched political interests. When a new technology enters the marketplace, it is especially vulnerable to competition from established energy technologies. This problem should be examined early in the technology development process. Technologies should be evaluated based upon the businesses and markets that might produce and employ them, and potential political resistance that they might encounter. Investigating these non-technical issues early is important, and will allow development of technology policies that cater to the context of particular technologies.

2. **Pursue multiple innovation pathways**. Just as no one technology will be able to solve the energy-climate problem, no one institution is capable of solving it either. A diverse ensemble of technologies should be pursued, recognizing that successful innovation is never certain and there will always be successes and failures. A successful innovation system will encourage technologies that will mature at a variety of short- to long-term timeframes: near-term, readily available technologies should not overwhelm and crowdout potential new technologies. Further, Richard Lester of MIT also argues for a diverse "system of innovation institutions," with different institutions having their own specializations.

(a) ***Encourage intra-governmental competition***. The Department of Energy has historically been focused toward basic research, and is not optimally equipped to work on more applied development projects. Encouraging multiple federal agencies, such as the Department of Defense, the National Aeronautics and Space Administration, and National Science Foundation to take a greater part in energy innovation can create competition that can help each agency better support innovation. Some successful examples of government sponsored innovation, including information technology, aircraft, and to an extent agricultural technology, reflect competition among a variety of government programs.

(b) ***Catalyze linkages between government, academia and the private sector, at multiple geographical scales***. Encouraging use-oriented research is a complex problem, and one way to do it is by linking public and private researchers at particular geographic scales, as is suggested by the Brookings Institution report. Their report focuses on innovation in metropolitan areas, as opposed to emphasizing national and international scales. These proposed innovation hubs would focus on solving problems that are relevant for that particular region, which provides a framework and context that can encourage innovation. The Harvard ERD3 reports reviewed principles that can apply within an individual research institute, with advice on managing innovation and balancing competition and collaboration amongst different sectors.

3. **Recognize $CO_2$ reduction as a public good, and pursue energy innovation through a public works model.** The market currently does not price the negative societal effects of climate change into the costs of carbon-intensive technologies, which means that some needed technologies that are not cost competitive may not develop in the current system. The public works model would justify the government's support for these technologies, essentially making the government into a customer (CSPO/CATF 2009). The burden of supporting energy innovation could be shared among multiple levels of government (federal, state, local), which could mirror the shared responsibilities for other public works projects like environmental protection. The following two principles are important in their own right, but they also represent two ways to pursue energy innovation as a public works project.

   (a) ***Stimulate demand using public procurement and regulatory mechanisms (including performance standards and carbon pricing) to encourage private sector innovation.*** Without a reliable demand for new energy technologies, firms will not aggressively pursue energy technology innovation. In the United States, most attempts to create demand for low-carbon energy technologies have focused on the establishment of a carbon cap or price. While this approach will push some innovation in the long run, carbon prices are likely to be low and unstable for an extended period, weakening their power. By contrast, direct government procurement is one of the most powerful ways that the Federal government has stimulated demand for innovation in past technological revolutions. Certain agencies, such as the Department of Defense, have uniquely powerful purchasing capabilities due to their large size. Procurement can be used to drive performance standards, and shows private industry that there will be a growing and sustained market, which in turn stimulates competition and innovation. In addition, direct technology-forcing regulatory mandates such as coal plant carbon performance standards are likely to move innovation in a shorter time scale.

   (b) ***Support late-stage development and demonstration projects.*** Some energy technologies can be well understood in the laboratory, but demonstrating technologies at a large commercial scale can reveal and create new, unforeseen problems. Successful demonstrations reduce uncertainty in a new technology, which can enable adequate technologies to develop and receive more investment. However, economic and structural biases often make it too risky for private corporations to undertake some demonstration projects, which prevents innovation. Governments should help provide financing and incentives to encourage these demonstration projects. Finding the right mechanism and balance of funding with private industry is critical, and various authors have discussed creating a publicly-funded Energy Technology Corporation that would invest in new demonstration projects.

4. **Encourage collaboration on energy innovation with rapidly industrializing countries**. While there may be political opposition to collaborating with countries like China and India, significant action on climate change may be impossible without them. Literature on innovation in rapidly industrializing countries like China and India shows that simply transferring technologies from developed countries to industrializing countries does not accelerate innovation. Industrializing economies need to develop their own innovation capacity and can best benefit from incremental improvements made in their industrial processes. Increased international collaboration may accelerate innovation, and as a result the United States can benefit from increased innovation capacities that exist in other countries.

## 29.5 Conclusion

The linear account of engineering knowledge seems unrelated to much of what actually happens in complex innovation systems. Cartwright's 'dappled' epistemology gives suggestions about the complexities that an adequate account of engineering must address, complexities which are explored in the above principles of innovation. Even beyond Bunge, the current literature on the philosophy of engineering may not yet be useful in characterizing the broader processes in engineering and innovation (Wimsatt 2007). The above principles hint at a better account of engineering, which could better define the role of engineering as a force for positive societal outcomes. To do this, we need better descriptions of *learning by doing* in complex, interconnected innovation systems and of the importance of getting technology built up to the proper scale.

In summary: while a decades-long scientific debate about anthropogenic climate change was followed by a rejection of comprehensive climate legislation, the United States has not yet focused on engineering a technological fix to the climate problem. Past efforts at the Department of Energy aimed at developing lower-cost, low-carbon-intensity energy technologies have been hampered in part because of an adherence to the linear model of science. While part of this problem stems from a poor connection between government and industry, the linear model is in part perpetuated by a conceptual error that identifies innovation as primarily related to science and research, with the developmental work of the engineer being ignored. A richer conceptual understanding of engineering and innovation could contribute to better innovation policies.

**Acknowledgments** I am grateful to Arizona State University's Consortium for Science, Policy and Outcomes and the Clean Air Task Force for the chance to briefly work on the still ongoing energy innovation project. I owe an intellectual debt to Daniel Sarewitz, John Alic, Armond Cohen, Joseph Chaisson, Robert Horner, and Travis Doom. All faults with the argument presented here are my own.

## References

Alexander, X. (2009). *An energy future transformed.* Climate Policy Center of Clean Air Cool Planet.

American Energy Innovation Council. (2010). *A business plan for America's energy future.* Washington, DC: American Energy Innovation Council.

Anadon, L. D., et al. (2009). *Tackling U.S. energy challenges and opportunities: Preliminary policy recommendations for enhancing energy innovation in the United States.* Cambridge, MA: Energy Research, Development, Demonstration & Deployment Policy Project, Energy Technology Innovation Policy Group, Harvard University.

Boon, M. (2011). In defense of engineering sciences: On the epistemological relations between science and technology. *Techné, 15*(1), 49–71.

Bunge, M. (1967). *Scientific research I: The search for system* (Vol. 3, Part 1). New York: Springer.

Bunge, M. (1983). Toward a philosophy of technology. In C. Mitcham & R. Mackey (Eds.), *Philosophy and technology: Readings in the philosophical problems of technology* (pp. 62–76). New York: The Free Press.

Bunge, M. (1985). *Treatise on basic philosophy* (Vol. 7, Part 2). New York: Springer.

Bunge, M. (2007). Interview with Mario Bunge. In J. K. Berg Olsen & E. Selinger (Eds.), *Philosophy of technology* (pp. 17–30). Copenhagen: Automatic Press/VIP.

Bush, V. (1945). *Science: The endless frontier.* Report to the President. Available at: http://www.nsf.gov/od/lpa/nsf50/vbush1945.htm

Cartwright, N. (1983). *How the laws of physics lie.* Oxford: Oxford University Press.

Cartwright, N. (1999). *The dappled world: A study of the boundaries of science.* Cambridge: Cambridge University Press.

Clean Air Task Force. (2009). *Coal without carbon: An investment plan for federal action.* Boston: Clean Air Task Force.

Cohen, L., & Noll, R. (1991). *The technology pork barrel.* Washington, DC: Brookings Institution.

Consortium for Science, Policy and Outcomes and the Clean Air Task Force. (2009). *Innovation policy for climate change.* Available at: http://www.cspo.org/projects/eisbu/report.pdf

Consortium for Science, Policy and Outcomes and the Clean Air Task Force. (2010). *Four policy principles for energy innovation: A synthesis.* Available at: http://www.catf.us/resources/publications/view/125

Duderstadt, J., et al. (2009). *Energy discovery innovation institutes: A step toward America's energy sustainability.* Washington, DC: Brookings Institution.

Gallagher, K. S., & Anadon, L. D. (2010). *DOE budget authority for energy research, development, & demonstration database.* Cambridge, MA: Energy Technology Innovation Policy research group, Belfer Center for Science and International Affairs, Harvard Kennedy School.

Hayward, S., et al. (2010). *Post-partisan power.* Oakland: The Breakthrough Institute. http://thebreakthrough.org/blog/Post-Partisan%20Power.pdf

Henderson, R., & Newell, R. (2011). *Accelerating energy innovation: Insights from multiple sectors.* Chicago: The University of Chicago Press.

Kealey, T. (2008). *Sex, science and profits: How people evolved to make money.* London: Vintage Press, Heineman. http://www.amazon.co.uk/Sex-Science-Profits-Terence-Kealey/dp/0434008249

Kevles, D. (1995). *The physicists: The history of a scientific community in modern America* (2nd ed.). New York: Knopf, Harvard. http://www.amazon.com/The-Physicists-History-Scientific-Community/dp/0674666569
Layton, E. (1986). *The revolt of the engineers: Social responsibility and the American engineering profession*. Baltimore: Johns Hopkins University Press.
Lester, R. (2009). *America's energy innovation problem (and how to fix it)*. Cambridge, MA: Report from the Energy Innovation Project, Massachusetts Institute of Technology.
Morone, J., & Woodhouse, E. (1989). *The demise of nuclear energy? Lessons for democratic control of technology*. New Haven: Yale University Press.
Mowery, D., Nelson, R., & Martin, B. (2009). *Technology policy and global warming*. London: UK, NESTA.
Narayanamurti, V., Anadon, L. D., & Sagar, A. D. (2009). Transforming energy innovation. *Issues in Science & Technology, 26* (Fall), 57–64.
National Commission on Energy Policy (NCEP). (2009). *Clean energy technology pathways*.
Pielke, R. A., Jr. (2010). *The climate fix: What scientists and politicians won't tell you about global warming*. New York: Basic Books.
President's Council of Advisors on Science and Technology (PCAST). (2010). *Report to the president on accelerating the pace of change in energy technologies through an integrated federal energy policy*. Washington, DC: Executive Office of the President's Council of Advisors on Science and Technology.
Sarewitz, D. (2004). How science makes environmental controversies worse. *Environmental Science and Policy, 7*, 385–403.
Sarewitz, D., & Nelson, R. (2008). Three rules for technological fixes. *Nature, 456*, 871–872.
Weiss, C., & Bonvillian, W. (2009). *Structuring an energy technology revolution*. Cambridge, MA: The MIT Press.
Wimsatt, W. (2007). *Re-engineering philosophy for limited beings: Piecewise approximations to reality*. Cambridge, MA: Harvard University Press.

# Chapter 30
# Is Engineering Philosophically Weak?

**David E. Goldberg**

**Abstract** In 2008, Carl Mitcham presented a paper to a gathering of philosophers and engineers entitled "*The Philosophical Weakness of Engineering as a Profession*," urging engineers to find ways to be more like *philosophically strong* professions such as law and medicine, strong in Mitcham's estimation because those professions aspire to the good-in-themselves ideals of justice and medicine. This chapter reflects on Mitcham's original argument from the standpoint of engineering practice, offering both an analysis of the distinction between philosophical weakness and strength and then considering the aspirational and institutional settings of the different professions or occupations that Mitcham compares and contrasts. It argues that (1) engineering is philosophically weak, but in a different sense of that term than Mitcham originally argued and that (2) Mitcham's ethical or aspirational urgings are made difficult by both the multiobjective nature of general technological invention and implementation and the usual institutional embeddings of engineers. In particular, simple aspirational ideals are largely impossible for engineers and their simple aggregation founders on the shoals of Arrow's impossibility result, and even if simple ideals were possible, the very institutional complexity and teamwork necessary in engineering work means that the simplified client-practitioner relationship of medicine and law are oftentimes inappropriate for engineering work in much the same way that individual soldiers cannot be free agents to pursue peace as individual actors.

The chapter concludes with a warning against overly simple professional and aspirational yearnings along Mitcham's line of argument as well as those that have recently arisen in calls for engineering education reform and various grand

D.E. Goldberg (✉)
ThreeJoy Associates, Inc., Douglas, MI, USA
e-mail: deg@threejoy.com

D.P. Michelfelder et al. (eds.), *Philosophy and Engineering: Reflections on Practice, Principles and Process*, Philosophy of Engineering and Technology 15,
DOI 10.1007/978-94-007-7762-0_30, 

challenges. The results of this chapter suggest that overly simple yearnings are likely to be frustrated because of the irreducible aspirational and institutional complexity of engineering effort explored herein.

**Keywords** Philosophical weakness of engineering • Ethics and complexity • Institutional organization of engineering • Changing engineering institutions

## 30.1 Introduction

As philosophers well know, interesting things start to happen—indeed philosophy starts to happen—when those with different perspectives and views engage in the practice of dialectic. This chapter traces its origin to 2008 when noted philosopher of technology Carl Mitcham presented an invited keynote paper in London to a packed hall of engineers and philosophers at the Royal Academy of Engineering at the 2008 Workshop on Philosophy and Engineering (WPE-2008). That paper, entitled *The Philosophical Weakness of Engineering as a Profession* (Mitcham 2008), was intended and served as a provocation to the assembled engineers, urging them to find ways to be more like *philosophically strong* professions such as law and medicine, strong in Mitcham's estimation because those professions aspire to the good-in-themselves ideals of justice and medicine.

The purpose of this chapter is to reflect on Mitcham's original argument[1] from a number of points of view. The orientation of the reflection is largely from the standpoint of engineering practice, but the chapter does offer conceptual analyses of some of the language used in and the institutional implications of the original paper.

In particular, the chapter starts with a brief abstract of a portion of Mitcham's argument in which five occupations—medicine, law, engineering, the military, and business—are considered *philosophically weak* or *philosophically strong* depending upon whether they lead toward the achievement of some good-in-itself ideal. Taking these terms literally leads into a different categorization scheme than Mitcham's, one that also categorizes engineering as philosophically weak, but for reasons different than those of Mitcham. The chapter returns to Mitcham's distinction, recovering it by choosing to distinguish different occupations along the lines of their *aspirational* strength or weakness. After briefly wondering whether medicine and law are as strong in practice as they are made out to be in theory, the chapter moves beyond the level of the individual practitioner and considers the *institutional* setting of Mitcham's five occupations. The move to an institutional level allows us to understand that Mitcham's strong occupations enjoy a presumption of *global ethical alignment* between working in their client's interest and working in societal interest. The chapter considers whether such realignment can be achieved for engineering, either by concocting a simple aspirational ideal or a different kind

[1] A subsequently published paper (Mitcham 2009) took a somewhat different tack, and the present paper largely argues from the version presented at WPE-2008.

of institutional arrangement for engineering, answering in the negative in both cases. Two bounding models of institutional rearrangement are then considered, neither one proves particularly practical or desirable, and the status quo of pursuing engineering within a regulated market framework is found to be intermediate between the two bounding forms. This leads to considering whether the institutional hypothesis stands up in practice by considering historical cases from engineering and recent developments in law and medicine. Those brief reflections support the institutional connection, and the chapter concludes by examining recent aspirational yearnings from within the engineering community, suggesting that caution is warranted along the lines of the arguments contained herein.

## 30.2 The Mitcham Five, a Criterion, and a Classification

Mitcham highlights five occupations[2] (Mitcham 2008)—law, medicine, business, the military, and engineering—and distinguishes between those that are *philosophically strong* (PS) and *philosophically weak* (PW)[3] according to whether or not the occupation aspires to a good-in-itself ideal.

By this criterion, medicine and law are found to be PS because they aspire to the good-in-themselves ideals of health and justice, whereas business, the military, and engineering are found to be PW, at least in part because they have no such ideal for individual practitioners to guide their conduct. Up to this point, the argument is largely descriptive, but Mitcham then makes a move from *is* to *ought* by suggesting that, or at least asking whether, engineering would be improved as an occupation or a profession if it, too, aspired to some higher ideal.

## 30.3 Preliminary Concerns

The juxtaposition of the five occupations is a useful conceptual framing, and the notion of philosophical strength and weakness is a stimulating distinction, one that we will pursue in a moment in two different ways. But first, we shouldn't let certain assumptions of the argument to go by unnoticed. In an age where Greek philosophical notions live side by side with more pragmatic and even postmodern approaches, the Platonic notion of an ideal, and the notion of something being good-in-itself can be challenged, especially by those with a pragmatic perspective (Pitt 2000). Moreover, the uncritical acceptance of Greek values from a slaveholding society 2,500 years ago can be particularly troubling to engineers for whom the idea that

[2] Mitcham uses the term *profession*, but we will use the less restrictive term *occupation* to avoid unnecessary discussion whether all five are professions and related matters.

[3] This terminology was changed in the published paper, but we will stay with the terminology of the original paper to follow where exploring those terms leads.

the pursuit of pure knowledge is somehow to be automatically elevated above pursuit of the applied is not something that can pass without question. Nonetheless, this paper accepts these assumptions for the sake of argument and proceeds within the framework Mitcham has created. An interesting question—one that we shall return to—is whether philosophy itself is PS or PW according to Mitcham's categorization, but we shall set aside any preliminary concerns—and this auxiliary question—and proceed.

## 30.4 Critical Examination of the PS/PW Distinction

Mitcham's notion of PS and PW turns on a concern for *ethics*, in particular virtue ethics, but ethics is but one of the five major divisions within philosophy: metaphysics (ontology), epistemology, ethics, politics, and aesthetics. When I first heard the title of the talk, I thought it was a much needed philosophical wake up call for engineering to become more broadly philosophically aware, and I was surprised that the talk focused more narrowly on ethical or aspirational concerns.

We will return to Mitcham's distinction in a moment, but we pause to critically examine the notions of philosophical strength and weakness more literally. In particular, we will say that an occupation is philosophical strong in this second sense (PS′) if it is broadly concerned with its ontology, epistemology, and ethics and philosophically weak in this second sense if it is not.

These are broad brush strokes, to be sure. First, we are casting aside issues connected with two branches of philosophy (aesthetics and politics) without comment or particular justification, except to say that some of what follows touches on issues of politics fairly directly. Second, we are going to use the broad category of metaphysics as the label, although our main concerns are ontological. This labeling will connect directly with the main categories of philosophy, even if the underlying categorization is fairly crude and incomplete.

To make this operational in the practical setting of real-world occupations, define the following three terms. An occupation is

1. **Metaphysically reflective** if it considers its history.
2. **Epistemologically reflective** if it consciously transmits its knowledge in forms appropriate to the practice of the occupation.
3. **Ethically reflective** if it has a code of ethics.

As a matter of simplicity we will say that an occupation is philosophically reflective (or philosophically strong in the second sense, PS′) if it is reflective on two of the three categories.

Many objections can be raised to this simple scheme. For example, history is a crude stand in for ontology, and using the label "metaphysically reflective" compounds the problem by taking concern for ontology as a concern for all of metaphysics; similar objections can be raised to the definitions of "epistemologically reflective" and "ethically reflective." Moreover, why should majority rule in the

**Table 30.1** Comparative analysis of five occupations

| | Metaphysics | Epistemology | Ethics |
|---|---|---|---|
| Medicine | Weak | Strong | Strong |
| Law | Strong | Strong | Strong |
| Military | Strong | Strong | Strong |
| Business | Weak | Strong | Weak |
| Engineering | Weak | Weak | Strong |

definition of "philosophically reflective?" What justifies the assumption that the three types of reflection are equally important? Clearly rough-and-ready practical measures have been used here in the mode of engineering reasoning to generate a simple model as a spur to deeper reflection. Of course, such a crude instrument cannot be the end of the story, but let's pursue the simple scheme somewhat further and see where it leads. Table 30.1 categorizes the Mitcham 5 according to these definitions, and the results are interesting.

Medicine, business, and engineering are metaphysically weak because they are not reflective on their nature through study of their history. Engineering is alone in its epistemological weakness, because it is blind to the ways in which its models are distinguished from those of math and science, to the importance of case histories, and to the role of language, more generally, in the capture of what it knows (Vincenti 1990). Business is alone in its ethical weakness because there are no generally accepted ethical standards for the conduct of business generally, although certain specialties (accounting, for example) do have codes of ethics.

Using these operational definitions designed for real-world professions, it is interesting to note that philosophy itself would be PS′, but with only a 2/3 score because it does not have a code of ethics. Of course, this result ignores that we are intentionally using the existence of a code of ethics as a stand-in for a deeper kind of reflection, reflection that is the heart and soul of philosophy, but the lack of a code of ethics for philosophy arises somewhat later in a more germane setting, and is, thus, mentioned in passing.

The more important point is that Mitcham's distinction between PS and PW downgrades engineering on ethical grounds, but it seems a bit unfair to take engineering to task in the one area in which it is philosophically reflective. Mitcham's ethical distinction, of course, values the aspiration to a higher ideal, something different than merely having a code of ethics; however, the larger point here is that engineering is philosophically weak (in the second sense), not because it doesn't pay attention to ethics, but because it doesn't pay sufficient attention to its nature or its knowledge.

As engineers approach philosophy for the first time, they will find a rather extensive discussion of ethics but only a meager discussion of epistemology and ontology. This, of course, is changing, but the plain usage of the terms philosophical strength and weakness, seems better reserved for this larger distinction than the one made solely on the basis of aspirational ethics. In the remainder of this paper, the terms philosophical strength and weakness will be reserved for the second kind (PS′/PW′) as defined in this section.

## 30.5 Recovering Mitcham's Distinction and a Concern

Having said this, Mitcham's distinction is a useful one and here we recover it by dividing the occupations almost tautologically—some might complain prescriptively—along the lines of their being *end-in-themselves* (or *aspirational*) or *instrumental* occupations. In this way, we immediately get back to Mitcham's division in which medicine and law aspire—in some sense—to a good-in-itself ideal and the others do not.

But before proceeding, it is useful to wonder how well aspirational occupations really work in practice. In some large sense, it is true that the medical and legal systems do aspire to health and justice, but looking under the hood at individual behavior suggests that the mechanism by which these aspirations are actually approached is messier than they might otherwise appear at first glance.

First, neither medicine nor law reach anything very close to an ideal form of health or justice. The medical system is largely concerned with disease and is largely unconcerned with wellness; doctors fix disease and administer medicine and ignore improvements in nutrition, exercise, or social engagement. Law is largely concerned with following its own rules regardless where they lead; lawyers work to acquit individuals who they know have committed crimes and they not infrequently pursue the deepest pockets in civil cases regardless of actual liability.

Second, the behavior of the individual doctors and lawyers is not the result of some selfless or abstract pursuit of the good. Doctors fix disease and administer medicine because they are compensated for so doing, and the ignore nutrition, exercise, and the wellness benefits of social interaction, because those things don't generate sufficient marginal income. Similarly, lawyers pursue work doggedly in the interest of their clients and are generally remunerated handsomely for so doing, even when the result is only a very rough approximation to anything remotely approaching justice.

Nonetheless, in the aggregate, it *can* be claimed that both medicine and law do approach their companion ideals, but what the preceding argument has made clear is that the good here cannot be attributed to high ideals of the individual physician or lawyer alone. No, the approximation to health and justice achieved—when it is—must be seen as coming from higher order institutional arrangements that promote an approximation to the desired ideal, something considered in the next section.

## 30.6 Institutions and Their Discontents

Institutions are more the domain of sociology and economics than philosophy, but we appeal to those other disciplines because the individualistic nature of philosophical thinking without augmentation might otherwise mislead us in the present case. First, we acknowledge that unlike the pensive philosopher thinking grand thoughts on his or her own, members of occupations are playing more-or-less a team sport within sophisticated institutional frameworks. Institutions arise to make certain things easier—to reduce certain transaction costs (Coase 1937) in economic terms—and it is an irony of free markets that using the free market is not itself generally without cost. It is these costs that largely shape the form and type of institutions that evolve over time.

A comparison of the institutional settings of the Mitcham 5 is instructive:

**Lawyer:** Officer of court, monopoly on practice by state. Work in private firms and in government.

**Doctor:** Member of regulated profession, in regulated institutions, monopoly on practice of medicine by state. Works in private practice, HMO, or government.

**Military person:** State has monopoly on force, military members are employees/ conscripts of state, and follow direct orders of civilian leaders through chain of command.

**Businessperson:** Free agent to contract with others, obeying laws of the state. Works in private enterprise.

**Engineer:** Free agent to contract with others, obeying laws of the state. Some licensed for some types of work. Work in free enterprise or public sector.

To this list, we might add the following:

**Philosopher:** Free agent to contract with others obeying laws of the state. Academic practitioners require PhD for tenure-track position. Work in private or public sector.

These listings are helpful, but the critical distinction comes with a bit more digging.

Doctors and lawyers work on behalf of their individual clients, doing their utmost to help the client, and this pursuit of the client's interest is assumed—in the larger institutional setting of medical and legal practice—to align with the good of society. Here we label the first of these conditions—when the practitioner's work aligns with the client's interest—*local ethical alignment*, and we label the second condition the *presumption of global ethical alignment.* Thus, we see that the fortuitous circumstance that makes medicine and law good-in-themselves occupations is when local ethical alignment leads to the presumption of global ethical alignment.

I call this circumstance "fortuitous" because it is this and largely only this that makes law and medicine "aspirational" professions. Doctors and lawyers do not need to be saintly figures pursuing society's bidding contrary to their own interests. They merely need to pursue the interest of their client, largely in alignment with their own interest, and the approximation to the aspirational ideals that is believed to exist comes about. Seen in this way, rather than calling these occupations aspirational, it might make more sense to call them merely *ethically simple.* Doctors and lawyers don't have to try very hard to be good in Mitcham's aspirational sense.

Despite the foregoing deconstruction of the need for individual goodness on the part of any given doctor or lawyer, it still makes sense to ask whether engineering can be made more like law or medicine in this regard. There appear to be two paths to so doing. First, can we imagine a simple aspirational ideal for engineers in general? Second, can we imagine a rearrangement of the institutional structure of engineering that permits local ethical alignment to lead to the presumption of global ethical alignment, in other words, to ethical simplicity? These two questions are pursued in what follows.

## 30.7 Quest for an Ideal: Engineering Version

In reflecting on technological artifacts and systems, it is difficult to imagine a simple ideal for engineering that can function like justice and health in law and medicine. Part of the problem is that technology satisfies many positive values from survival to εὐδαιμονία and a host of values in between. Moreover, different individuals will weight different values differently and Arrow's impossibility theorem (Arrow 1950) suggests immediately that there is no way to satisfy them all.

But even if there were a single value that we could all agree upon for all technological artifacts—something like sustainability or the like (and the paper is not conceding that sustainability is not without a host of problems in this regard)—it is reasonably straightforward to show that we still can't make engineering aspirational or ethically simple like medicine or law.

Without a concrete ideal to help make the point, however, this is hard to show in the context of engineering, so we temporarily shift our attention to a military context in which such an ideal suggests itself more readily.

## 30.8 Quest for an Ideal: Military Version

Since the multiple objectives that swirl around engineering make the search for a single unproblematic ideal so difficult, let's shift from engineering to one of the other instrumental occupations, in particular the military, and ask the following question: is there a single ideal that a benevolent military might aspire to? Survival doesn't seem strong enough, but perhaps *peace* is the ticket. Yes, peace can serve as the aspirational ideal for our ideal military and then we can ask how individual military personnel might go about acting to achieve that ideal.

Of course, as soon as we set out on this course of reflection, we run into some tough sledding. In particular, we have four problems:

1. Predictability problem
2. Effectiveness detection difficulty
3. Individual decomposition problem
4. Social effectiveness problem

Consider each briefly in turn.

The first difficulty comes from the unpredictable nature of warfare. How do we know that a particular military action will have the desired outcome and therefore the desired effect on the systemic state of peace?

The second difficulty comes from our inability to know whether any given outcome is actually contributing to peace in the long run.

The third problem is related to the first two and suggests that the individual is largely unable to know whether his or her contribution to the effort is adding to or detracting from the chances for peace.

Finally, the fourth problem is that if individual soldiers were to follow their own judgments about peace it would destroy the social cohesion needed to create effective military action. In other words, soldiers pursuing peace individually would not in any real sense be an army.

Put in this way, the picture of individual soldiers pursuing peace on their own terms is almost too absurd to contemplate, and thinking about engineers universally pursuing some larger societal goal on their own is strikingly analogous. Goldman has used the term *socially captive* (Goldman 1991) to suggest that engineers are often working at the direction of someone else, but the term "captive" might be interpreted by some to suggest that if only engineers could break out and do the right thing, that society would be better off, but the four problems above suggest strongly that this is not the case, that engineers are often necessarily part of large social aggregations and that the goals for those larger aggregations are determined at the top of a hierarchy and largely followed by those within it. This is not to suggest that engineers have no duty to oppose unlawful or unethical orders—no one is suggesting that soldiers should blindly follow unlawful orders under all circumstances, either. It is, however, fairly clear from the foregoing that the institutional complexity of these situations makes it difficult for those in instrumental occupations to pursue the good autonomously and that the ethical complexity of such situations deserves more respect than it often gets.

## 30.9 Ideal Quest Through Institutional Redesign

So even if we were able to come up with a simple ideal for engineering, the institutional/organizational setting of engineering would, like that of an army, thwart the ability of an individual engineer to depart from the role dictated by the needs of the institution/organization.

Given this chain of reasoning, it would seem that there remains one final mode for achieving what Mitcham suggests by considering ways in which the institutional setting might be redesigned to permit the kind of ethical simplicity that doctors and lawyers have, the ethical simplicity in which local ethical alignment leads to the presumption of global ethical alignment.

We consider two models for achieving these ends, what we here term the *absolute control model* and the *absolute fail-safe model.*

## 30.10 Absolute Control Model

In the absolute control model, we assume that engineers have some simple ideal and we construct an institutional framework that allows them to achieve it. Of course, engineering is an iterative process, so the engineer's orders must be obeyed in space and time to assure good outcomes eventually according to the engineer's vision of the good. This limits choice by others, of course, and in its extreme form engineers are given control over the state, so that their vision of the good can be achieved.

Of course, we have a name for this sort of thing, and it is called *technocracy*, and its extreme forms are approximated in authoritarian regimes.

From a philosophical perspective, these ideas have a long history. Plato envisioned something akin to this in *The Republic*, except instead of putting engineers in charge he made philosophers kings. Of course, if we're going to seriously entertain the notion of philosopher-kings, our earlier questions whether philosophers need a code of ethics becomes germane. It is one thing to have ideas as a matter of mere insight, but if people put ideas to practice in the world—if ideas become instrumental as they often do—the suggestions that engineers be held responsible for the technology applies to philosophers. This analogy would require philosophers to be held responsible for the consequences of their ideas in exactly the same way that Mitcham would have engineers be held more fully responsible for the consequences of their technology.

## 30.11 Fail-Safe Action Model

So perhaps we're all a bit nervous about giving engineers the absolute control they would need in the previous model to help ensure that local alignment results in the presumption of global alignment. Let's go to the other end of the spectrum and assume that engineers are free agents in something like a market economy and see if we can do some institutional redesign to permit their doing no harm.

Of course, the marketplace itself is insufficient institutionally to ensure ethical simplicity, so let's add an institutional constraint to prevent engineers from ever inventing, making, or sustaining any artefact that causes harm, ever! Think of this as a *precautionary principle* on steroids. If such a thing could be enforced, it would certainly solve the problem of creating anything bad (anything detrimental to the ideal), but it would almost certainly prevent anything very good from ever happening.

In game theoretic terms, this would require something like a minimax strategy in which the engineer would minimize harm subject to the actions of an adversary who would maximize misuse and mischief. Such a conservative strategy would essentially be the death knell for engineering as we know it. As Petroski (1982) has pointed out, engineering depends on error for its advancement. To prevent actions that would allow error to take place is tantamount to preventing engineering creativity and innovation from occurring at all.

## 30.12 Making in the Middle

Of course, the solutions of absolute control and fail-safe action are bounding models that live at the extremes, but this is exactly the point. Engineers do not (usually) run the whole show. Nor do we require them to never make a mistake. Instead we permit engineers to work in self-organizing institutions in the larger institution of a regulated market. In this way, engineering as practiced today in the real world is an intermediate tradeoff between accepting no harm (with no overall control) and accepting absolute control (with one engineer's idea of the good). This tradeoff is analogous to the one discussed in *The Calculus of Consent* (Buchanan and Tullock 1962) in which the compromise of democratic governance is seen as intermediate between dictatorship and individual veto.

And while democracy does not respect or veto individual choice, it does permit a messy advance of ideas that benefit the median voter most of the time. Analogously, letting engineers engage in regulated markets has similar benefits to the intermediate choice of democratic rule with the added benefit that markets do not require a single decision to be reached as in the political case. The plurality of markets permits many products and services to blossom within regulatory constraints in ways that offer a variety of choices to a variety of consumers who do not share thoughts about a single good.

## 30.13 Through an Institutional Lens: HMOs, Large Legal, and the Revolt Revisited

The ethical-institutional hypothesis of this paper, that occupations are increasingly ethically complex as their institutional complexity increases, may be examined with historical trends from medicine, law, and engineering.

Medical practice has become increasingly institutionally complex in the United States as of late with the rise of health maintenance organizations or HMOs, largely supplanting the individual small private practices of years ago. Our ethical-institutional modeling would suggest that the rise of institutional complexity would tend to make medicine increasingly ethically complex and move it away from Mitcham's aspirational ideal, and this is what we generally find. The interposition of account managers and other staff between doctor and patient, makes it increasingly difficult for physicians to act solely in their patients' interests.

Similarly the rise of large legal practices paid by a phalanx of third-parties (largely insurance companies) also should increase the ethical complexity of legal practice, thereby moving law away from the Mitchamian ideal, and this, too, is what we find. Here, the complexity comes—as in the case with medicine and HMOs—from the addition of third-party decision makers we find the purity of client advocacy somewhat corrupted by a variety of other considerations.

In the case of engineering, we can turn as many other have to Layton's (1971) account of the rise of the first professional societies in the United States during the latter part of the nineteenth and early part of the twentieth century. Our ethical-institutional tether should be expected to predict that engineering disciplines whose institutional arrangements permit something closer to the ideal of individual client-practitioner service should be more like Mitcham's aspirational ideal and those in which the nature of the product-service provided requires increasing institutional complexity should be increasingly distant from it. Examining the civil engineers, mining engineers, and the electrical engineers using Layton's historical narrative is enlightening in this regard.

From the beginning, civil engineering was closest to the ideal of individual legal and medical practice and the professional society the American Society of Civil Engineers (ASCE) reflected this in requiring high standards of professional knowledge, relatively long years of experience, and strong commitment to public safety and ethics.

By contrast, the mining engineers professional society, the American Institute of Mining Engineers (AIME), placed greater emphasis on the importance to serve the mining company and it was much less strict in its enforcement of educational or public service requirements.

Electrical engineering is interesting as something of a mixed case, a case that went through a transition from the founding of the discipline to its "captivity" within a small number of relatively large corporations. In the early days, individual engineers were rock stars—not unlike entrepreneurial software engineers of our own time—and the professional society emphasized professional knowledge and professional autonomy, but as large corporate interests became prevalent throughout electrical and radio industries, electrical engineers became increasingly captive to the interests of the company and less like the aspirational ideal of occupational practice.

Taken together, these cases suggest a coherence to the ethical-institutional connection discussed herein, one that should, at the very least, not be ignored when recommending a transformation of the aspirations of any occupation.

## 30.14 Educating Engineers and the Grand Challenges

Ethical urgings for engineers to aspire to the greater good are not isolated to Professor Mitcham's paper. The National Academy of Engineering has put out its list of Grand Challenges (NAE 2010) and a recent report sponsored by the Carnegie Foundation (Sheppard et al. 2008) recommends a kind of *neoprofessionalism* as the way toward a transformation of engineering education in better alignment with our times and societal needs. Detailed examination of these reports is beyond the scope of this chapter, but the Grand Challenges put forward a list of societal needs assembled by a team of elites and challenges engineers and engineering education to go forward and meet them, and the Sheppard report puts professionalism forward as an organizing principle to overcome the limitations of the cold war engineering curricular consensus. Both are aligned with the zeitgeist, but neither suggestion is particularly institutionally or ethically sophisticated, and both are likely to face difficulty unless accompanied by institutional rearrangement.

In particular, the analysis of this chapter would suggest that while both efforts may have significant public relations value, neither one is likely to have lasting occupational or professional effect unless accompanied by significant institutional reconfiguration. Such reorganization is not impossible, and the absolute control and fail-safe models of this chapter are suggestive of the directions in which the institutional arrangements can currently be changed to affect the ideals desired. To create something closer to a command-and-control economy or a more pervasive regulatory regime may or may not be desirable, but it should be clear that such decisions are political ones. Regardless of where one stands, the advocacy of one path or another should, at the very least, be recognized as being substantially value laden, not something objective or value neutral. Whether new forms of institutional arrangement, perhaps augmented by information technology, are possible is beyond the scope of this paper, but breaking out of the current limited array of institutional options would be desirable to effect practical change along the lines of Mitcham's urgings.

## 30.15 Conclusions

This chapter has briefly considered Carl Mitcham's provocative paper, *The Philosophical Weakness of Engineering as a Profession*, a paper presented to a gathering of engineers and philosophers at the 2008 Workshop on Philosophy and Engineering. Mitcham's distinction of philosophical strength and weakness depending upon whether a profession or occupation has well articulated aspirational ideals or not is questioned and replaced with a different scheme that uses epistemological, ontological, and ethical concerns in place of occupational aspirations as the classification criterion.

This chapter then shifts to consider whether the ethical urgings of Mitcham's paper are practical. By starting with the cases of medicine and law, the chapter suggests that those occupations achieve an approximation to their ideals, not by the goodness or aspirations of individual practitioners, but rather through the institutional arrangements of medicine and law in which local ethical alignment of practitioner and client is assumed to lead to global ethical alignment through the actions of institutional constraints. In other words, the "strength" of these professions comes more from their institutional arrangement and their resulting ethical simplicity than from the direct practice of aspirational ideals by individual practitioners. In a certain sense, this chapter accepts the initial categorization of Mitcham's basic argument, but explains it quite differently, thereafter justifying that different explanation with a quick examination of current events in medicine and law.

The chapter then considers whether a suitable ideal might be found for engineering analogous to health and justice for medicine and law. The difficulty in so doing leads to a shift in concern to the military and the aspirational ideal of peace is found to be problematic and a challenge to the very nature of what it means to be a military organization. Four difficulties are considered, and the chapter concludes that even if there were a simple ideal for engineering, the nature of engineering organization would, by analogy, make the situation for individual engineers necessarily and irreducibly ethically complex.

This leads to considering two bounding models of institutional rearrangement: absolute control and fail-safe action. Both are problematic, and the current norm of free market performance within a regulated economy is found to be intermediate between the two pure forms. Although the analysis does not rule out institutional invention that will overcome the limitations of the current state of institutional art, it does not appear that exploring the extremes or other points along the implied continuum is likely to provide substantially different qualitative outcomes.

Finally, the chapter briefly examines recent calls for higher aspiration in engineering practice in the NAE Grand Challenges and in the reform of engineering education. Although the examination is brief, both cases ask engineers to aspire to greater good in their work and in their educations. While these calls are noble, the results of this chapter suggest that simple ethical urgings or other calls for engineers to simply do better unaccompanied by calls for institutional redesign are unlikely to change very much. Institutions make certain things easy for human beings, and they make other things more difficult.

When medicine and law achieve approximations to health and justice and when engineering, business, and the military struggle against their tendency to be instrumental to the purposes of the powerful, both types of case are shaped by the institutional form of the occupations themselves. The wider spread recognition of this institutional shaping is important to all occupations, and such recognition may encourage creativity and innovation to find new—as yet undiscovered—institutional forms. Technology itself, in particular information technology, may be especially helpful in this latter creative endeavor, and innovations that make it easier for currently ethically complex occupations to become ethically simpler are desirable; however, until these innovations arise, the ethical complexity of the instrumental occupations is likely to remain a challenge to their practitioners and to society alike.

**Acknowledgments** The paper has benefited from discussions with Carl Mitcham, Joe Pitt, and Diane Michelfelder, and was originally presented at SPT 2009. A powerpoint of the original presentation is available at www.slideshare.net/deg511. The author is grateful for support from the Illinois Foundry for Innovation in Engineering Education (www.ifoundry.illinois.edu) and the National University of Singapore.

## References

Arrow, K. J. (1950). A difficulty in the concept of social welfare. *Journal of Political Economy, 58*(4), 328–346.

Buchanan, J. M., & Tullock, G. (1962). *The calculus of consent*. Ann Arbor: University of Michigan Press.

Coase, R. (1937). The nature of the firm. *Economica, 4*(16), 386–405.

Goldman, S. L. (1991). The social captivity of engineering. In P. T. Durbin (Ed.), *Critical perspectives on non academic science and engineering* (pp. 121–145). Bethlehem: Lehigh University Press.

Layton, E. T. (1971). *The revolt of the engineers: Social responsibility and the American engineering profession*. Cleveland: Press of Case Western Reserve University.

Mitcham, C. (2008). The philosophical weakness of engineering as a profession. In D. E. Goldberg & N. McCarthy (Eds.), *Abstracts of the 2008 workshop on philosophy and engineering (WPE-2008)* (p. 6). www.philengtech.org

Mitcham, C. (2009). A philosophical inadequacy of engineering. *The Monist, 92*(3), 339–356.
National Academy of Engineering. (2010). *Grand challenges for engineering*. Website: www.engineeringchallenges.org
Petroski, H. (1982). *To engineer is human: The role of failure in successful design*. New York: St. Martins Press.
Pitt, J. (2000). *Thinking about technology*. New York: Seven Bridges Press.
Sheppard, S. D., Macatangay, K., Colby, A., & Sullivan, W. M. (2008). *Educating engineers: Designing for the future of the field*. San Francisco: Jossey-Bass.
Vincenti, W. (1990). *What engineers know and how they know it*. Baltimore: Johns Hopkins University Press.

# Contributors

**M.H. Abolkheir** is an industrial inventor who is a firm believer in the unique opportunities which are present in applying the abstract tools of philosophy to the examination of industrialised inventions, as the outcome can be of considerable practical use, and the theoretical insights which emerge from such a philosophy of technology can have wider benefits in the philosophy of science. He is currently completing a Ph.D. in the philosophy of technology at the University of Bristol. His contribution to this volume is part of a long-term research programme to increase the synergy between technological practices and philosophical investigation.

**Sarah Bell** is Senior Lecturer at University College London in the Department of Civil, Environmental and Geomatic Engineering. Her research interests lie in the relationships between engineering, technology and society as they impact on sustainability, particularly in relation to water systems. This includes work on water efficiency, the public acceptability of water reuse and water sensitive urban design. She works in collaboration with partners including Thames Water, Waterwise, AECOM and Arup. She is a Chartered Engineer and holds a Ph.D. in Sustainability and Technology Policy from Murdoch University, Western Australia. She is a co-director of the UCL Urban Laboratory, former co-Director of the UCL Environment Institute, and led the EPSRC funded Bridging the Gaps: Sustainable Urban Spaces project at UCL which provided support for new research collaborations across 26 departments, involving 63 researchers.

**W. Richard Bowen** is a Professor and Fellow of the UK Royal Academy of Engineering. His long-term commitment to ethical issues includes membership of the Academy's Engineering Ethics Working Group, the UK Academies' Human Rights Committee and the Advisory Board of the Centre for Emerging Technologies and Bioethics at St Mary's University College London. Recent publications include *Engineering Ethics: Outline of an Aspirational Approach* (London: Springer, 2009) and *Peace Engineering*, 2nd edition, edited with P. Aarne Vesilind (Woodsville: Lakeshore Press, 2013).

**William M. Bulleit** is a Professor of Structural Engineering in the Department of Civil and Environmental Engineering at Michigan Tech. He received B.S.C.E. and

D.P. Michelfelder et al. (eds.), *Philosophy and Engineering: Reflections on Practice, Principles and Process*, Philosophy of Engineering and Technology 15,
DOI 10.1007/978-94-007-7762-0, 

M.S.C.E. degrees from Purdue in 1974 and 1975, going on to design submersibles before obtaining a Ph.D. in Engineering Science from Washington State University in 1980. He worked on the design of bridges and a soft-earth tunnel before going to Michigan Tech in 1981. His research is primarily in structural reliability, which led to his interest in the philosophy of engineering. He has published about 100 technical papers and two books.

**Andrew Chilvers** is a Doctor of Engineering from the Department of Civil, Environmental & Geomatic Engineering, University College London. Here he has developed research in partnership with the global engineering and design consultancy, Arup, on how values are appropriated in and through engineering practice. This employs a range of ethnographic techniques to develop empirical insight on the mechanisms that normatively shape engineering practice and its outcomes. Insights cover the role of professional intermediaries such as design codes and guidance, the role of organizational history and discourse, and the values and perspectives introduced by individual engineers. Andrew's work seeks to improve the way engineers engage with the normative shaping of their practice.

**Mandar M. Dewoolkar** (PE) is Associate Professor in the School of Engineering at the University of Vermont. He holds bachelor's and master's degrees from Mumbai, India, and from the University of Colorado at Boulder, all in Civil Engineering. He has worked on incorporating educational modules on laboratory and computational research into undergraduate engineering courses as well as service learning in the curricula. His other interests are in applying experimental and analytical methods to address geotechnical and geoenvironmental engineering problems.

**Priyan Dias** is a Senior Professor in Civil Engineering at the University of Moratuwa, a fellow of the National Academy of Sciences and member of the National Research Council in Sri Lanka. His doctorate from Imperial College London (1986) is in the area of concrete structures and technology, but his current research interests also include engineering philosophy and systems approaches for engineering. He is an Associate Editor of the international journal *Civil Engineering & Environmental Systems.* He has held Commonwealth and Fulbright Research Fellowships at the universities of Bristol (1992/93) and Carnegie Mellon (2000/01) respectively.

**PAN Enrong** is Associate Professor of Marxist Studies at Zhejiang University. Previously he studied as a visiting graduate student at Delft University of Technology. His current research mainly focuses on philosophy of engineering design, technological methodology, and engineering ethics.

**Rick Evans** is a Senior Lecturer and the Robert Noyes Director of the Engineering Communications Program in the College of Engineering at Cornell University. He is a sociolinguist with research interests in literacy studies, language and disciplinary socialization, communication/writing in the disciplines, and the scholarship of teaching and learning. Among his recently published works are articles on performance theory and its role in engineering education, and the role of "micro-genres" in disciplinary socialization.

**Scott Forschler** received his Ph.D. in Philosophy from the University of Minnesota in 2004. He has published articles on ethical rationalism, the principle of moral universalizability (especially as found in Hare and Kant), consequentialism, and poetic justice. His current projects involve the relationship between supervenience, the concept of a reason, higher-order valuation, and the foundations of ethics.

**David E. Goldberg** is President of ThreeJoy Associates, Inc. a training, consulting, and coaching firm and Founder of Big Beacon, a global movement for the transformation of engineering education. Before this, Dave was the Jerry S. Dobrovolny Distinguished Professor in Entrepreneurial Engineering at the University of Illinois at Urbana-Champaign, where he was known for his path-breaking research in genetic algorithms, for his role in starting ShareThis, for his role in starting the Workshop on Philosophy and Engineering and the Forum on Philosophy, Engineering, and Technology, as well as his work in starting the Illinois Foundry for Innovation in Engineering Education (iFoundry).

**William Grimson** received his B.A. and B.A.I. from Trinity College Dublin and his M.A.Sc. from the University of Toronto. He is a Chartered Engineer and a Fellow and current Vice-President of Engineers Ireland. Now retired, he worked as a Research & Development Engineer for Ferranti Ltd before joining the academic staff of the Dublin Institute of Technology. His academic output was and remains eclectic, ranging from publications in areas as diverse as plasma physics, clinical information systems, philosophy of engineering, and development issues.

**WANG Guoyu** is Professor of Philosophy and Director of the Institute of Philosophy and Center for German Philosophy of Technology at Dalian University of Technology, China. She earned degrees from both the FU Berlin and TU Dalian and has served as a research scholar in the Universities of Magdeburg and Stuttgart. Her research focuses on the ethics of science, technology, and engineering. She has initiated ten research projects on technical ethics and coordinates two national projects in these areas.

**Charles E. (Ed) Harris, Jr.** has an undergraduate major in biology and minor in chemistry and a Ph.D. in philosophy. He has published numerous articles in applied ethics, most of them in engineering ethics. He is also author of *Applying Moral Theories* and co-author of *Engineering Ethics: Concepts and Cases.*

**Nancy J. Hayden** (PE) is Associate Professor Emeritus in the School of Engineering at the University of Vermont. She has been working on innovative educational opportunities for students in civil and environmental engineering for 22 years. Her interests include sustainable water and waste issues. She has a Ph.D. and an M.S. in Environmental Engineering from Michigan State University, a B.S. in Forest Biology from the Environmental Science and Forestry School at Syracuse, B.A. degrees in English and Studio Art from the University of Vermont, and an M.F.A. in creative writing from the Stonecoast Program at the University of Southern Maine.

**Wybo Houkes** is an Associate Professor in Philosophy of Science and Technology at Eindhoven University of Technology. He is the co-author (with Pieter Vermaas) of *Technical Functions* (Springer, 2010); and one of the associate editors of the *Handbook of Philosophy of Technology and Engineering Sciences* (Elsevier, 2009). His research interests include artefacts and their functions, the role of intentions in design and use, technological knowledge, and cultural-evolutionary models of technology.

**SU Junbin** is Associate Professor of Journalism at the School of Journalism and Communication, Xiamen University. He received his Ph.D. in STS from Tsinghua University in Beijing. His research interests are mainly focused on the social implications of media technologies. He participated in the Seventh Framework Programme project "China-EU Information Technology Standard Research" from 2008 to 2010. Funded by the China-US Education Trust, in 2011 SU Junbin was a visiting professor in Washington, D.C. in the School of Communication at American University.

**Billy Vaughn Koen**, Professor Emeritus of Mechanical Engineering, The University of Texas at Austin, is the author of *Definition of the Engineering Method* (1985) and *Discussion of the Method: Conducting the Engineer's Approach to Problem Solving* (2003). He is a Fellow of the American Society for Engineering Education and the American Nuclear Society.

**Russell Korte** is an Assistant Professor in Human Resource Development and a Fellow with the Illinois Foundry for Innovation in Engineering Education at the University of Illinois at Urbana-Champaign. He has been a research assistant for the Center for the Advancement of Engineering Education. His research investigates how professionals navigate their education and how they transition into the workplace—specifically studying how they learn the social norms of work and navigate the social and political systems in organizations. Research interests include theory, philosophy, workplace learning and performance, socialization, adult and professional education, social psychology, and organization studies.

**Natasha McCarthy** is Head of Policy at The Royal Academy of Engineering. She leads a team engaging with policy issues across the whole spectrum of engineering research and practice, from energy systems to medical implants. Natasha has particular interest in engineering ethics, professional responsibilities and technologies at the interface with society – from surveillance technologies to critical infrastructure. Natasha's background is in philosophy and history of science and technology and she previously lectured at the University of St Andrews. She authored the book *Engineering: A Beginner's Guide*, which explores the impact of engineering on society and culture and the nature of engineering knowledge and practice.

**Diane P. Michelfelder** is Professor of Philosophy at Macalester College in Saint Paul, MN, USA. She received her degrees in philosophy from Bryn Mawr College (B.A.) and The University of Texas at Austin (Ph.D.). One of the first philosophers in the USA to teach a course on the Internet, her primary areas of research addresses ethical aspects of the Internet and Internet-"infused" technologies, engineering ethics, and the relationship between engineering and education in the liberal arts. The co-chair of fPET-2010, Diane has also been president of the Society for

Philosophy and Technology and has served as an editor of its journal *Techné: Research in the Philosophy of Technology.*

**HU Mingyan** received her B.A. in Philosophy from Nanjing University, and a Ph.D. in Philosophy of Science and Technology from Tsinghua University. From 2009 to 2010, she was a visiting scholar in the Philosophy Department of Delft University of Technology. At present, she is an Assistant Professor in the Philosophy Department of the Central Party School of the Communist Party of China. Her research interests are in the ethics of newly emerging sciences and technologies, and in Science and Technology Studies.

**Carl Mitcham** is Professor of Liberal Arts and International Studies at the Colorado School of Mines, where he also directs the Hennebach Program in the Humanities and co-directs an Ethics Across Campus program. His publications include *Thinking through Technology: The Path between Engineering and Philosophy* (1994) and he is editor of the four volume *Encyclopedia of Science, Technology, and Ethics* (2005).

**Mike Murphy** received an H. Dip. from the Dublin Institute of Technology, a B.Sc. in engineering from Trinity College Dublin, and an M.Eng. and Ph.D. from Stevens Institute of Technology. He is a Fellow of Engineers Ireland and a member of the IEEE. He worked in Bell Labs and later at Bell Communications Research before returning to academia in 2002. Presently he is the Director of the Dublin Institute of Technology and Dean of the College of Engineering & Built Environment. A member of SEFI Administrative Council and the European Engineering Deans Council, his area of special interest is engineering and technology education.

**CAO Nanyan** received her B.S. in Psychology from Peking University, and her M.A. in Philosophy and History of Science from the Graduate University of Chinese Academy of Sciences. She has been a visiting scholar at Massachusetts Institute of Technology (1993–1994 and 1999), and the University of Toronto (1998). Currently, she is Professor of Social Studies of Science and Technology at Tsinghua University. She has published several books including *Cognitive Learning Theory* (1991) and *Car Culture* (1997). Her research and teaching areas are in the ethics of scientific research, engineering ethics, and science and technology studies.

**Byron Newberry** holds a B.S. and M.S. in Aerospace Engineering from the University of Alabama and Iowa State University, respectively, and a Ph.D. in Engineering Mechanics from Iowa State University. He is Professor of Mechanical Engineering at Baylor University, a Baylor Fellow for teaching, and a Professional Engineer (PE) in Texas, USA. His research interests are in engineering design, engineering ethics, philosophy of engineering and technology. He is also an aircraft structural engineering consultant, an executive board member of the National Institute for Engineering Ethics, and an editor of Springer's *Philosophy of Engineering and Technology* book series.

**Rune Nydal** is Associate Professor at the Program for Applied Ethics, Department of Philosophy, Norwegian University of Science and Technology (NTNU). He is interested in the normative character of technological and scientific research activities. He has focused on large scale scientific programs and the institutional integration of ethics within them. His Ph.D. draws on a study of the emergence of the field of functional genomics in Norway. In recent years this interest has been extended into a study of other emerging fields such as systems biology and nanotechnology.

**Kieron O'Hara** is a senior research fellow in Electronics and Computer Science at the University of Southampton. His research interests are trust, privacy, memory and the politics of computing (including transparency and open data), with a focus on the World Wide Web and the Semantic Web. He is the author of several books, chairs the transparency sector panel for open data releases in crime and criminal justice for the UK Home Office and Ministry of Justice, and writes the Digital Citizen column for IEEE Internet Computing.

**Zachary Pirtle** is an engineer at the National Aeronautics and Space Administration headquarters in Washington, D.C. and an affiliate of the Consortium for Science, Policy and Outcomes at Arizona State University. His interests include robustness analysis, the management and governance of complex technological systems, and the philosophy of engineering and science. From ASU, Zach earned his B.S. in Mechanical Engineering, B.A. in Philosophy, and M.S. in Environmental Engineering. Previously, Zach was a Fulbright Scholar in Mexico and a Mirzayan Fellow at the National Academy of Engineering. He recently began a Ph.D. in Systems Engineering at George Washington University.

**Joseph C. Pitt** is Professor of Philosophy and of Science and Technology Studies at Virginia Tech. He is the author of three books, *Pictures, Images and Conceptual Change; Galileo, Human Knowledge and the Book of Nature; Thinking About Technology,* edited 11 additional books and published over 100 articles and book reviews in scholarly journals. He is the founding editor of the interdisciplinary journal *Perspectives on Science; Historical, Philosophical, Social,* published by MIT Press. He is currently Editor-in-Chief of *Techné: Research in Philosophy and Technology, The Journal of the Society for Philosophy and Technology* of which he has served as President.

**Auke J.K. Pols** is a postdoctoral researcher in the Department of Philosophy and Ethics at Eindhoven University of Technology. He studied Cognitive Artificial Intelligence at Utrecht University and got his Ph.D. at Eindhoven University of Technology, where he wrote a dissertation on how to conceptualise what actions with artefacts are, and how artefacts influence agents and agency. He currently works on the Netherlands Organization for Scientific Research Responsible Innovation research project 'Biofuels: sustainable innovation or gold rush?', where he investigates how to understand the notion of 'sustainability' in the context of biofuel production and use, and what this implies for biofuel policy recommendations.

**Hans Poser** studied Mathematics and Physics (Staatsexamen, equivalent to M.A.) as well as Philosophy (Ph.D. and Habilitation). He was Professor of Philosophy at Technische Universitaet Berlin since 1972 and retired in 2008. He was President of the German Society of Philosophy and is Honorary Director of the Institute of Philosophy of Dalian University of Technology in China. As a Visiting Professor he has taught at many international universities, among them Rice University at Houston/USA. His main areas of research are philosophy of science and technology as well as the history of modern philosophy (Descartes, Leibniz, Kant).

**Raymond L. Price** holds the William H. Severns Chair of Human Behavior in the College of Engineering at the University of Illinois at Urbana-Champaign and is Director of the Illinois Foundry for Innovation in Engineering Education. Prior to joining the University, he held management positions at Allergan, Boeing, and Hewlett-Packard. At Allergan he was Vice President of Human Resources for the North America Region. He earned a Ph.D. in Organizational Behavior from Stanford University. Professor Price's research is on innovation, creativity, and new product development. He collaborated with Charles House as authors of The HP Phenomenon: Innovation and Business Transformation.

**Donna M. Rizzo** is Associate Professor in the University of Vermont's School of Engineering. She holds undergraduate degrees in Civil Engineering from the University of Connecticut and in Fine Arts from the University of Florence, Italy, and an M.S. and Ph.D. from the University of California, Irvine, and University of Vermont, respectively. She works on a variety of activities designed to recruit students from underrepresented groups and promote interdisciplinary undergraduate research experiences through mentoring, laboratory and field research. Her research focuses on the development of new computational tools to improve the understanding of human-induced changes on natural systems and the way decisions about natural resources are made.

**Wade L. Robison** is the Ezra A. Hale Professor of Applied Ethics at the Rochester Institute of Technology. He has published extensively in practical and professional ethics, philosophy of law, and the work of David Hume, writing or co-editing books on Hume, business and professional ethics, health-care reform, David Hume, and Adam Smith. His book *Decisions in Doubt: The Environment and Public Policy* won the Nelson A. Rockefeller Prize in Social Science and Public Policy. Besides directing an NEH seminar on David Hume, he has received a number of NEH fellowships and NSF grants and directed numerous conferences on Hume and practical and professional ethics.

**Viola Schiaffonati** is a researcher at the Dipartimento di Elettronica, Informazione e Bioingegneria of Politecnico di Milano (Italy), where she teaches philosophy. She got her doctorate in Philosophy of Science from Università di Genova and has been visiting scholar at the Department of Philosophy of the University of California at Berkeley and visiting researcher at the Suppes Center for the Interdisciplinary Study of Science and Technology of Stanford University. Her research interests include: the philosophy of computer science and the foundational issues of artificial intelligence, formal approaches to the philosophy of science, and the epistemological issues of computational science.

**Jon Alan Schmidt** is an associate structural engineer at Burns & McDonnell in Kansas City, Missouri. He chairs the editorial board for *STRUCTURE* magazine and writes a bimonthly column that often explores the interface between philosophy and engineering. He is also founding chair of the Engineering Philosophy Committee within the Structural Engineering Institute of the American Society of Civil Engineers.

**Peter Simons** holds the 1837 Chair of Philosophy at Trinity College Dublin. His main research focus is metaphysics, pure and applied. In pure metaphysics he authored the definitive monograph on mereology (part–whole theory), *Parts* (1987, 2000), and has also investigated other key metaphysical concepts. His interest in the application of these concepts to various disciplines led to his engagement with the philosophy of engineering, via work on software design for knowledge systems applied to manufacturing.

**John P. Sullins** is an Associate Professor of Philosophy at Sonoma State University in California, where he has taught since 2004. He received his Ph.D. in 2002 from the Philosophy, Computers, and Cognitive Science program at Binghamton University in New York. His current research and publications involve the study of computer ethics, malware ethics, and the analysis of the ethical impacts of military and personal robotics technologies. He is the 2011 recipient of the Herbert Simon Excellence in Research award from the International Association of Computers and Philosophy. He lives in Sonoma County California with his wife and two daughters.

**Ibo van de Poel** is Anthonie van Leeuwenhoek Professor in Ethics and Technology at Delft University of Technology in the Netherlands. He has published on engineering ethics, the moral acceptability of technological risks, values and engineering design, moral responsibility in research networks, and ethics of new emerging technologies such as nanotechnology. His current research focuses on new technologies as social experiments and conditions for morally responsible experimentation.

**Bruce A. Vojak** is an Associate Dean of Engineering at the University of Illinois at Urbana-Champaign, and Adjunct Professor of Electrical and Computer Engineering and of Industrial and Enterprise Systems Engineering. Earlier he held positions at M.I.T. Lincoln Laboratory, Amoco, and Motorola, where he was Director of Advanced Technology. He holds a Ph.D. in Electrical Engineering from the University of Illinois at Urbana-Champaign and an M.B.A. from the University of Chicago's Booth School of Business. In addition to administrative responsibilities, he teaches and conducts research on innovation, serves on the Board of Directors of Midtronics, Inc., and consults for Proctor & Gamble.

# Name Index



D.P. Michelfelder et al. (eds.), *Philosophy and Engineering: Reflections on Practice, Principles and Process*, Philosophy of Engineering and Technology 15,
DOI 10.1007/978-94-007-7762-0, 

**W**



**Z**

# Subject Index




D.P. Michelfelder et al. (eds.), *Philosophy and Engineering: Reflections on Practice, Principles and Process*, Philosophy of Engineering and Technology 15,
DOI 10.1007/978-94-007-7762-0, 

**Q**



**R**

CPSIA information can be obtained at www.ICGtesting.com
Printed in the USA
LVOW10*0750010714

392453LV00007BA/44/P